Animal Cell Culture

The Practical Approach Series

SERIES EDITORS

D. RICKWOOD
Department of Biology, University of Essex
Wivenhoe Park, Colchester, Essex CO4 3SQ, UK

B. D. HAMES
Department of Biochemistry and Molecular Biology,
University of Leeds, Leeds LS2 9JT, UK

Affinity Chromatography
Anaerobic Microbiology
Animal Cell Culture (2nd Edition)
Animal Virus Pathogenesis
Antibodies I and II
Biochemical Toxicology
Biological Membranes
Biomechanics—Materials
Biomechanics—Structures
 and Systems
Biosensors
Carbohydrate Analysis
Cell Growth and Division
Cellular Calcium
Cellular Neurobiology
Centrifugation (2nd Edition)
Clinical Immunology
Computers in Microbiology
Crystallization of Nucleic Acids
 and Proteins
Cytokines
The Cytoskeleton
Diagnostic Molecular Pathology
 I and II
Directed Mutagenesis

DNA Cloning I, II, and III
Drosophila
Electron Microscopy in Biology
Electron Microscopy in
 Molecular Biology
Enzyme Assays
Essential Molecular Biology I
 and II
Fermentation
Flow Cytometry
Gel Electrophoresis of Nucleic
 Acids (2nd Edition)
Gel Electrophoresis of Proteins
 (2nd Edition)
Genome Analysis
HPLC of Macromolecules
HPLC of Small Molecules
Human Cytogenetics I and II
 (2nd Edition)
Human Genetic Diseases
Immobilised Cells and
 Enzymes
Iodinated Density Gradient
 Media
Light Microscopy in Biology
Lipid Analysis

Animal Cell Culture

A Practical Approach
Second Edition

Edited by

R. I. FRESHNEY

CRC Department of Medical Oncology,
University of Glasgow, Alexander Stone Building,
Garscube Estate, Switchback Road,
Bearsden, Glasgow G61 1BD, UK

OXFORD UNIVERSITY PRESS
Oxford New York Tokyo

Oxford University Press, Walton Street, Oxford OX2 6DP

Oxford   New York   Toronto
Delhi   Bombay   Calcutta   Madras   Karachi
Kuala Lumpur   Singapore   Hong Kong   Tokyo
Nairobi   Dar es Salaam   Cape Town
Melbourne   Auckland   Madrid

and associated companies in
Berlin   Ibadan

Oxford is a trade mark of Oxford University Press

A Practical Approach   is a registered trade mark
of the Chancellor, Masters, and Scholars of the University of Oxford
trading as Oxford University Press

Published in the United States
by Oxford University Press Inc., New York

A catalogue for this book is available from the British Library

Library of Congress Cataloging-in-Publication Data

Animal cell culture : a practical approach / edited by R. I. Freshney.—2nd ed.
(Practical approach series)
Includes bibliographical references and index.
1. Cell culture.   2. Tissue culture.   I. Freshney, R. Ian.
II. Series.
QH585.2.A55   1992   591.87′0724—dc20   92—4082
ISBN 0 19 963212 X (Hbk)
ISBN 0 19 963213 8 (Pbk)

Printed in Great Britain by Information Press Ltd, Eynsham, Oxford

Preface to the second edition

Like any other field in the biological sciences, cell culture is changing rapidly
in the way that it is being used by scientists and industrial biotechnologists and
I have attempted to make the changes between the first and second editions of
this book reflect this. The basic principles remain the same, so there is little
alteration to the first chapter, but there are major changes in other chapters,
particularly serum-free medium, scale-up, and flow cytometry, where significant advances have been made, not just in the development of these techniques
as tools for the cell biologist, but in their availability to the non-specialist. At
the time of the first edition reliable commercially available serum-free
medium was hard to obtain, but now a number of suppliers are providing
media such as the keratinocyte medium of the Clonetics Corporation and
numerous hybridoma media, which have been quality controlled with the
appropriate cells. Flow cytometry (Ch. 6), while still a very sophisticated
technique, is now available to the novice, with benchtop equipment that can
be used without employing a fulltime technician.

The reduction of the tetrazolium salt MTT (Ch. 8), as a colorimetric
indicator of viability, has made a considerable impact on cytotoxicity assay,
and DNA fingerprinting has revolutionized the identification of individual cell
strains (Ch. 4). These, and other developments in the techniques described in
this book, made a new edition essential. The emphasis remains on presenting
specialized techniques in a readily accessible form, with the emphasis on the
practical aspects and the presentation of easily followed protocols.

Since the first edition was written, one of the authors, Luciano Morasca,
who was both a friend and a scientific colleague, has died. He was well liked
and respected by those who knew him and worked with him, and it is for this,
and the many interesting discussions that I had with him about tissue culture
and its use in the study of cytotoxicity in tumour cells, that I would like to
dedicate this second edition of *Animal Cell Culture* to his memory.

Glasgow   I. F.
April 1992

Preface to the first edition

In recent years the use of animal cell culture has undergone a major expansion from being a purely experimental procedure to become an accepted technological component of many aspects of biological research and commercial exploitation. New developments have arisen to improve both versatility and standardization and it is the aim of this book to make these developments readily accessible to scientists and technologists in the field. These developments include major advances that have been made in serum-free culture and in the scaling up of animal cell culture to industrial and semi-industrial levels. The importance of cell line characterization, standardization, and banking are now widely recognized and information is presented to allow the reader both to perform their own banking and to take advantage of central repositories and data banks.

Analytical and preparative fractionation of cell populations is dealt with in two chapters on flow cytometry and centrifugal elutriation, and analysis at the cellular and subcellular level is described in a chapter on *in situ* hybridization. The importance of cell interaction and retention of histological structure is stressed in a chapter on organ culture, where most of the standard techniques are described. Finally, a chapter on viability and cytotoxicity measurements has been included to cover the requirements of *in vitro* toxicology and anti-neoplastic drug screening, as well as routine monitoring of culture viability.

I. F.

Contents

3. Scaling-up of animal cell cultures 47
Bryan Griffiths

Contributors

D. CONKIE
Beatson Institute for Cancer Research, Garscube Estate, Switchback Road, Bearsden, Glasgow G61 1BD, UK.

EUGENIO ERBA
Istituto di Ricerche Farmacologiche 'Mario Negri', Via Eritrea 62, I-20157 Milano, Italy.

R. I. FRESHNEY
CRC Department of Medical Oncology, University of Glasgow, Alexander Stone Building, Garscube Estate, Switchback Road, Bearsden, Glasgow G61 1BD, UK.

BRYAN GRIFFITHS
Division of Biologics, PHLS CAMR, Porton Down, Salisbury, Wilts SP4 0JG, UK.

R. J. HAY
American Type Culture Collection, 12301 Parklawn Drive, Rockville, MD 20852, USA.

S. M. LANG
Department of Pathology, Medical School, University of Edinburgh, Teviot Place, Edinburgh EH8 9AG, UK.

ILSE LASNITZKI
Strangeways Laboratory, Cambridge (correspondence to 11 Porson Road, Cambridge CB2 2ET, UK).

H. R. MAURER
Freie Universität Berlin, Fachbereich Pharmazie, Institut für Pharmazie (WE1), Kelchstrasse 31, 0-1000 Berlin 41, Germany.

JAMES V. WATSON
MRC Clinical Oncology and Therapeutics Unit, Medical Research Council Centre, Hills Road, Cambridge CB2 2QH, UK.

Contributors

ANNE P. WILSON

Oncology Research Laboratory, Derby City Hospital, Uttoxeter Road, Derby DE3 3NE, UK.

A. H. WYLLIE

Department of Pathology, Medical School, University of Edinburgh, Teviot Place, Edinburgh EH8 9AG, UK.

Abbreviations

ACP	acid phosphatase
ACTH	adenocorticotrophic hormone
ADA	adenosine deaminase
AK	adenylate kinase
ATCC	American Type Culture Collection
bFGF	basic fibroblast growth factor
BHK	baby hamster kidney
BME	Eagle's basal medium
BPE	bovine pituitary extract
BrdU	bromodeoxyuridine
BSA	bovine serum albumin
BSS	balanced salt solution
CAM	chorioallantoic membrane
CCNSS	Cancer Chemotherapy National Service Centre
CPE	cytopathogenic effect
CT	cholera toxin
DAPI	2-diamidinophenylindole
DEPC	diethyl pyracarbonate
Dex	dexamethasone
DHT	5α-dihydrotestosterone
DIP	dipeptidase
DME or DMEM	Dulbecco's modified Eagle's medium
DMSO	dimethyl sulphoxide
E-2	oestradiol
EA	ethanolamine
ECGF	endothelial cell growth factor
ECM	extracellular matrix
EDTA	ethylenediamine tetra-acetic acid
EGF	epidermal growth factor
EGTA	ethyleneglycol-bis(β-aminoethyl)N,N,N',N'-tetra-acetic acid
EPO	erythropoietin
ES-D	esterase D
E-2	estradiol

FAF	fatty acid free
FALS	forward angle light scatter
FCS	fetal calf serum
FGF	fibroblast growth factor
FITC	fluorescein isothiocyanate
FN	fibronectin
FRAME	Fund for Replacement of Animals in Medical Experiments
FSH	follicle stimulating hormone
FUCA	α-fucosidase
G-CSF	granulocyte colony stimulating factor
G6PD	glucose-6-phosphate dehydrogenase
GLO	glyoxalase
Glu	glutamine
GOT	glutamic-oxaloacetic transaminase
GPI	glucose-phosphate isomerase
GPK	glucose-phosphate kinase
H-E	haematoxylin-eosin
HBS	hepatitis B surface antigen
HBSS	Hanks' balanced salt solution
HC	hydrocortisone (=cortisol)
HDL	high-density lipoprotein
HGH	human growth hormone
HLA	human leukocyte antigens
HPAP	human placental alkaline phosphatase
HS	high salt
IF	intermediate filament
IFN	interferon
IGF	insulin-like growth factor
IL	interleukin
IMDM	Iscove's modified Dulbecco's medium
INT	2-(*p*-iodophenyl)-(*p*-nitrophenyl)-5-phenyl tetrazolium chloride
ISH	*in situ* hybridization
LA	linoleic acid
LDH	lactate dehydrogenase
LDL	low-density lipoprotein
LSB	liposome B
ME2	malate dehydrogenase
5-ME	5-mercaptoethanol
MEM	minimal essential medium
MTT	dimethylthiazol diphenyltetrazolium
NaPyr	sodium pyruvate
NGF	nerve growth factor
NP	nucleoside phosphorylase
OD	optical density

ODR	oxidation-reduction (redox) potential
OTR	oxygen transfer rate
PA	prostate antigen
PAS	periodic acid—Schiff
PBMEM	phosphate-buffered Eagle's MEM
PBS	phosphate-buffered saline (Dulbecco)
PBSA	phosphate-buffered saline without Ca^{2+} and Mg^{2+}
PDGF	platelet-derived growth factor
PE	phosphatidylethanolamine
PEP-D	peptidase D
PG	prostaglandin
PGD	6-phosphogluconate dehydrogenase
PGE	prostaglandin E
PGF	prostaglandin $F_{2\alpha}$
PGM	phosphoglucomutase
PMS	phenazine methosulphate
PRL	prolactin
Pyr	pyruvate
RBC	red blood cell
SDS	sodium dodecyl sulphate
SEM	scanning electron microscopy
SF	serum-free
SG	specific gravity
SRB	sulphorhodamine B
SSC	standard saline citrate
T_3	tri-iodothyronine
TAE	Tris-acetate-EDTA
TC	Tris-citrate
TCA	trichloroacetic acid
TEB	Tris-EDTA-borate
TEM	transmission electron microscopy
TGF	transforming growth factor
tPA	tissue plasminogen activator
TSH	thyroid stimulating hormone
VLDL	very-low-density lipoprotein

1

Introduction to basic principles

R. I. FRESHNEY

1. Introduction

The culture of animal cells and tissues is now a widely used technique in many different disciplines from the basic sciences of cell and molecular biology to the rapidly evolving applied field of biotechnology. An introduction to the basic procedures is available in many laboratories and frequently features as an integral part of undergraduate study in the biological sciences. Several textbooks are already available (1–3) to assist the complete novice during his or her introduction to the basic principles of preparation, sterilization, and cell propagation, so this book will concentrate on certain specialized aspects, many of which are essential for complete understanding and correct utilization of the technique.

This chapter will review some of the general aspects of cultured cells, their biology, derivation, and characterization, and set out some of the basic assumptions and definitions.

2. Biology of cells in culture

2.1 Origin and characterization

The list of different cell types which can now be grown in culture is quite extensive, and includes connective tissue elements such as fibroblasts, skeletal tissue (bone and cartilage), skeletal, cardiac, and smooth muscle, epithelial tissues (e.g. liver, lung, breast, skin, bladder, and kidney), neural cells (glia and neurones, though neurones do not proliferate), endocrine cells (adrenal, pituitary, pancreatic islet cells), melanocytes, and many different types of tumour. (For further information see ref. 4.)

The use of markers that are cell type specific (see Chapter 4) has made it possible to determine the lineage from which many of these cultures were derived, but what is not entirely clear, in many cases, is the position of the cells within the lineage. In a propagated cell line, a precursor cell type will predominate, rather than a fully differentiated cell, which would not normally proliferate. Consequently the cell line may be heterogeneous as some cultures,

e.g. epidermal keratinocytes, contain stem cells, precursor cells, and keratin-
ized squames. There is constant renewal from the stem cells, proliferation and
maturation in the precursor compartment, and terminal and irreversible differentiation releasing squames into the culture medium. Other cultures, such
as fibroblasts, contain a fairly uniform population of proliferating cells at low
cell densities (about 10^4 cells/cm^2) and an equally uniform, more differentiated, non-proliferating population at high cell densities (10^5 cells/cm^2). This
high-density population of fibrocyte-like cells can re-enter the cell cycle if the
cells are trypsinized or scraped to reduce the cell density or create a free edge.

Culture heterogeneity also results from multiple lineages being present in
the cell line. The only unifying factors are the selective conditions of the
medium and substrate, and the predominance of the cell type (or types) which
have the maximum growth rate. This tends to select a common phenotype,
but, due to the cell interactive nature of growth control, may obscure the fact
that the population may contain several distinct phenotypes only detectable
by cloning.

Nutritional factors like serum or Ca^{2+} ions (5), hormones, cell (6) and
matrix interactions (7), in addition to the density of the culture (8), can all
affect differentiation and cell proliferation, often inversely. Hence, it is not
only essential to define the lineage of cells being used, but also to characterize
and stabilize the stage of differentiation, by controlling cell density and the
nutritional and hormonal environment, to obtain a uniform population of
cells.

Because the dynamic properties of cell culture are sometimes difficult to
control, and the appropriate cell interactions found *in vivo* difficult to recreate *in vitro*, many people have forsaken the idea of serial propagation in
favour of retaining the structural integrity of the original tissue. Such a
system is called histotypic or organ culture and is dealt with in Chapter 7.
Attempts have also been made to recreate tissue-like structures *in vitro* by re-
aggregating different cell types and culturing at high density as multicellular
spheroids (9), perfused multilayers on glass or plastic substrates (10) or
floating cultures on collagen (11) or synthetic microporous filters (12).

2.2 **Differentiation**

As propagation of cell lines requires that the cell number increases continually,
culture conditions which have evolved over the years have been selected to
favour maximal cell proliferation. It is not surprising that these conditions are
often not conducive to cell differentiation where cell growth is severely limited
or completely abolished. Those conditions which favour cell proliferation are
low cell density, low Ca^{2+} concentration (13) (100–600 μM) and the presence
of growth factors such as epidermal growth factor (EGF), fibroblast growth
factor (FGF), and platelet derived growth factor (PDGF). High cell density
($> 10^5$ cells/cm^2), high Ca^{2+} concentration (300–1500 μM) and the presence

of differentiation inducers (hormones such as hydrocortisone (14), glia maturation factor (15), nerve growth factor (16), retinoids (17), and polar solvents, such as dimethyl sulphoxide (18), will favour cytostasis and differentiation.

The role of serum in differentiation is not entirely clear and depends on the cell type and medium used. While a low serum concentration promotes differentiation in oligodendrocytes (19), a high serum concentration causes squamous differentiation in bronchial epithelium (20). In the latter case, the active principle is a molecule closely resembling, or identical to, transforming growth factor β (TGF-β) isolated from platelets. The use of defined media will hopefully help to resolve this question (see Chapter 2).

The establishment of the correct polarity and cell shape may also be important, particularly in epithelium. Many workers have shown that growing cells to high density on a floating collagen gel allows matrix interaction, access to medium on both sides (or, perhaps more important, to the basal surface), the possibility of establishing correct polarity with respect to the basement membrane and the adoption of the correct cell shape due to the plasticity of the substrate (21, 22).

Different conditions are required, therefore, for propagation and differentiation and hence an experimental protocol may require a growth phase to increase cell number and allow for replicate samples, followed by a non-growth maturation phase to allow for increased expression of differentiated functions.

3. Choice of materials

The primary determinant in selecting a tissue or cell line for further study is the nature of the observations that have to be carried out. General cellular processes such as DNA synthesis, membrane permeability, or determination of cytotoxicity may be feasible with any cell type, while the study of specialized properties such as myotube fusion, antibody production, or regulation of urea cycle enzymes will require cell types expressing these specific functions.

3.1 Organ culture or cell culture?

Originally, tissue culture was regarded as the culture of whole fragments of explanted tissue with the assumption that histological integrity was maintained, at least in part. Now 'tissue culture' has become a generic term and encompasses organ culture, where a small fragment of tissue or whole embryonic organ is explanted to retain tissue architecture, and cell culture where the tissue is dispersed mechanically or enzymatically, or by spontaneous migration from an explant, and may be propagated as a cell suspension or attached monolayer.

In adopting a particular type of culture the following points should be taken

Table 1. Respective advantages of cell and organ culture

Organ culture	Cell culture
Histology	Propagation and expansion
Differentiation	Cloning, selection, and purification
Cell interaction	
(homotypic and heterotypic)	Characterization and preservation
Matrix interaction	Replicate sampling and quantitation

into account (*Table 1*). Organ culture (see Chapter 7) will preserve cell interaction, retain histological and biochemical differentiation for longer, and, after the initial trauma of explantation and some central necrosis, will generally remain in a non-growing steady state for a period of several days and even weeks. They cannot be propagated, however, generally incur greater experimental variation between replicates, and tend to be more difficult to use for quantitative determinations due to minor variations in geometry and constitution.

Cell cultures on the other hand, are usually devoid of structural organization, have lost their histotypic architecture and often the biochemical properties associated with it, and generally do not achieve a steady state unless special conditions are employed. They can, however, be propagated and hence expanded and divided into identical replicates, they can be characterized and a defined cell population preserved by freezing, and they can be purified by growth in selective media (Chapter 2), physical cell separation (Chapters 5 and 6) or cloning (Chapters 4 and 8) to give a characterized cell strain with considerable uniformity.

3.2 Source of tissue

3.2.1 Embryo or adult?

In general, cultures derived from embryonic tissues will survive and proliferate better than those from the adult. This presumably reflects the lower level of specialization and presence of replicating precursor or stem cells in the embryo. Adult tissues will usually have a lower growth fraction and a high proportion of non-replicating specialized cells, often within a more structured, and less readily disaggregated, extracellular matrix. Initiation and propagation are more difficult, and the lifespan of the culture often shorter.

Embryonic or fetal tissue has many practical advantages, but it must always be remembered that in some instances the cells will be different from adult cells and it cannot be assumed that they will mature into adult-type cells unless this can be confirmed by appropriate characterization.

Examples of widely used embryonic cell lines are the various 3T3 lines (mouse embryo fibroblasts) and WI-38, MRC-5 and other human fetal lung fibroblasts. Mesodermally derived cells (fibroblasts, endothelium, myoblasts) are on the whole easier to culture than epithelium, neurones, or endocrine tissue but this may reflect the extensive use of fibroblast cultures during the early years of the development of culture media together with the response of mesodermally-derived cells to mitogenic factors present in serum. A number of new selective media have now been designed for epithelial and other cell types (see Chapter 2) and with some of these it has been shown that serum is inhibitory to growth and may promote differentiation (20). Primary cultures from epithelial tissues, such as skin, lung, and mammary gland, are now routine procedures in some laboratories (21).

3.2.2 Normal or neoplastic?

Normal tissue usually gives rise to cultures with a finite lifespan while cultures from tumours can give continuous cell lines (see below), although there are several examples of continuous cell lines (MDCK dog kidney, 3T3 fibroblasts) which have been derived from normal tissues and which are non-tumorigenic (see Section 3.3.1 below).

Normal cells will generally grow as an undifferentiated stem cell or precursor cell, and the onset of differentiation is accompanied by a cessation in cell proliferation which may be permanent. Some normal cells, e.g. fibrocytes or endothelium, are able to differentiate and still de-differentiate and resume proliferation and in turn re-differentiate, while others, e.g. squamous epithelium, skeletal muscle, and neurones, once committed to differentiate, are incapable of resuming proliferation.

Cells cultured from neoplasms, such as B16 mouse melanoma, can express at least partial differentiation, while retaining the capacity to divide. Many studies of differentiation have taken advantage of this fact and used differentiated tumours such as the minimal deviation hepatomas of the rat (23) and human and rodent neuroblastomas (24), although whether this can be taken as normal differentiation is always in doubt.

Tumour tissue can often be passaged in the syngeneic host, providing a cheap and simple method of producing large numbers of cells, albeit with lower purity. Where the natural host is not available, tumours can also be propagated in athymic mice with greater difficulty but similar advantages.

Many other differences between normal and neoplastic cells are similar to those between finite and continuous cell lines (see below) and indeed the importance of immortalization in neoplastic transformation has been recognized (25).

3.3 Subculture

Freshly isolated cultures are known as *primary cultures* until they are passaged or subcultured. They are usually heterogeneous, and have a low growth fraction, but are more representative of the cell types in the tissue from which they were derived and in the expression of tissue specific properties. Sub-culture allows the expansion of the culture (it is now known as a *cell line*), the possibility of cloning (see Chapters 4 and 8), characterization and preservation (Chapter 4), and greater uniformity, but may cause a loss of specialized cells and differentiated properties unless care is taken to select out the correct lineage and preserve or re-induce differentiated properties (see below).

The greatest advantage of subculturing a primary culture into a cell line is the provision of large amounts of consistent material suitable for prolonged use.

3.3.1 Finite or continuous cell lines?

After several subcultures a cell line will either die out (*finite cell line*) or 'transform' to become a *continuous cell line*. It is not clear in all cases whether the stem line of a continuous culture pre-exists masked by the finite population or arises during serial propagation. Because of the time taken for such cell lines to appear (often several months) and the differences in their properties, it has been assumed that a mutational event occurs, but the pre-existence of immortalized cells, particularly in cultures from neoplasms, cannot be excluded. Complementation analysis has shown that senescence is a dominant trait, and immortalization the result of one or more gene deletions (26).

The appearance of a continuous cell line is usually marked by an alteration in cytomorphology (smaller cell size, less adherent, more rounded, higher ratio of nucleus to cytoplasm), an increase in growth rate (population doubling time decreases from 36–48 h to 12–36 h), a reduction in serum dependence, an increase in cloning efficiency, a reduction in anchorage dependence (i.e. an increased ability to proliferate in suspension as a liquid culture or cloned in agar), an increase in heteroploidy (chromosomal variation among cells) and aneuploidy (divergence from the donor, euploid, karyotype) and an increase in tumorigenicity. The resemblance between spontaneous *in vitro* transformation and malignant transformation is obvious but nevertheless the two are not necessarily identical although they have much in common. Normal cells can 'transform' to become continuous cell lines without becoming malignant, and malignant tumours can give rise to cultures which 'transform' and become more (or even less) tumorigenic but acquire the other properties listed above.

In vitro transformation is primarily the acquisition of an infinite lifespan. Alterations in growth control which may occur simultaneously, or subsequently, are under distinct positive and negative control genes (oncogenes and tumour suppressor genes, refs 27 and 28).

The advantages of continuous cell lines are their greater growth rates to higher cell densities and resultant higher yield, their lower serum requirement and general ease of maintenance in simple media, and their ability to grow in suspension. Their disadvantages include greater chromosomal instability, divergence from the donor phenotype, and loss of tissue-specific markers.

3.3.2 Propagation in suspension

Most cultures are propagated as a monolayer, anchored to a glass or plastic substrate, but some, principally transformed cells, haemopoietic cells, and ascites tumours, can be propagated in suspension. This has the advantage of simpler propagation (subculture only requires dilution, no trypsinization), no requirement for increasing surface area with increasing bulk, ease of harvesting and the possibility of achieving a 'steady-state' culture if required (see Chapter 3).

3.4 Selection of medium

Regrettably, the choice of medium is still often quite empirical. What was used previously by others for the same cells, or what is currently being used in your own laboratory for different cells, often dictates the choice of medium and serum. For continuous cell lines it may not matter a great deal as long as the conditions are consistent, but for specialized cell types, primary cultures, and growth in the absence of serum, the choice is more critical. This subject is dealt with in more detail in Chapter 2.

There are two major advantages of using more sophisticated media in the absence of serum: they may be selective for particular types of cell, and the isolation of purified products is easier in the absence of serum. Nevertheless, culture in the presence of serum is still easier and often, surprisingly, no more expensive, though undoubtedly less controlled. Two major determinants still regulate the use of serum-free media: (a) they are only gradually becoming available commercially (many people do not have the time, facilities, or inclination to make up their own), (b) requirements for serum-free media are more cell-type specific. Serum will cover many inadequacies revealed in its absence. Furthermore, because of their selectivity, a different medium may be required for each type of cell line. This problem may be particularly acute when culturing tumour cells where cell line variability may require modifications for cell lines from individual tumours.

In the final analysis the choice is often still empirical: read the literature and determine which medium has been used previously. If several different media have been used (and this is often the case) test them all for yourself, with a few others added to the series if you wish. Test growth (doubling time and saturation density) (see below), cloning efficiency (Chapter 8), and expression of specific properties (differentiation, viral propagation, cell products, etc.). The choice of medium may not be the same in each case, e.g. differentiation of lung epithelium will proceed in serum, propagation is better without. If possible, include one or more serum-free media in your panel, supplemented with growth factors, hormones, and trace elements as required (see Chapter 2).

Once a type of medium has been selected, try to keep this constant for as long as possible. Similarly, if serum has to be used, select a batch by testing samples from commercial suppliers and reserve enough to last 6 months to one year, before replacing it with another pre-tested batch. Testing procedures are as described above for media selection.

3.5 Gas phase

The gas phase is determined by

- the type of medium
- whether the culture vessel is open (Petri dishes, multiwell plates) or sealed (flasks, bottles)
- the amount of buffering required

Several variables are in play, but one major rule predominates and three basic conditions can be described. The rule is that the bicarbonate concentration and carbon dioxide tension must be in equilibrium. The three conditions are summarized in *Table 2*. It should be remembered that carbon dioxide/bicarbonate is essential to most cells, so a flask or dish cannot be vented without providing carbon dioxide in the atmosphere. Prepare medium to about pH 7.1 or 7.2 at room temperature, incubate with the correct carbon dioxide tension for at least 0.5 h in a shallow dish, and check that the pH stabilizes at pH 7.4. Adjust with sterile 1 M HCl or 1 M NaOH if necessary.

Table 2. Composition of gas phase

Buffering capacity	[HCO_3^-]	Gas phase	Hepes
Low (sealed flask)	4 mM	Air	—
Moderate (open or sealed vessel)	23 mM	5% CO_2	—
High (open or sealed vessel)	8 mM	2% CO_2	20 mM

Oxygen tension is usually maintained at atmospheric but variations have been described, e.g. elevated for organ culture (see Chapter 7) and reduced for cloning melanoma (see Chapter 8).

3.6 Substrate

The nature of substrate is determined largely by the type of cell and the use to which it will be put. Polystyrene which has been treated to make it wettable and give it a net negative charge is now used almost universally. In special cases (culture of neurones, muscle cells, capillary endothelial cells, and some epithelial cultures) the plastic is pre-coated with gelatin, collagen, or polylysine to give a net positive charge. Glass may also be used, but must be washed carefully with a non-toxic detergent (see below).

Culture vessels vary in size from Terasaki multiwell plates (1 mm^2 surface area, 5–10 μl medium volume) and microtitration plates (30 mm^2, 100–200 μl) up through a range of dishes and flasks to 180 cm^2, and roller bottles and multisurface propagators (see Chapter 3) for large-scale culture. The major determinants are the number of cells required (5×10^4–10^5/cm^2 maximum for most untransformed cells, 10^5–10^6/cm^2 for transformed), the number of replicates (96 in a microtitration plate), and the times of sampling: a 24-well plate is good for a large number of replicates for simultaneous sampling, but individual tubes or bottles are preferable where sampling is carried out at different times. Petri dishes are cheaper than flasks and good for subsequent processing, e.g. staining or extractions. Flasks can be sealed, do not need a humid carbon dioxide incubator, and give better protection against contamination. For suspension cultures, volume is the main determinant. Sparging and agitation may be necessary as the depth increases (see Chapter 3).

4. Preparation

4.1 Substrate

Many laboratories now utilize disposable plastics as substrates for tissue culture. They are optically clear, prepared for tissue culture use by modification of the plastic to make it wettable and suitable for cell attachment, and come already sterilized for use. On the whole they are convenient and provide a source of reproducible vessels for both routine and experimental work. However, they are disposable and thus are more expensive than glass.

4.1.1 Washing glassware

For routine passage of continuous cell lines and many finite cell lines, glass is perfectly adequate and cheaper to use provided you have the staff and glass washing facilities and can control the quality of washing (1). A non-toxic detergent must be used, the glassware given a thorough soak, preferably overnight, and then thoroughly rinsed in tap water followed by deionized or distilled water. It is assumed that all glassware specified in each chapter meets or exceeds this specification (see also Chapter 7).

4.1.2 Sterilizing glassware

Although plastics come already sterilized, glassware must be sterilized before use. All the protocols described in subsequent chapters assume that all glassware is sterile. Many laboratories routinely sterilize all glassware, regardless of whether it will be used directly for culture or not, to avoid possible confusion between sterile and non-sterile stocks.

Sterilization is best done by dry heat (160°C for 1 h), in containers or foil covered, with screw caps autoclaved separately.

Periodically it may be necessary to check the quality of your glassware, the wash-up process, or the choice of detergent. This is best done by washing in the usual way, checking the glassware by eye, drying and sterilizing the glass and cloning monolayer cells on the glass (see Chapters 4 and 8), preferably in a low serum concentration (2–5%) or serum free (see Chapter 2).

4.2 Medium

Most of the commonly used media are available commercially (see Appendix A1) but for special formulations (see Chapter 4) or additions it may be necessary to prepare and sterilize your own. In general, stable solutions (water, salts, and media supplements such as tryptose or peptone) may be autoclaved at 121°C (100 kPa or 1 atm above ambient) for 20 min, while labile solutions (media, trypsin, and serum) must be filtered through a 0.2 μm porosity membrane filter (Millipore, Sartorius, Gelman, Pall). Sterility testing (see Chapter 4) should be carried out on samples of each filtrate.

Where an automatic autoclave is used care must be taken to ensure that the timing of the run is determined by the temperature of the centre of the load and not just by drain or chamber temperature or pressure which will rise much faster than the load. The recorder probe should be placed in a package or bottle of fluid similar to the load and centrally located.

4.3 Cells

4.3.1 Primary culture

When cells are isolated from a tissue, grown *in vitro*, and before subculture, they are regarded as a primary culture. Primary cultures lack many of the cells present in the original tissue, which were unable to attach and survive *in vitro*. Furthermore, if the culture proliferates, then any non-dividing or slow-growing cells will be diluted out. Hence it may be necessary at this stage to select specific cell types by cloning, selective culture (Chapter 2) or physical cell separation (Chapters 5 and 6).

The first step in preparing a primary culture is sterile dissection followed by mechanical or enzymatic disaggregation. The tissue may simply be chopped to around 1 mm^3 and the pieces attached to a dish by their own adhesiveness or by scratching the dish, or by using clotted plasma (1). In these cases cells will grow out from the fragment and may be used directly or subcultured. The fragment of tissue, or explant as it is called, may be transferred to a fresh dish or the outgrowth trypsinized to leave the explant and a new outgrowth generated.

The trypsinized cells from the outgrowth are re-seeded in a fresh vessel and become a secondary culture, and the culture is now technically a cell line.

Primary cultures can also be generated by disaggregating tissue in enzymes such as trypsin (0.25% crude or 0.01–0.05% pure) or collagenase (200–2000 units/ml, crude) and the cell suspension allowed to settle on to, adhere, and spread out on a glass or plastic substrate (1). This type of culture gives a higher yield of cells though it can be more selective as only certain cells will survive dissociation. In practice, many successful primary cultures are generated using enzymes such as collagenase to reduce the tissue, particularly epithelium, to small clusters of cells which are then allowed to attach and grow out.

For methods of primary culture see (1).

4.3.2 Subculture

A monolayer culture may be transferred to a second vessel and diluted by dissociating the cells of the monolayer in trypsin (suspension cultures need only be diluted). This is best done by rinsing the monolayer with PBS or PBS containing 1 mM EDTA, and removing the rinse, adding cold trypsin (0.25% crude or 0.01–0.05% pure) for 30 s, removing the trypsin and incubating in the residue for approximately 15 min. Cells are then re-suspended in medium, counted, and re-seeded.

4.3.3 Growth curve

When cells are seeded into a flask they enter a lag period of 2–24 h, followed by a period of exponential growth (the 'log phase') and finally enter a period of reduced or zero growth after they become confluent ('plateau phase') (*Figure 1*). These phases are characteristic for each cell line and give rise to reproducible measurements—the length of the lag period, the population doubling time in mid-log phase, and the saturation density at plateau, given that the environmental conditions are kept constant.

The determination of the growth cycle is important in designing routine subculture and experimental protocols. Cell behaviour and biochemistry changes significantly at each phase and it is therefore essential to control the stage of the growth cycle when drugs or reagents are added and cells harvested. The shape of the growth curve can also give information on the reproductive potential of the culture (*Figure 1*) but it is generally recognized that the analysis of clonal growth is easier (29), and less prone to ambiguity and misinterpretation.

Further discussion of cell growth and viability are presented in Chapter 8.

4.3.4 Feeding

Some rapidly growing cultures, e.g. transformed cell lines like HeLa, will require a medium change after 3–4 days in a seven-day subculture cycle. This is usually indicated by an increase in acidity to below pH 7.0.

4.3.5 Contamination

The problem of microbial contamination has been greatly reduced by the use of antibiotics and laminar airflow cabinets. However, cultures should be maintained in the absence of antibiotics whenever possible so that chronic, cryptic contaminations are not harboured.

Check frequently for contamination by looking for a rapid change in pH (usually a fall, but some fungi can increase the pH), cloudiness in the

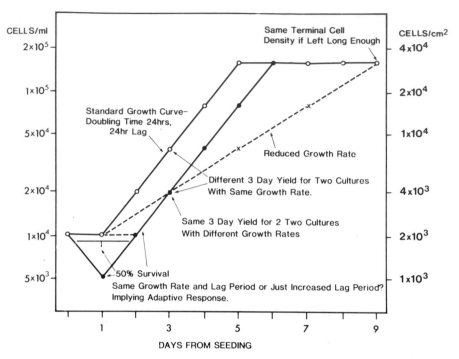

Figure 1. Hypothetical growth curves, calculated from a seeding concentration of 1×10^4 cells/ml, emphasizing the need to examine the entire curve and not just count cells at one timepoint.

medium, extracellular granularity under the microscope, or any unidentified material floating in the medium. If a contamination is detected, discard the flask unopened and autoclave.

If in doubt, remove a sample and examine by phase microscopy, Gram's stain, or standard microbiological techniques (see Chapter 4).

Mycoplasma

Cultures can become infected by mycoplasma from media, sera, trypsin, or the operator. They are not visible to the naked eye, and, while they can affect cell growth, their presence is often not obvious. It is important to test for mycoplasma at regular intervals (every 1–3 months) as they can seriously affect cellular biochemistry, antigenicity, and growth characteristics. Several tests have been proposed, but the fluorescent DNA stain technique of Chen is the most widely used (see Chapter 4).

4.3.6 Cross-contamination

This problem is dealt with in detail in Chapter 4. Its severity is often underrated, and consequently it still occurs with a higher frequency than many people admit. To avoid cross-contamination, do not share bottles of media or reagents among cell lines and do not return a pipette, which has been in a flask or bottle with cells, back to the medium bottle; use a fresh pipette. Do not share medium among operatives.

Animal cell culture is a precise discipline. Beware of those who say it is not, or it is 'magic', or due to 'green fingers'; they are not controlling all the variables. While this may not be fully attainable as yet, consistency can be achieved, and the following chapters are intended to indicate how best to control conditions within the present limitations of our knowledge. By adhering to their advice, cell culture, by the benefit of these authors' experience, and the addition of yours, will become an even more exact science.

References

1. Freshney, R. I. (1987). *Culture of Animal Cells: A Manual of Basic Technique*, 2nd edn. Alan R. Liss, New York.
2. Paul, J. (1975). *Cell and Tissue Culture*, 5th edn. Livingstone, Edinburgh.
3. Adams, R. L. P. (1980). *Cell Culture for Biochemists. Laboratory Techniques in Biochemistry and Molecular Biology*. (Series ed. T. S. Work and R. H. Burdon). Elsevier, Amsterdam.
4. Harris, C. C., Trump, B. F., and Stoner, G. D. (1980). *Normal Human Tissue and Cell Culture*. Methods in Cell Biology 21, Academic Press, New York.
5. Wakelam, M. J. O. (1985). *Biochem. J., 228*, 1.
6. Cunha, G. R. and Chung, L. W. K. (1981). *J. Steroid Biochem.*, **14**, 1317.
7. Kemp, R. B. and Hinchliffe, J. R. (1984). *Matrices and Cell Differentiation*. Progress in Clinical and Biological Research 15, Alan R. Liss, New York.
8. Frame, M. C., Freshney, R. I., Vaughan, P. F. T., Graham, D. I., and Shaw, R. (1984). *Br. J. Cancer, 49*, 269.
9. Freyer, J. P. and Sutherland, R. M. (1980). *Cancer Res., 40*, 3956.
10. Kruse, P. F., Jr and Miedema, E. (1965). *J. Cell Biol., 27*, 273.
11. Michalopoulos, G. and Pitot, H. C. (1975). *Fed. Proc., 34*, 826.
12. Chambard, M., Verrier, B., Gabrion, J., and Mauchamp, J. (1983). *J. Cell Biol., 96*, 1172.
13. Peehl, D. M. and Ham, R. G. (1980). *In Vitro, 16*, 526.
14. Piddington, R. and Moscona, A. A. (1967). *Biochim. Biophys. Acta, 141*, 429.
15. Kato, T., Fukui, Y., Turriff, D. E., Nakagawa, S., Lim, R., Arnason, B. G. W., and Tanaka, R. (1981). *Brain Res., 212*, 393.
16. Levi-Montalcini, R. (1964). *Science, 143*, 105.
17. Sporn, M. B. and Roberts, A. B. (1983). *Cancer Res., 43*, 3034.
18. Friend, C., Scher, W., Holland, J. G., and Sato, T. (1971). *Proc. Natl. Acad. Sci. USA, 68*, 378.
19. Raff, M. C., Miller, R. H., and Noble M. (1983). *Nature, 303*, 390.
20. Lechner, J. F., McClendon, I. A., Laveck, M. A., Shamsuddin, A. M., and Harris, C. C. (1983). *Cancer Res., 43*, 5915.
21. Freshney, R. I. (1992). *Culture of Epithelial Cells*. Wiley-Liss, New York.

22. Sattler, C. A., Michalopoulos, G., Sattler, G. L., and Pitot, H. C. (1978). *Cancer Res.*, **38,** 1539.
23. Pitot, H., Peraino, C., Morse, P., and Potter, V. R. (1964). *Natl. Cancer Inst. Monographs,* **13,** 229.
24. Littauer, U. Z., Giovanni, M. Y., and Glick, M. C. (1979). *Biochem. Biophys. Res. Commun.,* **88,** 933.
25. Land, H., Parada, L. F., and Wemberg, R. A. (1983). *Nature,* **304,** 596.
26. Periera-Smith, O. and Smith, J. (1988). *Proc. Natl. Acad. Sci. USA,* **85,** 6042.
27. Bishop, J. M. (1991). *Cell,* **64,** 235.
28. Marshall, C. J. (1991). *Cell,* **64,** 313.
29. Ham, R. G. and McKeehan, W. L. (1978). *In Vitro,* **14,** 11.

2

Towards serum-free, chemically defined media for mammalian cell culture

H. R. MAURER

1. Introduction

When a cell is removed from its original tissue or organism and placed in culture, the medium must provide all the environmental conditions that the cell has been exposed to *in vivo*; only then will it be able to survive, to proliferate, and to differentiate. The extracellular medium must meet the essential requirements for survival and growth (i.e. must provide nutritional, hormonal, and stromal factors). Among the biological fluids that have proved successful for culturing cells outside the body, serum has gained the most widespread significance. Although the requirements which lead to the development of complex, chemically defined media have been partially elucidated, 5–20% serum was, and still is, usually needed for optimum cell growth. For a variety of reasons it would be highly advantageous to eliminate the indefinable serum constituents, thus creating fully chemically defined media. Some of the various approaches and relevant problems have been described previously (1–7).

2. Role of serum in cell culture

2.1 Role of serum components

Serum is an extremely complex mixture of many small and large biomolecules with different, physiologically balanced growth-promoting and growth-inhibiting activities. Major serum components and their average concentrations are listed in *Table 1*, and further factors that have been found to support survival and growth of many mammalian cells in culture are listed in *Table 2*.

Serum is not simply the liquid formed by defibrination of plasma. In addition to various plasma proteins, peptides, lipids, carbohydrates, minerals, etc., it contains factors released during platelet aggregation (e.g. platelet derived growth factor (PDGF), transforming growth factor β (TGF-β))

Table 1. Major serum components and profile of fetal calf serum (Lindl and Bauer, ref. 12)

Component	Average concentration per litre	
Na^+	137	meq
K^+	11	meq
Cl^-	103	meq
Fe^{2+}, Zn^{2+}, Cu^{2+}, Mn^{2+}, Co^{2+}, VO_3^-, $Mo_7O_{24}^{6-}$	µg to ng	
SeO_3^{2+}	26	µg
Ca^{2+}	136	mg
Inorganic phosphorous	100	mg
Glucose	1250	mg
Nitrogen (urea)	160	mg
Total protein	38	g
Albumin	23	g
α-2-Macroglobulin	3	g
Fibronectin	35	mg
Uric acid	29	mg
Creatinine	31	mg
Haemoglobin	113	mg
Bilirubin (total)	4	mg
Alkaline phosphatase	255	U
Lactate dehydrogenase	860	U
Insulin	0.4	µg
TSH (thyroid stim. hormone)	1.2	µg
FSH (follicle stim. hormone)	9.5	µg
Bovine growth hormone	39	µg
Prolactin	17	µg
T_3 (triiodothyronine)	1.2	µg
Cholesterol	310	µg
Cortisone	0.5	µg
Testosterone	0.4	µg
Progesterone	80	ng
Prostaglandin E	6	µg
Prostaglandin F	12	µg
Vitamin A	90	µg
Vitamin E	1	mg
Endotoxin	0.35	µg

which are not present in plasma, and which induce the proliferation of some cells (e.g. fibroblasts) and may be cystostatic and induce differentiation of others (e.g. some epithelia).

While the nutritional role of serum is largely supplied by the formulation of the medium, there remain several functions which must, potentially, be replaced if serum is removed.

Table 2. Further serum components essential for cell survival and growth *in vitro*

Proteins
  Fibronectin
  α_2-Macroglobin
  Fetuin
  Transferrin
Growth factors
  Insulin-like growth factors I and II (IGF)
  Somatomedin A and C
  Multiplication stimulating activity
  Platelet-derived growth factor (PDGF)
  Epidermal growth factor (EGF)
  Fibroblast growth factor (FGF)
  Endothelial cell growth factor (ECGF)
Amines
  Amino acids
  Polyamines (spermine, spermidine)
Peptide
  Glutathione
Lipids
  Linoleic acid
  Phospholipids

Some of the major functions of serum are to provide the following:

- basic nutrients, in solution and bound to proteins
- hormones and growth factors stimulating cell growth and specialized functions
- attachment and spreading factors
- binding proteins (albumin, transferrin) carrying hormones, vitamins, minerals, lipids, etc.
- non-specific protection factors against mechanical damage; viscosity (shear forces during agitation of cell suspensions)
- protease inhibitors
- pH buffer

2.1.1 Growth factors

Most growth factors are present in serum in concentrations of nanograms/ml or below. Some of them are specific for cells at distinct differentiation stages, like the haematopoietic growth factors (colony stimulating factors). Others are less specific, even acting on different cell lineages. Epidermal growth factor (EGF), for example, promotes the proliferation of fibroblasts, epidermal and glial cells. Moreover, one cell type may be stimulated by different growth factors. Thus fibroblasts respond to fibroblast growth factor (FGF), EGF, PDGF, and somatomedins. Some of these growth factors may act synergistically. At present the physiological role of these numerous factors is not fully understood; for more on growth factors see refs 1 and 8.

2.1.2 Hormones

Among the hormones, insulin is essential for the growth of nearly all cells in culture. Due to its short half-life and sensitivity to inactivation by cysteine, it is usually added at a relatively high concentration and immediately before use. Glucocorticoids (hydrocortisone, dexamethasone) may stimulate or inhibit the proliferation of cells in culture depending on the cell type and cell density; they may moderate cell proliferation by altering responsiveness to growth factors. Some cell lines may require other steroid hormones (e.g. oestradiol, testosterone, progesterone). Thyroid hormones (mainly triiodothyronine (T_3) seem to support the growth of several cell lines (see *Table 5*).

2.1.3 Attachment and spreading factors

A major role of serum is to provide attachment and spreading factors, essential components of the extracellular matrix. Anchorage-dependent cells must first adhere to an adequate substrate, then spread out before they start proliferating and forming a monolayer. While these attachment factors are normally cellular in origin, trypsinized cells, not surprisingly, often lack the capacity to adhere following subculture. Serum provides a number of proteins which promote cell attachment. The role of these (e.g. fibronectin) is discussed in Section 3.3.8.

2.1.4 Binding proteins

Serum provides several so-called binding proteins the function of which is to carry essential factors of low molecular mass. Albumin carries vitamins, lipids (fatty acids, cholesterol) and hormones among others. Iron-saturated transferrin is essential for most cells in culture, and cells possess a specific transferrin receptor on their surface. While transferrin's conventional role is to supply essential iron to the cell in a non-basic form, it has been suggested that binding of transferrin to its receptor is, in itself, a mitogenic stimulus.

2.1.5 Lipids

Serum is a rich source for the various lipids cultured cells generally need for survival and particularly for growth. Cell lines differ in their requirements for essential fatty acids, phospholipids, lecithin, and cholesterol. For example, optimum fatty acid/cholesterol/albumin balance varies radically for different haemopoietic lineages and for different maturation stages within a given lineage. Prostaglandins E and $F_{2\alpha}$ have been found to be involved in cell proliferation in a synergistic manner with growth factors like EGF.

2.1.6 Minerals

The role of various inorganic tracer elements (Cu, Zn, Co, Mn, Mo, Va, Se) present in serum has been only partially elucidated, but many act as enzyme co-factors. For example, SeO_3^{2-} is required to activate distinct enzymes essential for metabolic detoxification. Selenium may also serve to inactivate free radicals.

2.1.7 Viscosity

Serum proteins contribute viscosity, which is no doubt important in protecting circulating cells from mechanical damage. In culture, particularly stirred suspension culture, and in the resuspension of trypsinized cells, it may perform a similar role.

2.1.8 Anti-proteases

Tissue cells, particularly endothelium and myeloid cells, may release proteases which require to be neutralized by serum anti-proteases. At first fortuitously, and now deliberately, serum is used to arrest proteolysis after trypsinization.

2.2 Advantages and disadvantages of the use of serum in cell culture

Anyone working on a usual laboratory scale with cell cultures and interested in controlling the culture conditions precisely may wish to avoid uncertainties imposed by batch variations of undefined serum components.

Two main reasons are particularly applicable to large-scale cultures, for example hybridomas producing monoclonal antibodies, fibroblasts producing interferon, or lymphomas producing lymphokines and other factors: the avoidance of foreign proteins which must otherwise be removed, and the costs of FCS.

2.2.1 Advantages of low-serum and serum-free media

There are several potential benefits from the use of low-serum or serum-free media:

(a) improved reproducibility between cultures, and avoidance of batch-to-batch variations of sera

(b) standardization of media formulations among different laboratories

(c) improved economy (particularly regarding FCS), although some of the constituents required to replace serum can partially or even completely negate this

(d) easier downstream purification of culture products

(e) less protein interference in bioassays

(f) avoidance of serum cytotoxicity

(g) no serum proteases that may degrade sensitive proteins

(h) prevention of fibroblast overgrowth in primary cultures

(i) selective culture of differentiated and functional cell types from heterogeneous populations of primary cultures

2.2.2 Disadvantages of using serum in cell culture

As a corollary to these advantages it should be noted that there are some pronounced disadvantages of using serum:

(a) For most cells, serum is not the physiological fluid which they contact in the original tissue except during wound healing and blood-clotting processes, where serum promotes fibroblast growth, but suppresses epidermal keratinocyte growth (1).

(b) The potential cytotoxicity of serum is often overlooked. Besides selective inhibitors, bacterial toxins and lipids, serum may contain polyamine oxidase which, upon reaction with polyamines (like spermine, spermidine) secreted from highly proliferative cells, forms cytotoxic polyaminoaldehydes (9). FCS and human pregnant serum, in contrast to horse serum, show a relatively high level of this enzyme. This may explain reports on the suppressive activity of these sera, and the preference of some workers for horse serum.

(c) The relatively large batch-to-batch variations of sera require extensive serum screening, which is time-consuming and costly, and even then can never ensure consistency.

(d) Serum may contain inadequate levels of cell specific growth factors, which have to be supplemented, and an overabundance of others which may be cytostatic.

(e) Risk of contamination (virus, fungi, mycoplasma).

(f) Periodic crises in the supply of serum.

(g) Sterilization problems associated with filtration of colloids and particulate content.

2.2.3 Potential disadvantages of serum-free media

Although the use of serum-free media has many advantages, there are also potential disadvantages:

(a) Although greater economy is possible with many cell lines, supplementing medium with hormones and growth factor can be as expensive as serum.

(b) Serum-free media are often highly specific to one cell type, so a different medium may be required for each cell line carried.

(c) Reliable serum-free media are, as yet, not readily available commercially, with a few exceptions.

(d) Serum-free conditions are generally less tolerant of physicochemical variations. Ultra-high-purity reagents and water are required, and a more critical control of pH and temperature.

(e) Cell growth is often slower, saturation density may be lower, and maximum generation number of finite cell line may be less.

3. Strategies for designing serum-free cultures and appropriate media

3.1 General procedures

Investigations to replace serum in cell cultures have followed three different strategies:

(a) *Analytical approach*. Isolation and identification of each of the various serum factors required for survival and growth; a laborious task, which proved unsuccessful in the long run, since most hormonal factors act synergistically and are present only in minute quantities in serum.

(b) *Synthetic approach* (Sato and co-workers, ref. 10). Supplementation of existing basal media by adding various combinations of growth factors performing serum functions: cell specific and non-cell specific hormones (including mitogens), binding proteins, attachment and spreading factors.

(c) *Limiting factor approach* (Ham and co-workers, ref. 11). Development of new formulations of existing nutritional media by lowering the serum concentration, until cell growth is limited, and then adjusting the concentration of each component of the nutrient medium (i.e. vitamins, amino acids, hormones, etc.) until growth resumes. A detailed procedure has been described by Ham (1).

3.2 Considerations prior to medium selection

3.2.1 Non-transformed and transformed cells

Some considerations are necessary prior to establishing serum-free cultures and selecting appropriate media. As a first step, one should specify to which category the cells to be cultured belong: normal finite lifespan, immortalized, or transformed (with autonomous growth control and malignant properties). Is the culture intended for growth (proliferation) and/or survival (with maintenance of viability and capacity to differentiate and produce desired proteins)? Is the cell population homogeneous or heterogeneous (e.g. are epithelial cells mixed with fibroblasts)? Usually primary explant cells from healthy tissues behave as *non-transformed cells* in the first generations and may absolutely require defined quantities of proteins, growth factors, and hormones, even in the most advanced media. When serum is added it is worthwhile considering whether FCS is absolutely necessary, or would new-born or calf serum serve as well? What is the minimum serum quantity with which the desired culture properties are still achieved?

Immortalized cells, spontaneously immortalized or transfected with viral sequences such as SV40-T antigen, adenovirus E1A, or whole EB virus, may be less dependent on serum, but still not fully autonomous, and will still require some serum growth factors.

Transformed cells, however, may grow without these added growth factors due to autocrine growth factor production, and in these cases added growth factors may even be inhibitory for growth. Yet some essential non-cell specific requirements may still have to be met, apart from those factors common to most media such as insulin, transferrin, selenite, lipids, and possibly some more specific growth factors and biomatrix proteins (see below).

3.2.2 Serum-free and chemically defined media: a distinction

Serum-free (SF) is not synonymous with 'chemically defined', as most SF media still contain some source of protein such as albumin or bovine pituitary extract. For most practical purposes albumin is a suitable high-molecular-mass protein, but its purity (absence of contaminants) may not be adequate (see Section 3.3.6). Completely chemically defined media have been developed, but in general cells tend to become 'fragile' and vulnerable during prolonged, agitated, culture unless protective biopolymers or synthetic polymers are added. Even media with apparently chemically well-defined constituents may reveal, upon careful analysis, trace amounts of contaminants (Cu^{2+} from NaCl, for example) (12). With the advent of gene cloning and production of functional mammalian proteins in bacterial or eukaryotic vectors it is now possible to obtain a pure protein additive, such as insulin or EGF, which is chemically defined. So there are now three classes of media: *serum-free* (but potentially otherwise protein supplemented), *protein-free* (no protein supplementation), and *chemically defined* (ultrapure with or without small molecular constituents, genetically engineered peptides, or proteins).

3.3 Factors and components required for serum-free media: roles and properties

3.3.1 Minerals and trace elements

The significance of various inorganic minerals and trace elements (Cu, Zn, Co, Mn, Mo, Va, Se, Ag, Al, Ba, Cr, Ge, Ti, Sn, Ni among others) for cell growth and function in serum-free media is poorly understood. Some of them may act as enzyme cofactors. For example, *selenite* is known to activate glutathione peroxidase, a key enzyme involved in the detoxifying metabolism of otherwise cytotoxic oxygen radicals. According to Hewlett (13) it is the only trace element with any obvious effect on cell growth, particularly human cells.

The role of trace elements should not be underestimated. Cleveland *et al.* (14) reported that they could even discount the addition of insulin, transferrin, albumin and liposomes for serum-free hybridoma cell culture, when they expanded the number of trace elements, thus creating a totally protein- and peptide-free medium. Besides the trace elements (in ionic form) Fe, Cu, Mn, Si, Mo, V, Ni, Sn, Zn, and Se they included Al, Ag, Ba, Br, Cd, Co, Cr, F, Ge, I, Rb, and Zr. They regarded the elimination of albumin, transferrin, and insulin as an important step since these substances are potential sources of artifacts in the use of monoclonal antibodies to identify cell surface antigens. More recently Darfler (15) refined the protein-free medium of Cleveland *et al.* (14) for the growth of hybridoma and other immune cells by complexing 18 trace elements with EDTA and substituting transferrin for sodium nitroprusside.

3.3.2 Vitamins

Ascorbic acid (vitamin C) and α-tocopherol (vitamin E) are included in some SF media, mainly as antioxidants (6). Other vitamins such as those of the B series (folic acid, biotin, pantothenate) are common components of most media.

3.3.3 Carbohydrates

Glucose generally provides the major energy supply and serves as the predominant carbohydrate nutrient. Lactate accumulation from high-density culture may acidify the medium. Substitution of mannose decreases the rate of hexose utilization, which correlates with decreased lactate formation (16). However, total hexose substitution may change the antibody glycosylation pattern in hybridomas (17), hence only partial substitution is recommended.

Pyruvate is added to some media as an immediate precursor for biosynthetic processes requiring acetyl-CoA.

3.3.4 Lipids

Many cells require an exogenous source of lipids for optimal *in vitro* growth, because biosynthesis of membranes from acetyl-CoA requires a lot of energy.

It is easier and more economical for cells to use intermediate lipid precursors such as cholesterol, long-chain fatty acids, and glycerides. However, an overload of lipids can inhibit growth, so the lipid composition must be carefully balanced. In fact, 'experience has shown that the lipid content of the medium requires the most careful optimization' (13).

Lipids present in serum are mainly either bound to albumin or complexed with proteins to form lipoproteins. Since lipids are generally only slightly soluble or even insoluble in water, special procedures are necessary when adding them to culture media. Albumin may carry undefined lipids, qualitatively and quantitatively, and should therefore be used in a defatted form. However, the procedure of defatting should be standardized for reproducible results; unfortunately, this is not usually the case. Methods for adding lipids to media without albumin involve

(a) solubilization in water-miscible organic solvents such as ethanol, acetone, DMSO

(b) liposome preparations (with phospholipids) in aqueous suspensions

(c) microemulsions in water-soluble suspensions

(d) cyclodextrine preparations (see Section 3.3.6)

The disadvantages of the first method are that the solvents might be cytotoxic and that dilution with the aqueous medium might precipitate the lipids. For the other two methods, reproducibility of preparation and long-term stability (oxidation, aggregation) of pre-formed liposomes or microemulsions can present problems.

Soybean lipids, containing about 40% triglycerides, 60% phosphatides (lecithin, phosphatidylethanolamine, phosphatidylinositol), and sterol traces, are prepared as liposomes at 2 mg lipid/ml aqueous dispersion or mixed with BSA at 20 mg/ml for stabilization. These preparations, commercially available, are supplements for Iscove's modified Dulbeccos's medium (IMDM), which has been successfully used for the culture of hybridoma, lymphoid, and other haemopoietic cells (18). However, soybean lipids usually tend to aggregate on storage and may lose components during sterile filtration.

More recently, water-soluble *mammalian lipid* preparations (known as Ex-Cyte, Miles Inc.) presenting a balanced lipid composition of cholesterol, triglycerides, fatty acids, phosphatides, and lipoproteins have been tested and found not only to support antibody production in bioreactors (19) but also interleukin-1 production by A 431 human keratinocytes (13) and growth of L 929 mouse fibroblasts (20).

A *protein-free lipid microemulsion* has been prepared from purified synthetic lipids to yield a homogeneous, water-soluble and stable mixture that can be sterile-filtered (21). The sonicate essentially contains cholesterol and esters thereof, phosphatidylcholine diglycerides, sphingomyelin, α-tocopherol, and Tween 80. This supplement proved suitable for serum and protein-free cultures of CHO cells, murine hybridomas, transformed T-lymphoblasts, and human skin keratinocytes.

Known *lipoproteins* (HDL and LDL) have also been tested for cell growth promotion in SF media, since they can deliver cholesterol to the cells (6, 22). Concentration-dependent stimulatory and inhibitory effects and an apparent species specificity of lipoproteins from non-human sources have so far complicated their application. Individual protein and lipid fractions still await further characterization (23).

Oxidation of lipids in medium may be a problem (13), although addition of α-tocopherol and selenite (see below) can help to prevent oxidative processes and eliminate cytotoxic peroxides and oxygen radicals, respectively.

3.3.5 Amino acids

Amino acid accumulation and metabolism during prolonged cultures of hybridomas require careful attention. While alanine, glycine, and glutamic acid were found to accumulate in mouse–mouse hybridoma cultures, glutamine, leucine, isoleucine, valine and tryptophan rapidly disappeared from the culture media (24). Obviously culture conditions (batch, chemostat, or perfusion bioreactor system (see Chapter 3)) influence the amino acid consumption rates. In general, glutamine, methionine, and serine are the amino acids that pose relatively high nutrient limitations (25). Rapid exhaustion of *glutamine* was found to be responsible for diminished cell viability and productivity, although the glutamine deamidation rate could be reduced by adding glycyl-glutamine (16). In any case a careful amino acid analysis of both the input and spent medium is recommended to determine the requirements for cell growth and productivity. It is important to recognize that such an amino acid utilization profile is generally unique for each cell type, culture conditions, and biological product (16).

3.3.6 Proteins and other polymers with multiple functions

i. *Albumin*

This major serum protein serves various functions *in vivo* as well as *in vitro* in SF media, such as:

- buffer for pH changes
- carrier for hormones, growth factors, lipids (cholesterol, fatty acids, and many others) lipophilic vitamins, drugs, minerals, etc.
- osmotic pressure regulator
- protecting agent against mechanical damage (including shear forces)

The quality of commercial albumin preparations varies considerably, although it usually meets the requirements of the European Pharmacopeia (including absence of hepatitis virus). Bovine serum albumin (BSA) is available as crude fraction V globin free, fatty acid free, and globin and fatty acid free. Special procedures have been developed to strip essential lipids from BSA. When tested on rat tracheal epithelial cells, BSA was found to stimulate or to inhibit cell proliferation, depending on the type and concentration of BSA used and on the presence of cholera toxin (26).

ii. *Synthetic polymers as albumin substitutes*

The frequent requirement for albumin has been mainly attributed to a non-specific protective effect of the biopolymer on the cells (27). In attempts to substitute other polymers for albumin, polyethylene glycol 20 000 or carboxymethyl cellulose has been included in suitable supplemented media (19, 28). Furthermore, polyvinylpyrrolidone (29) and pluronic polyols (30, 31) have proved able to promote growth of various mammalian and even insect cells. Thus, in long-term hybridoma cell cultures the addition of 0.01% pluronic F-68 yielded relatively high antibody productivity (32).

Biopolymers and synthetic polymers apparently protect cells from mechanical damage and provide optimal colloidal osmotic pressure and viscosity (33, 34).

α-Cyclodextrin (35) and β-cyclodextrin (36) have been introduced as albumin substitutes, particularily for lipid transportation.

iii. Transferrin

This serum β-globulin serves as an iron (Fe^{3+})-transporter and, like albumin, as a detoxifying protein. Transferrin is essential for the growth of almost every cell line in long-term SF cultures. It may be replaced by 0.05–0.5 mM Fe^{3+} salts following complexing with chelating agents such as citrate, ascorbic acid, iminodiacetic acid, glycylglycine (32), or phosphate salts (37).

It appears that only Fe^{3+} requires transferrin for transport into the cells, while Fe^{2+} is taken up independently (13). Human transferrin is effective for all hybridoma types, although a species-specificity of transferrin has been noted (38). It should be mentioned that the quality of transferrin, like that of albumin, varies considerably according to its preparation. It should also be tested for virus contamination.

iv. α2-Macroglobulin

α2-Macroglobulin (725 kDa) is a protease inhibitor that inactivates enzymes such as trypsin. Quantitatively it is one of the major serum proteins, at concentrations in healthy individuals of 2–4 g/l. It is also an outstanding transporter for various cytokines (39).

3.3.7 Hormones and growth factors

The requirement for hormones such as estradiol, thyroxine, cortisol (hydrocortisone), dexamethasone (synthetic fluoridated hydrocartisone), progesterone, glucagon, and last but not least, insulin, varies with the cell type cultured. The optimal concentration of the growth and/or differentiation promoting factors must be determined for each cell line separately. For example, cortisol was found to be stimulatory for WI-38 cell growth, but inhibitory to BHK cell proliferation (16).

Proliferation and differentiation of many mammalian cells in culture are dependent on the interaction of *polypeptide growth factors* with their corresponding cell membrane receptors. The availability of highly purified and recombinant growth factors facilitates the formulation of tailored SF media with these essential growth factors. Thus recent SF media include recom-

Table 3. Growth factors for serum-free media

Growth factor (GF)	Target cells (general)	Target cells (special)	Recommended concentration
EGF	Ectodermal, mesodermal cells	Keratinocytes, fibroblasts, chondrocytes, etc.	1–20 ng/ml
bFGF	Mesodermal, neuroectodermal cells	Endothelial cells, fibroblasts, chondrocytes, myoblasts, etc.	0.5–10 ng/ml
FGF	Mesodermal, neuroectodermal cells	Fibroblasts, vascular endothelial cells	1–100 ng/ml
ECGF (= acidic FGF)	Endothelial cells		1–3 mg/ml
Insulin	Most cells		1–10 µg/ml
IGF-I (= Somatomedin C)	Most cells		1–50 ng/ml
PDGF	Mesenchymal cells	Fibroblasts, muscle, glial cells	1–50 ng/ml
NGF (7S, −β, 2.5S)	Sensory, sympathic cells	Neurones (Glial cells)	5–100 ng/ml
TGF-α	As for EGF		1–10 ng/ml
TGF-β	As for EGF	Stimulatory for mesenchymal, inhibitory for ectodermal cells	0.1–3 ng/ml

binant human epidermal growth factor (EGF) for human epidermal keratino-cyte and normal human endothelial cell cultures, basic fibroblast growth factor (bFGF) for normal human dermal fibroblast and epidermal melanocyte cultures. *Table 3* summarizes the growth factors and their target cells. Some so-called growth factors may be inhibitory, however, like EGF toward various lung carcinoma cell lines (40), so it is more useful to think of positively acting and negatively acting growth factors, particularly as the same factor may act in different ways on different target cells. It is beyond the scope of this review to cover all the more or less characterized growth factors that have been applied in SF media.

Some extracts upon purification may reveal factors that are already known: for instance, bovine pituitary extract (BPE) has been found necessary for mammary and prostatic epithelial cell growth. Prostatrophin has been iso-lated from BPE as a protein factor very similar to acidic FGF (41) which, in addition, is a growth factor for a mouse keratinocyte cell line (42). bFGF can also replace BPE for growth of human breast epithelial cells in SF medium (43).

Insulin is a general supplement to SF media, although in concentrations about 1000-fold higher than the levels found in human serum. It has been suggested that insulin is rapidly degraded in culture media at 37°C (13) and that it may be mimicking insulin-like growth factors (IGFs, somatomedins). Some cell lines do not require insulin, e.g. mouse–mouse hybridomas (32).

3.3.8 Cell–substrate adhesion, extracellular matrix (biomatrix), attachment, and spreading factors

Most non-transformed cell types must attach to a solid substrate in order to proliferate, and the interaction of these cells with the substrate or extra-cellular matrix (ECM) determines many of their properties. Apart from haemopoietic cells, only transformed cells show the capacity to multiply without attachment.

Cells do not adhere directly to glass or specially treated plastic (so-called tissue culture plastic with modified surface charges), but rather bind via specific cell surface receptors to attachment proteins adsorbed to these surfaces. Serum generally provides some of these adhesion factors that are, otherwise, essential components of the ECM.

Cells may produce their ECM proteins themselves. Alternatively the growth surface may be coated with the ECM isolated as biomatrix from growing cultures, or with the ECM component proteins, before cell seeding. Prominent ECM proteins are *fibronectin, laminin, vitronectin*, and different *collagen* types (*Table 4*). The receptor binding domain of fibronectin has been localized to a tripeptide (L-Arg–Gly–Asp (RGD)), but the most active peptide in different assays is a fibronectin pentapeptide. In addition to the natural attachment and spreading factors contained in the ECM, *gelatin* and *poly-D-lysine* have also been shown to act in a non-specific manner. Some of these factors are available in precoated dishes.

The ECM plays an important role on the proliferation and differentiation of many epithelial cells. Laminin and fibronectin are intrinsic mitogens and differentiation factors for the maintenance of normal cellular functions. The addition of these factors to SF media permits the establishment of an *in vitro* environment that induces or supports the survival, differentiation and matur-ation of cells for study of their morphogenetic or biochemical functions. Thus, while isolated hepatocytes will only survive on plastic for one week, they may survive up to one month on collagen, and on ECM (biomatrix) for more than 6 months. Cells on plastic tend to de-differentiate: epithelial cells assume fibroblastoid cell shape and make fibroblast-specific proteins; culture on col-lagen or biomatrix maintains the start of differentiation, however, and re-duces the necessity to add various growth factors or serum.

Fibronectin mainly promotes attachment and differentiation of mesenchyme-derived cells including fibroblasts, sarcoma, granulocytes, kidney epithelium, adrenal cortex epithelium, CHO cells, myoblasts etc., whereas laminin mainly enhances attachment of cells of ectodermal and endodermal origin, such as epidermal, bronchial and secretory epithelial cells, and neuronal cells, hepato-

Table 4. Attachment and spreading factors for monolayer cells

Factor	Source	Target cells	Recommended coating concentration
Collagen type I	Rat tail	Most normal and transformed mammalian cells, e.g. fibroblasts, hepatocytes, epithelial, endothelial, muscle, lung, neuronal cells, etc.	$5-10\,\mu g/cm^2$
Collagen type I	Bovine skin	As above, particularly for primary cultures of epitheloid cells	$5-10\,\mu g/cm^2$
Collagen type IV (Matrigel)	Mouse EHS sarcoma cell line	All epithelial, endothelial, muscle and neuronal cells	$5-10\,\mu g/cm^2$
Fibronectin	Human bovine, rat plasma	Mesenchymal cells: fibroblasts, sarcoma; endothelial, epithelial, muscle, neuromal cells etc.	$1-5\,\mu g/cm^2$
Laminin (Matrigel)	Mouse EHS sarcoma cell line	Ectodermal and endodermal cells: epithelial, endothelial, muscle, neuronal cells, hepatocytes	$1-2\,\mu g/cm^2$
Gelatin	Bovine porcine skin	Many cell types	$100-200\,\mu g/cm^2$
Poly-D-lysine	Synthetic	Many cell types	$2.5-5\,\mu g/cm^2$

cytes, etc. In fact, laminin suppresses fibroblast overgrowth in primary cultures of epithclioid cells.

For procedures to coat tissue culturc plastic with poly-D-lysine, collagen, or fibronectin, see Section 6.

3.4 Assays to determine the best media

The complex growth requirements of many mammalian cells impose several problems for the adequacy of assay systems for the determination of optimal growth media. The reader is referred to a thorough review by Ham (1). Generally, a long-term (several days) cell multiplication assay is to be preferred to a short-term assay using [^{3}H]-thymidine uptake (for several hours) as the sole means to measure cell proliferation, since several artifacts are known that may simulate cell proliferation (44). Nevertheless [^{3}H]thymidine methods are often used for screening, but the results should be cross-checked by a clonal growth assay or at least by cell growth curve analysis.

3.5 Media for low-serum and serum-free cell cultures

A variety of defined media have been derived from Eagle's minimal essential medium (MEM) during the last three decades, including Dulbecco's enriched modification (DME), and rather complex media such as Ham's F12, CMRL 1066, RPMI 1640, McCoy's 5A, and Iscove's modified Dulbecco's medium (IMDM). All are now commercially available, and their compositions can be found in the commercial catalogues. For serum-free formulations often a 1:1 (v/v) mixture of DME and F12 is used to combine the qualitative richness of F12 (trace elements, more vitamins) with the higher nutrient concentration of DME. For haemopoietic cells, including hybridoma cultures, IMDM proved valuable when supplemented with transferrin, BSA, and several lipids. Other blood cells including leukaemic cells have been successfully grown in RPMI 1640. CMRL 1066 is particularly rich in nucleosides and some vitamins. The MCDB series was developed by Ham and co-workers especially to meet the different requirements of non-transformed cells.

The most frequently used media may be categorized (11) as:

- *media for permanent lines with some serum or protein supplementation*: MEM, DME, α-MEM, McCoy's 5A, RPMI 1640

- *media for permanent lines with purified protein or hormone supplementation*: F10, F12, DME

- *media for permanent lines in monolayer without protein supplementation*: CMRL 1066, MCDB 411, DME

- *media for clonal growth of permanent lines without protein supplementation*: F12, MCDB 301, DME, IMDM

- *media for non-transformed cells*: DME, IMDM, MCDB 104, 105, 202, 401, and 501

Studies with various mammalian cells have documented the tissue- and species-specificity of their requirements for hormones and nutrients. Chondrocytes and fibroblasts require a lipid supplement but epithelial cells do not; chondrocytes require FGF; keratinocytes, bronchial, mammary and prostatic epithelial cells require EGF but not FGF; ethanolamine is needed by keratinocytes and bronchial epithelial cells, etc. (1). Some media have been found generally useful for cells from several different species (e.g. 199, DME, F12K, and MCDB 202) but the following preferences have been suggested for individual species:

- *for human and monkey cells*: 199, 5A, RPMI 1640, CMRL 1969, MCDB 104, and MCDB 202

- *for rat and rabbit cells*: 5A, F12, and MCDB 104

- *for non-transformed mouse cells*: DME, CMRL 1415, MCDB 202, and MCDB 401

- *for chicken cells*: 199, DME, F12K, and MCDB 202

3.6 Selection of hormones and growth factors

It follows from the above discussion and from an inspection of *Table 5* that most serum-free media include insulin, transferrin, and selenite. Insulin is added at relatively high concentrations (1–10 µg/ml), since it rapidly loses bioactivity in the usual serum-free culture media. Insulin may also mimic the activity of related peptides due to peptide sequence homologies with insulin-like growth factors and somatomedins. Transferrin should be added Fe^{2+}-saturated. Among the trace elements, selenite seems to be important for the activation of glutathione peroxidase, a key enzyme involved in oxygen metabolism and detoxification. Glucocorticoids improve cloning efficiencies of glial cells, fibroblasts, chondrocytes, keratinocytes, bronchial, mammary, and prostate epithelial cells. Exocrine cells may also require their specific hormones (e.g. oestrogens, prolactin, progesterone for mammary cells). Some cells benefit from supplementation with T_3, and EGF is often added to many different types of epithelial cells.

In general, no clear rules can be established that lead to a generally applicable serum-free medium. So, in many cases trial and error may be the only method to determine the best medium and to select the appropriate supplements. A well-supplemented starting medium for monolayer cultures may contain in DME/F12 (1:1, v/v): insulin 5 µg/ml, transferrin 5 µg/ml, selenite 30 nM, EGF 25 ng/ml, albumin (fatty acid- and cholesterol-free) 1 mg/ml, and fibronectin 1 µg/ml. Systematic omission of each component may identify the essential factor(s). The first-limiting-factor approach by Ham and co-workers (1) may eventually lead to optimized concentrations of the essential components. However, although straightforward and intellectually satisfying, this approach takes a lot of time and effort.

To help the beginner, a list of media for various serum-free cell cultures with various supplements has been compiled (*Table 5*). To start with, a medium may be selected that is most closely related to one already found to 'work' with similar cells. In general, media formulated for one cell type will support growth of other lines and primary cultures of independent origin if they are of the same cell type. For instance, for leukaemic cells try medium RPMI 1640 or CMRL 1066, possibly mixed with F12 (1:1, v/v).

However the reader wishing to set up a SF medium should, in any case, consult the original literature for further details not shown in *Table 5*.

3.7 Designing serum-free and chemically defined media for producer cells in large-scale cultures

3.7.1 Hybridoma cell cultures for monoclonal antibodies

Before designing an SF medium, one has to decide whether the medium should be optimized mainly for cell proliferation or for productivity. Most basal media, generally used for animal cell cultures, are optimized for cell

Table 5. Serum-free and chemically defined media for mammalian cell cultures

Mammalian cell type	Reference	Basal medium	Insulin ($\mu g/ml$)	Transferrin ($\mu g/ml$)	Selenite	Additional supplements (per ml or final conc.)
A. *Normal, non-transformed cells*						
Human oral epithelial cells	57	MCDB 153	5	–	–	HC 0.5 µg EGF 10 ng Ca^{2+} 0.15 mM
Human granulocyte colony-forming cells	58	IMDM	2 U	300	–	rh-G-CSF 20 ng BSA 10 mg Cholesterol 48 µg Phosphatidyl choline 80 µg (HC 50 µM)
Murine thymic epithelial cells	59	DME/F12 1:1	3	25	–	Ca^{++} 21 µg final EGF 100 ng HC 0.5 µg CT 10 ng
Mouse prostatic epithelial cells	60	DME/F12 1:1	10	10	–	BSA V 10 mg EGF 10 ng CT 10 ng DHT 1 µg PRL 5 µg Putrescine 10 µg bFGF 200 µg Trace elements
Human endothelial	61	IMDM	5	5	60 nM	ECM coat FGF 1 ng 5-ME 50 µM
Dog kidney cells MDCK	62	DME/F12 1:1	5	5	50 nM	HC 50 nM T_3 5 pM PGE_2 25 ng

No.	Cell type	Medium				Supplements
63	Rat mammary epithelial cells	DME/F12	10	5	–	PRL 1 µg, Progesterone 1 µg, HC 1 µg, EGF 10 ng, BSA-FAF 1 mg, Vit. C 5 µM
64	Human placental trophoblasts	HB 102	15	10	–	BSA 0.7 mg, LDL 2 µg
65	Human malignant and non-malignant squamous cells	PF 86–1	–	–	+	Vit. C/0.15 mg, Calciferol 80 ng, Trace elements
66	Human foreskin fibroblasts Rabbit aortic smooth muscle cells	DME	10	–	–	BSA 1 mg, casein 1 mg, PDGF } various conc., ECGF } various conc.
67	Rat cardiac muscle cells	MEM Eagle, Earles salts	+	+	+	EGF 10 ng, bFGF 50 ng, Norepinephrine 10 nM, T_3 10 nM, BSA 1 mg, Vit. C/20 µg
68	Human normal and cystidfibrosis epithelial cells	LHC-8e				FN coat
69	Bovine follicular thecal cells	DME/F12 1:1	10	5	–	Collagen gel matrix, BSA-FAF 5 mg, EGF 10 ng, T_3 2.5 ng, Dex. 40 ng
70	Hamster fibroblasts	DME/F12 1:1		+		

Table 5. *Continued*

Mammalian cell type	Reference	Basal medium	Insulin (μg/ml)	Transferrin (μg/ml)	Selenite	Additional supplements (per ml or final conc.)
B. Transformed and tumour cells						
Human ALL CD4$^+$ cells	71	RPMI 1640	5	5	5 ng	BSA 0.1 mg
Human leukaemic CFU (AML)	72	ASF 101 (DME deriv.)	5	5	–	BSA deionized 1 mg, Cholesterol 7.8 μg, rh-G-CSF 100 U, rh-GM-CSF 10 ng, rh-IL-3 20 U
Human multiple myeloma CFU	73	DCCM 1	–	25	–	BSA V 2 mg
Human ovarian carcinoma cells	74	KSLMS	1	1	–	BSA 0.5 mg
Human hepatoma cells	75	RPMI 1640	–	–	30 nM	–
Human colon carcinoma cells	76	McCoy's 5 A	20	4	10 nM	EGF 10 ng, HC 2 μg, T$_3$ 0.4 nM
Human prostate epithelial cells	77	P4	5	–	30 nM	Ca^{2+} 0.5 mM, HC 0.28 μM, PE 0.5 μM, BSA 250 μg, BPE 0.5%, CT 0.1 nM
Rat renal glomerular cells	78	DME/F12 1:1 (DHFs)	5	5	5 ng	BSA-FAF 1 mg, LA 5 μg, EGF 10 ng, Collagen I/IV pre-coat

	Cell type	Medium				Supplements
79	Human lung adenocarcinoma cell lines	RPMI 1640 (ACL 3)	20	10	25 nM	EGF 10 ng HC 50 nM BSA 5 mg T_3 100 pM Na Pyr 0.5 mM Glu 2 mM Collagen and FN pre-coat
80	Human primary non-small cell lung carcinoma cells	RPMI 1640 (ACL 4)	20	10	25 nM	EGF 1 ng HC 50 nM T_3 100 pM BSA 2 mg Na Pyr 0.5 mM Glu 2 mM EA 10 μM PE 10 μM Collagen pre-coat
81	Human mammary carinoma cells (hormone dependent and independent)	DME/F12 1:1 (HBCA)	3	25	—	Dex 10 nM E–2 10 nM T_3 10 nM PGF 2α 100 ng FN 100 ng Vitrogen pre-coat
82	Rat mammary carcinoma cells	MCDB 202 (54–24)	5	5	—	EGF 10 ng HC 1 μg PRL 10 ng LSB 5 μg (including Vitamin E and other lipids)

Table 5. *Continued*

Mammalian cell type	Reference	Basal medium	Insulin (µg/ml)	Transferrin (µg/ml)	Selenite	Additional supplements (per ml or final conc.)
C. Hybridoma cells						
Human and murine hybridomas T-lymphoblastoid cells, normal T-lymphocytes	15	ABC	–	–	+	18 trace elements EDTA 5 µg MOPS 3.1 mg Nitroprusside 6 µg Progesterone 6 ng Defin. lipids 1 µg Tocopherol 0.15 µg
Mouse–human hybridomas	83	RPMI 1640/ DME/F12 2:1:1	10	10	20 nm	EA 50 µM VLDL 2.5 µg BSA 5 mg
Mouse–mouse hybridomas	84	DME	10	10	1 nM	2-ME 50 µM EA 20 µM Vit. C 20 µg
Mouse–mouse hybridomas	32	MEM (amino acid modif.)	–	–	100 nM	Dex. 1 ng Vitamin E 20 µM Pluronic F 68 0.025% EA 10 µM 2-ME 10 µM Fe$_2$(SO$_4$)$_3$ 100 µM Glycylglycin 11 mM Vitamin C 20 µM pO$_2$ 10%

BPE: bovine pituitary extract; BSA: bovine serum albumin; bFGF: basic fibroblast growth factor; CT: cholera toxin; Dex: dexamethasone; DHT: 5α-dihydrotestosterone; EA: ethanolamine; ECM: ECGF: endothelial cell growth factor; extracellular matrix; EGF: epidermal growth factor; E-2: oestradiol; FAF: fatty acid free; FN: fibronectin; Glu: glutamine; HC: hydrocortisone = cortisol; LA: linoleic acid; LDL: low density lipoprotein; LSB: liposome B; 5-ME: 5-mercaptoethanol; NaPyr: sodium pyruvate; PDGF: platelet derived growth factor; PE: phosphatidylethanolamine; PGE: prostaglandine E; PGF: prostaglandine F$_{2α}$; PRL: prolactin; T$_3$: triiodothyronine; VLDL: very-density lipoprotein
See Appendix A2 for list of suppliers of growth factors and serum-free-medium.

proliferation, e.g. DME + Hams' F 12, 1 : 1 (DF medium) or RPMI 1640 + DME + Ham's F 12, 2 : 1 : 1 (RDF medium). This is still true after enrichment (E) with amino acids, vitamins (ERDF medium) and growth factors such as insulin (I), transferrin (T), ethanolamine (E), and selenite (S) to give the ITES-ERDF medium for SF-culture of human–human hybridomas (23). Transferrin stimulates proliferation of various types of cells, whereas lactoferrin (L) specifically promotes proliferation and productivity of lymphocytic cells (45). Since hybridoma cells are not necessarily dependent on insulin, the modified ITES-ERDF medium was found to yield best results for proliferation and IgM production of human hybridoma cells (46). Murakami (23) pointed out that serum contains both stimulatory and inhibitory factors for proliferation and Ig production of hybridomas.

SF media are clearly preferable to serum-supplemented media in order to optimize proliferation and productivity of hybridoma cultures (47), and simplify downstream purification of proteins produced.

The monoclonal antibody productivity of high-density human hybridoma cultures can be significantly improved by the addition of various exogenous proteins, such as very-low-density lipoproteins from chicken egg yolk or caseins. Moreover, cellular Ig production stimulating factors (IPSFs), isolated and purified from the culture supernatants of lymphoblastoid cells, were found to promote proliferation of several hybridomas including IgG and IgM producers (48, 49).

Medium buffering with HEPES can be substituted by lowering the Ca^{2+} concentration to 0.1 mM and increasing the phosphate concentration to 10 mM (32). 14 mM phosphate will also increase antibody productivity of human–human hybridomas (50).

Merten and Litwin (32) have compiled and discussed, in detail, the various SF media and homogeneous and heterogenous, immobilized bioreactor systems developed for optimized *hybridoma cell cultures*. They concluded that

(a) Only four basal media are generally used either singly or in combination: DMEM, IMDM, F-12 and RPMI 1640.

(b) Supplements to these media almost always include insulin, transferrin, ethanolamine, and often BSA.

(c) Additional supplements such as hormones, growth and differentiation factors may be required for many hybridomas, the proliferation and productivity of which is still not yet satisfactory.

In addition, it must be presumed that there are a number of formulations used by the biotechnology industry and subject to commercial secrecy.

3.7.2 Cell cultures for the production of recombinant proteins (biopharmaceuticals)

During the last few years we have seen a tremendous development in the biotechnological manufacture of recombinant biopharmaceuticals from mammalian cells in large-scale cultures. Various production processes (batch, chemostat, perfusion culture on micro carrier beads, etc.) have been comparatively evaluated (see Chapter 3). At present fed-batch suspension cultures in totally serum-free media seem to be favoured, provided that cell densities and nutrient feedings are optimized (51). Thus a couple of already indispensable protein drugs have been produced on a large scale mainly in the Chinese hamster ovary line CHO, the baby hamster kidney line, BHK, the mouse mammary tumour line, C 127, and rodent myeloma lines. These drugs include, among others, tissue plasminogen activator (tPA), human growth hormone (HGH), erythropoietin (EPO), chimeric monoclonal antibodies, blood clotting factor VIII, interferons (IFN), interleukin 2 (Il-2), and hepatitis B surface antigen (HBS). The problems involved in large-scale serum-free cultures have been discussed by Broad *et al.* (51), Jayme (16), Merten and Litwin (32).

4. Practical suggestions for the preparation and handling of serum-free media

4.1 General rules and precautions

4.1.1 Quality of water, chemicals, and biologicals

The highest possible purity should be attained, since serum proteins that could bind toxic contaminants are not present in the medium. For example, 0.1 p.p.b. cadmium ions and 1 p.p.m. humic acid can be found as inhibitory contaminants in double-distilled water (52).

4.1.2 Bacterial endotoxin

Generally the lowest possible levels should be attained.

4.1.3 Quantity of basal medium at any one time

Generally a 2-week supply kept at 4 °C is recommended to avoid decomposition of constituents otherwise protected by the detoxifying components of serum. Plastic flasks may be preferred for storage to avoid adsorbance of protein or peptide factors to glass. Alternatively, siliconized glass may be used.

4.1.4 Storage and addition of growth factors and other labile components

Store, sterile, at the manufacturer's recommended temperature in small aliquots (e.g. 20 μl/ml basal medium to be prepared) thaw and add to medium immediately before use. Make sure that handling and sterilization of your stock solutions and media does not remove or inactivate essential nutrients and growth factors. Specifically:

(a) *Transferrin* and some *hormones* were found to be inactive in fully prepared media after 48 h even at 4 °C (53).

(b) Fe^{2+} may be oxidized to Fe^{3+}, may precipitate, and will consequently be removed during sterile filtration. Hence add Fe^{2+} just before medium sterilization.

(c) Concentrated Ca^{2+} may deteriorate due to the precipitation of Ca_3 $(PO_4)_2$. Ca^{2+} solutions should be stored phosphate-free below pH 7.0.

(d) High *cysteine* concentrations may inactivate insulin, so the cysteine concentration should be reduced to 10 μM. Alternatively, add a different reducing agent (e.g. dithiothreitol or thioglycerol).

(e) Always prepare *microemulsions* or *liposomes* freshly. Unsaturated fatty acids may be oxidized; microemulsions and liposomes may be partly lost during sterile filtration. Check cytotoxicity of liposome stabilizer.

4.1.5 Proteases and protease inhibitors

Cells may release proteases during the subcultivation procedures. A mixture of protease inhibitors like aprotinin and soybean trypsin inhibitor may be advisable, especially after trypsinization, in order to prevent damage to the cells or medium components by trypsin or serine proteases.

4.1.6 Antibiotics

Some common antibiotics may prove to be cytotoxic, even at high serum concentrations (54). Moreover, longterm use of antibiotics may select resistant microorganisms or induce cryptic contamination. The use of antibiotics should therefore be kept to a minimum.

4.1.7 Sterile filtration of stock solutions

This may reduce the growth factor concentration and, if it is necessary, it should be done with filters of low dead space and minimal protein binding, e.g. Millipore GS.

4.1.8 Stock solutions

For some complex media (e.g. the MCDB series) *separate* stock solutions must be prepared and combined just before filtration. *Unstable* stock solutions (e.g. glutamine, Fe^{2+}) should be stored frozen and added immediately before use. Otherwise most sterile working-strength media may be stored at 4°C for several weeks.

Care should be taken when using *concentrated* stock solutions that all constituent remain in solution during sterile filtration. In addition, if several stock concentrations are to be combined they should be diluted as they are combined, not mixed as concentrates.

4.2 Adaptation of cells from serum-containing to serum-free and chemically defined cultures

Although some transformed cells can grow in SF media without prolongation of the lag phase of growth following seeding, most cells show diminished survival by 50% and more within 2–3 days, unless they are gradually adapted to SF media. Adaptation may have to be done slowly and gradually, weaning off serum and changing to a new medium formulation, to avoid selection of an undesirable subpopulation of cells and loss of the functionally active cells. Strategies have been developed for weaning from serum-containing to scrum-free media (13): *Protocol 1* may provide a general guide-line.

Protocol 1. Strategy for the adaptation of cells to serum-free culture

1. Select an assay for cellular behaviour (cell proliferation or productivity).
2. Reduce serum concentration gradually, for example from 10% to 3% to 1% to 0.5% to 0.1% to 0%. At one step cell growth may cease, so always preserve previous stages in parallel cultures and in frozen stocks.
3. For adherent cells: provide a substrate for the cells to attach by coating the vessel surface with poly-D-lysine, collagen, or fibronectin (see Section 6).
4. Select basal medium. A medium recommended for a closely related cell type may be tried first. If such a medium is not available, try a mixture of DME/F 12 (1:1, v/v) and add a cocktail (ITS) of insulin (I, low endotoxin), transferrin (T, 90% Fe^{2+} saturated), and selenite (S) at final concentrations of 10 μg/ml, 10 μg/ml, and 100 nM, respectively.
5. In case of poor growth, add BSA (fatty-acid-free) at 1–5 mg/ml and/or a cocktail of trace elements prior to adding the ITS cocktail (see step 4 above).
6. Try to reduce the BSA concentration to about 0.2 mg/ml while supplementing with hormones and/or growth factors (e.g. EGF, IGF I, glucocorticoids) and/or more trace elements (see *Table 3*).
7. If growth is still poor, check lipid requirements of your cells; provide microemulsions or liposomes containing soya bean lipids, cholesterol (1–30 μg/ml), linoleic acid, phosphoethanolamine.
8. Control cell detachment method prior to cell seeding. Use pure trypsin, carefully check concentration, time, and temperature (even down to 4°C) of enzyme treatment using the microscope. Stop reaction with added trypsin inhibitor (as little as possible). Alternatively use Dispase (Boehringer Mannheim) instead of trypsin.
9. Determine optimum inoculum (i.e. cell density at seeding), usually between 10^4 and 10^5 cells/ml. Inoculum of cells, from mid-log phase, should be 2–5 fold higher than in serum-containing medium.
10. Test cell behaviour in selected assays. Cell proliferation parameters: lag phase, doubling time, maximal cell concentration from a growth kinetic curve, i.e. hours vs. $\log_{10}$ cell number; cell productivity, e.g. Ig quantity/ 10^6 hybridoma cells.

11. Avoid harsh subcultivation handling (centrifugation, vigorous pipetting, over-trypsinization).

12. Increase or decrease oxygen delivery, increase surface/volume ratio of culture medium by inclining the culture vessels and gently agitate to facilitate adaptation.

During the adaptation process various changes of the cells may occur, such as morphology, karyotype, cell surface markers, capacity to metabolize drugs, etc. Appropriate cell properties should therefore be tested before and after adaptation in order to determine whether a specific cell mutant has been selected or true adaptation has occurred. Although this process has been called adaptation and can in fact result from metabolic adaptation over a relatively short period (<100 h), adaptation to serum-free conditions over a prolonged period (weeks or months) may result from selection of a subline or mutant clone. It is in these cases, particularly, that caution must be exercised to ensure the subline is still active.

It is essential to monitor productivity or other specific properties during the selection process, as there have been many mistakes when the adaptation to serum-free conditions has caused a loss of specialized properties. Again, samples should be stored at intervals, and, if necessary (and if possible) the cells should be cloned under serum-free conditions, using a feeder layer if necessary.

Finally, the investigator should determine which ultimate goal is required with the cells to be cultured. Generally, cultured cells should reflect their origin, should show essential properties of the tissue they are derived from, and should reveal the characteristics of differentiation and proliferation they exert *in situ*. While this goal may be reached with some cells, it is important to realize the limitations of present tissue culture methods for other cells. It is also possible that cultured cells will not proliferate and express differentiated functions simultaneously. Hence a period of propagation and amplification may be required, followed by maintenance at high cell density ($>10^5/cm^2$) to allow differentiation to occur.

5. Procedures for coating of culture surface with biomatrix

Protocol 2. Poly-D-lysine coating of culture surface

For anchorage-dependent cells coating the culture surface with a positively charged polymer such as poly-D-lysine has proved to be useful (11).

1. Dissolve 0.1 mg/ml poly-D-lysine HBr (M_r 3–7 $\times$ 10^4) (Sigma) in distilled water and sterilize (cellulose acetate filter, 0.22 μm).

Protocol 2. *Continued*

2. Pipette 0.5 ml into each 60 × 15 mm Petri dish.

3. Rock the dish gently to spread the solution uniformly over the entire culture surface.

4. After 5 min remove the solution completely with a Pasteur pipette.

5. Wash the dish with 1.5 ml of distilled water, rock gently, and remove by aspiration.

6. It is essential to remove all solutions completely, since free polylysine may be cytotoxic.

7. The dishes can be used immediately or allowed to dry and used later.

8. Dishes coated under non-sterile conditions may be sterilized by u.v. irradiation.

Protocol 3. Collagen isolation and coating of culture surface

Collagen type I can be obtained according to Strom and Michalopoulos (55) from the tendons that attach to the vertebrae of the rat tail.

1. Cut fresh or thawed tails longitudinally from the base to the tip.

2. Pull the skin off the tail and remove vessels.

3. Remove the white collagen fibres and sterilize with u.v. light.

4. Stir 1 g of collagen fibres (from 3–4 tails) into 300 ml of 0.1% acetic acid for 48 h at 4°C.

5. Filter the solution through 2–3 sterile gauze layers.

6. Dilute the stock solution 1:20 with double-distilled water.

7. Cover the bottom of a plastic dish with this solution and air dry (max. 37°C, 1–2 days).

8. The dishes can be stored in a humid atmosphere at 4°C for several months. Alternatively, the stock solution may be neutralized and diluted in culture medium whereupon the collagen will gel due to the increased pH and salt concentration. Denatured (air-dried) collagen enhances cell attachment (e.g. for endothelial cells and skeletal muscle), but native (undenatured) collagen gel may be required for correct phenotype expression.

Protocol 4. Fibronectin coating of culture surface

Fibronectin may be prepared according to Öbrink (56) from fresh human plasma, avoiding heparin and proteolytic degradation (by means of phenylmethysulphonyl fluoride) or obtained commercially from BRL, Collaborative

Research, or Gibco. Fibronectin is used at 1–5 μg/cm^2 growth area of the culture dish.

1. Dissolve 20 μg of fibronectin lyophilizate gently in 1 ml of double-distilled water.

2. Cover the bottom of the dish (35 mm diameter) with solution and air-dry at 40°C.

3. Rinse extensively with distilled water and air-dry again.

4. Coated dishes can be stored at room temperature for several months.

References

1. Ham, R. G. (1981). In *Tissue Growth Factors* (ed. R. Baserga), p. 13. Handbook of Experimental Pharmacology **57**.
2. Sato, G. H., Pardee, A., and Sirbasku, D. A. (ed.) (1982). *Growth of Cells in Hormonally Defined Media*. Cold Spring Harbor Conf. on Cell Proliferation, Vol. 9, Books A and B. Cold Spring Harbor Press, Cold Spring Harbor, New York.
3. Fischer, G. and Wieser, R. J. (ed.) (1983). *Hormonally Defined Media*. Springer-Verlag, Berlin.
4. Barnes, D. W., Sirbasku, D. A., and Sato, G. H. (ed.) (1984). *Cell Culture Methods for Molecular and Cell Biology*. Alan R. Liss, New York. Vol. 1: *Methods for Preparation of Media, Supplements, and Substrate for Serum-free Animal Cell Culture*. Vol. 2: *Methods for Serum-free Culture of Cells of the Endocrine System*. Vol. 3: *Methods for Serum-free Culture of Epithelial and Fibroblast Cells*. Vol. 4: *Methods for Serum-free Culture of Neuronal and Lymphoid Cells*.
5. Mather, J. P. (1984). *Mammalian Cell Culture: The Use of Serum-free, Hormone-supplemented Media*. Plenum, New York.
6. Barnes, D. W. (1987). *BioTechniques*, **5**, 534.
7. *Cytotechnology* (1991) **5**, No. 1, 1–94.
8. Habenicht, A. (ed.) (1990). *Growth Factors, Differentiation Factors and Cytokines*. Springer-Verlag, Berlin.
9. Ali-Osman, F. and Maurer, H. R. (1983). *J. Cancer Res. Clin. Oncol.*, **106**, 17.
10. Barnes, D. and Sato, G. (1980). *Anal. Biochem.*, **102**, 255.
11. Ham, R. G. and McKeehan, W. L. (1979). *Meth. Enzymol.*, **53**, 44.
12. Lindl, T. and Bauer, J. (1989). *Zell-und Gewebekultur*. G. Fischer Verlag, Stuttgart.
13. Hewlett, G. (1991). *Cytotechnology*, **5**, 3–14.
14. Cleveland, W. L., Wood, I., and Erlanger, B. F. (1983). *J. Immunol. Methods*, **56**, 221.
15. Darfler, F. J. (1990). *In Vitro Cell. Dev. Biol.*, **26**, 769.
16. Jayme, D. W. (1991). *Cytotechnology*, **5**, 15–30.
17. Moellering, W. J., Tedesco, J. L., Townsend, R. R., Hardy, M. R., Scott, R. W., and Prior, C. P. (1990). *Biopharmacology*, **3**, 30.
18. Iscove, N. N. (1984). In *Cell Culture Methods for Molecular and Cell Biology* (ed. D. W. Barnes, D. A. Sirbasku, and G. H. Sato), Vol. 4, p. 169. Alan R. Liss, New York.
19. Blasey, H. D. and Winzer, U. (1989). *Biotechnol. Lett.*, **11**, 455.

20. Hewlett, G., Duvinski, M. S., and Montalto, J. G. (1989). *Miles Sci. J.*, **11**, 9.
21. Darfler, F. J. (1990). *In Vitro Cell. Dev. Biol.*, **26**, 779.
22. Gospodarowicz, D. (1984). In *Cell Culture Methods for Molecular and Cell Biology* (ed. D. W. Barnes, D. A. Sirbasku, and G. H. Sato), Vol. 1, p. 69. Alan R. Liss, New York.
23. Mukarami, H., Yamada, K., and Shirahata, S. (1991). *Cytotechnology*, **5**, 83–94.
24. Geaugey, V., Duval, D., Geahel, I., Marc, A., and Engasser, J. M. (1989). *Cytotechnology*, **2**, 119.
25. Büntemeyer, H., Lütkemeyer, D., and Lehmann, J. (1991). *Cytotechnology*, **5**, 57–67.
26. Thomassen, D. G. (1989). *In Vitro Cell. Dev. Biol.*, **25**, 1046.
27. Clark, J. M., Gebb, C., and Hirtenstein, M. D. (1981). *Develop. Biol. Standard*, **50**, 193.
28. Shintani, Y., Iwamoto, K., and Kitano, K. (1988). *Appl. Microbiol. Biotechnol.*, **27**, 533.
29. Katsuta, H., Takaoka, T., Hosaka, S., Hibino, M., Otsuki, I., Hattori, K., Suzuki, S., and Mitamura, K. (1959). *Jpn. J. Exp. Med.*, **29**, 45.
30. Mizrahi, A. (1975). *J. Clin. Microbiol.*, **2**, 11.
31. Murhammer, D. W. and Goochee, C. F. (1988). *BioTechnology*, **6**, 1411.
32. Merten, O. W. and Litwin, J. (1991). *Cytotechnology*, **5**, 69–82.
33. Handa-Corrigan, A., Emery, A. N., and Spier, R. E. (1989). *Enzyme Microb. Technol.*, **11**, 230.
34. Gürhan, S. I. and Özdural, N. (1990). *Cytotechnology*, **3**, 89.
35. Yamane, I., Kan, M., Minamoto, Y., and Amatsuji, Y. (1981). *Proc. Japan Acad.*, **57B**, 385.
36. Ohmori, H. (1988). *J. Immunol. Method.*, **112**, 227.
37. Rasmussen, L. and Toftlund, H. (1986). *In Vitro Cell. Dev. Biol.*, **22**, 177.
38. Ill, C. R., Brehm, T., Lyderson, B. K., Hernandez, R., and Burnett, K. G. (1988). *In Vitro Cell. Dev. Biol.*, **24**, 413.
39. James, K. (1990). *Immunol. Today*, **11**, 163.
40. Collodi, P., Rawson, C., and Barnes, D. (1991). *Cytotechnology*, **5**, 31–46.
41. Crabb, J. W., Armes, L. G., Carr, S. A., Johnson, C. M., Roberts, G. D., Bordoli, R. S., and McKeehan, W. L. (1986). *Biochemistry*, **25**, 4988.
42. Miller-Davis, S., McKeehan, W., and Carpenter, G. (1988). *Exp. Cell Res.*, **179**, 595.
43. Takahashi, K., Suzuki, K., Kawahara, S., and Ono, T. (1989). *Int. J. Cancer*, **43**, 870.
44. Maurer, H. R. (1981). *Cell Tissue Kinet.*, **14**, 111.
45. Hashizume, S., Kuroda, K., and Murakami, H. (1980). *Biochim. Biophys. Acta*, **763**, 377.
46. Yamada, K., Ikeda, I., Sugahara, T., Shirahata, S., and Murakami, H. (1990). *Cytotechnology*, **3**, 123.
47. Glassy, M. C., Peters, R. E., and Mikhalev, A. (1987). *In Vitro Cell. Dev. Biol.*, **23**, 745.
48. Yamada, K., Akiyoshi, K., Murakami, H., Sugahara, T., Ikeda, I., Toyoda, K., and Omura, H. (1989). *In Vitro Cell. Dev. Biol.*, **25**, 243.
49. Toyoda, K., Murakami, H., Inoue, K., Yamada, K., Shirahata, S., and Murakami, H. (1990). *Cytotechnology*, **3**, 189.

50. Sato, S., Murakami, H., Sugaharo, T., Ikcgami, T., Yamada, K., Omura, H., and Hashizume, S. (1989). *Cytotechnology*, **2**, 63.
51. Broad, D., Boraston, R., and Rhodes, M. (1991). *Cytotechnology*, **5**, 47–55.
52. Mather, J. (1986). *BioTechnology*, **4**, 56.
53. Avner, E. D., Sweney, W. E., Jr, and Ellis, D. (1984). In *Cell Culture Methods for Molecular and Cell Biology* (ed. D. W. Barnes, D. A. Sirbasku, and G. H. Sato), Vol. 3, p. 33. Alan R. Liss, New York.
54. Metzger, J. F., Fusilio, M. H., Cornman, I., and Kuhns, D. M. (1954). *Exp. Cell. Res.*, **6**, 337.
55. Strom, S. C. and Michalopoulos, G. (1982). *Meth. Enzymol.*, **82**, 544.
56. Öbrink, B. (1982). *Meth. Enzymol.*, **82**, 513.
57. Oda, D. and Watson, E. (1990). *In Vitro Cell. Dev. Biol.*, **26**, 589.
58. Ieki, R., Kudoh, S., Kimura, H., Ozawa, K., Asano, S., and Takaku, F. (1990). *Exp. Hematol.*, **18**, 883.
59. Ropke, C., van Deurs, B., and Petersen, O. W. (1990). *In Vitro Cell. Dev. Biol.*, **26**, 671.
60. Turner, T., Bern, H. A., Young, P., and Cunha, G. R. (1990). *In Vitro Cell. Dev. Biol.*, **26**, 722.
61. Weiss, T. L., Selleck, S. E., Reusch, M., and Wintroub, B. U. (1990). *In Vitro Cell. Dev. Biol.*, **26**, 759.
62. Taub, M., Chuman, L., Saier, M. H., Jr, and Sato, G. (1979). *Proc. Natl Acad. Sci. USA*, **76**, 3338.
63. Hahm, H. A., Ip, M. M., Darcy, K., Black, J. D., Shca, W. K., Forczck, S., Yoshimura, M., and Oka, T. (1990). *In Vitro Cell. Dev. Biol.*, **26**, 803.
64. Branchaud, C. L., Goodyer, C. G., Guyda, H. J., and Leferbve, Y. (1990). *In Vitro Cell. Dev. Biol.*, **26**, 865.
65. Rikimaru, K., Toda, H., Tachikawa, N., Kamata, N., and Enomoto, S. (1990). *In Vitro Cell. Dev. Biol.*, **26**, 849.
66. Anderson, S. N., Ruben, Z., and Fuller, G. C. (1990). *In Vitro Cell. Dev. Biol.*, **26**, 61.
67. Nag, A. C., Lee, M. L., and Kosiur, J. R. (1990). *In Vitro Cell. Dev. Biol.*, **26**, 455.
68. Gruenert, D. C., Basbaum, C. B., and Widdicombe, J. H. (1990). *In Vitro Cell. Dev. Biol.*, **26**, 411.
69. Ikeda, H. (1990). *In Vitro Cell. Dev. Biol.*, **26**, 193.
70. Hill, M., Hillova, J., and Mariage-Samson, R. (1990). *In Vitro Cell. Dev. Biol.*, **26**, 44.
71. Chilton, D. G., Johnson, B. H., Danel-Moore, L., Kawa, S., and Thompson, E. B. (1990). *In Vitro Cell. Dev. Biol.*, **26**, 561.
72. Takanashi, M., Motoji, T., Masuda, M., Oshimi, K., and Mizoguchi, H. (1990). *Exp. Hematol.*, **18**, 868.
73. Rhodes, E. G. H., Olive, C., and Flynn, M. P. (1990). *Exp. Hematol.*, **18**, 79.
74. Golombik, T., Dansey, R., Bezwoda, W. R., and Rosendorff, J. (1990). *In Vitro Cell. Dev. Biol.*, **26**, 441.
75. Teece, M. F. and Terrana, B. (1988). *Anal. Biochem.*, **169**, 306.
76. Boyd, D. D., Levine, A. E., Brattain, D. E., McKnight, M. K., and Brattain, M. G. (1988). *Cancer Res.*, **48**, 2469.
77. Kaighn, M. E., Reddel, R. R., Lechner, J. F., Peehl, D. M., Camalier, R. F., Brash, D. E., Saffiotti, U., and Harris, C. C. (1989). *Cancer Res.*, **49**, 3050.

78. Yamada, M., Kawaguchi, M., Takamiya, H., Wada, H., and Okigaki, T. (1988). *Cell Struct. Funct.*, **13,** 495.
79. Brower, M., Carney, D. N., Oie, H. K., Gazdar, A. F., and Minna, J. D. (1986). *Cancer Res.*, **46,** 798.
80. Gazdar, A. F. and Oie, H. K. (1986). *Cancer Res.*, **46,** 6011.
81. Calvo, F., Brower, M., and Carney, D. N. (1984). *Cancer Res.*, **44,** 4553.
82. van der Haegen, B. A., Ham, R. G., and Kano-Sueoka, T. (1989). *In Vitro Cell. Dev. Biol.*, **25,** 158.
83. Takazawa, Y., Tokashiki, M., Murakami, H., Yamada, K., and Omura, H. (1988). *Biotechnol. Bioeng.*, **31,** 168.
84. Tharakan, J. P. and Chan, P. C. (1986). *Biotech. Lett.*, **8,** 457.

3

Scaling-up of animal cell cultures

BRYAN GRIFFITHS

1. Introduction

Small cultures of cells, for example in flasks of up to 1 litre volume (175 cm^2 surface area), are the best means of establishing new cell lines in culture, for studying cell morphology, and for comparing the effects of different agents, or different concentrations of an agent, on growth and metabolism. However, there are many applications in which large numbers of cells are required, for example extraction of a cellular constituent (10^9 cells will provide 7 mg DNA); to produce viruses for vaccine production (typically 5×10^{10} cells per batch) or other cell products (interferon, plasminogen activator, interleukins, hormones, enzymes, erythropoietin, and antibodies); and to produce inocula for even larger cultures. Animal cell culture is now a widely used production process in biotechnology with systems in operation at scales up to 10 000 litres. This has been achieved by graduating from multiples of small cultures—which is tedious, labour-intensive, and expensive—to large 'unit process' systems. Although unit processes are more cost-effective and efficient, achieving the necessary scale-up has required a series of modifications to overcome limiting factors such as oxygen limitation, shear damage, and metabolite toxicity. One of the aims of this chapter is to describe these limitations, indicate at what scale they are likely to occur, and suggest possible solutions to the problems they create. This theme has to be applied to cells that grow in suspension and also to those that will only grow when attached to a substrate (anchorage-dependent cells). Another aspect of scale-up that will be discussed is increasing unit cell density 50–100 fold, particularly by the use of cell immobilization techniques.

Free suspension culture offers the easiest means of scale-up because a 1 litre vessel is conceptually very similar to a 1000 litre vessel. The changes concern the degree of environmental control and the means of maintaining the correct physiological conditions for cell growth, rather than significantly altering vessel design. Monolayer systems (for anchorage-dependent cells) are more difficult to scale-up in a unit system, as opposed to a multiple process, and consequently a wide range of diverse systems have evolved. The aim of increasing the surface area available to the cells in relation to both the medium and the total culture volume has been successfully achieved, and the most effective of these methods (microcarrier) will be described (Section 3.3.4).

The more generalized aspects of cell culture are first discussed, to provide a better understanding of the way in which these techniques operate.

2. General methods and culture parameters

Familiarity with certain biological concepts and methods is essential when understanding scale-up of a culture system. In small-scale cultures there is often some leeway for error. If the culture fails it is a nuisance, but not necessarily a disaster in terms of time and costs. As the size of the culture increases it represents an ever-increasing investment of resources. Culture failure is more serious, but at the same time the system increasingly demands that all conditions are more critically met. This section describes the basic ingredients of all cell culture systems which have to be attended to as the culture size gets larger and more diverse in design and operation.

2.1 Cell quantification

Measuring total cell numbers (by haemocytometer counts of whole cells or stained nuclei) and total cell mass (by determining protein or dry weight) is easily achieved. It is far more difficult to get a reliable measure of cell viability because the methods employed either stress the cells or use a specific, and not necessarily typical, parameter of cell physiology (see Chapter 8). An additional difficulty is that in many culture systems the cells cannot be sampled (most anchorage-dependent cultures), or visually examined, and an indirect measurement has to be made.

2.1.1 Cell viability

i. *Dye exclusion*

Dye exclusion. The dye exclusion test is based on the concept that viable cells do not take up certain dyes, whereas dead cells are permeable to these dyes. Trypan blue (0.4%) is the most commonly used dye, but has the disadvantage of staining soluble protein. In the presence of serum, therefore, erythrosine B (0.4%) is often preferred (see Chapter 4). Cells are enumerated in the standard manner using a haemocytometer. Some caution should be used when interpreting results as the uptake of the dye is pH- and concentration-dependent, and there are situations in which misleading results can be obtained. Two relevant examples are membrane leakiness caused by recent trypsinization and freezing and thawing in the presence of dimethylsulphoxide. A colorimetric assay using the MTT assay (see Chapter 8) is being increasingly used both to measure viability after release of cytoplasmic contents into the medium from artificially lysed cells, and for microscopic visualization within the attached cell (1). Further details on measuring cell viability are given in Chapters 4 and 8.

2.1.2 Indirect measurements

Indirect measurements of viability are based on metabolic activity. The most commonly used parameter is glucose utilization, but oxygen utilization, lactic or pyruvic acid production, or carbon dioxide production can also be used, as can the expression of a product, such as an enzyme. When cells are growing logarithmically, there is a very close correlation between nutrient utilization and cell numbers. However, during other growth phases, utilization rates, caused by maintenance rather than growth, can give misleading results. The measurements obtained can be expressed as a growth yield (Y) or specific utilization/respiration rate (Q):

$$\text{Growth yield } (Y) = \frac{\text{Change in biomass concentration } (dx)}{\text{Change in substrate concentration } (ds)}$$

Equation 1

Specific utilization/respiration rate (Q_A)

$$= \frac{\text{Change in substrate concentration } (ds)}{\text{Time } (dt) \times \text{cell mass/numbers } (dx)}$$

Equation 2

Typical values of growth yields for glucose (10^6 cells/g) are 385 (MRC-5), 620 (Vero), and 500 (BHK).

A method which is not so influenced by growth rate fluctuations is the lactate dehydrogenase (LDH) assay. LDH is measured in cell-free medium at 30°C by following the oxidation of NADH by the change in absorbance at 340 nm. The reaction is initiated by the addition of pyruvate (2). One unit of activity is defined as 1 μmol/min NADH consumed.

2.2 Equipment and reagents

2.2.1 Culture vessel and growth surfaces

The standard non-disposable material for growth of animal cells is glass, although this is replaced by stainless steel in larger cultures. It is preferable to use borosilicate glass (e.g. Pyrex) because it is less alkaline than soda glass and withstands handling and autoclaving better. The primary requirement is that glassware is thoroughly cleaned and rinsed at least three times in deionized water. Cells usually attach readily to glass, but, if necessary, attachment may be augmented by various surface treatments (see Section 3.1). In suspension culture, cell attachment has to be discouraged, and this is achieved by treatment of the culture vessel with a proprietary silicone preparation. Examples are Dow Corning 1107 (which has to be baked on) or dimethyldichlorosilane (Repelcote, Hopkins and Williams) which requires thorough washing of the vessel in distilled water to remove the trichloroethane solvent.

Complex systems use a combination of stainless steel and silicone tubing to connect various components of the system. Silicone tubing is very pervious to gases, and loss of dissolved carbon dioxide can be a problem during transfer of media. It is also liable to rapid wear when used in a peristaltic pump. Thick-walled tubing with additional strengthening (sleeve) should be used. When connecting silicone tubing to stainless steel, it should be secured with plastic ties to prevent blow-off during autoclaving or if back pressure builds up in the system during use. Also, silicone tubing tips should be used on the end of stainless steel tubes to prevent scratching the culture vessel, or causing mechanical damage to the cells. Custom-made connectors should be used to ensure good aseptic connections of additional vessels, etc., during process operation. (These are available from all fermenter supply companies.) (See Appendix A2.)

Safe removal of samples of the culture at frequent intervals is essential. An entry with a vaccine stopper through which a hypodermic syringe can be inserted provides a simple solution, but is only suitable for small cultures. Repeated piercing of the vaccine stopper can lead to a loss of culture integrity. The use of specialized sampling devices, also available from fermenter supply companies, is recommended. These automatically enable the line to be cleared of static medium containing dead cells, and thus avoiding the necessity of taking small initial samples which are then discarded, and increasing the chances of retaining sterility.

Air filters are required for the entry and exit of gases. Even if continuous gassing is not used, one filter entry is usually needed to equilibrate pressure and for forced input or withdrawal of medium. The filters should have a 0.22 μm rating and be non-wettable. Suitable examples are the Microflow 50 and Pall Ultipor.

2.2.2 Non-nutritional medium supplements

i.

Sodium carboxymethyl cellulose (15–20 c.p.s.) is often added to media (at 0.1%) to help minimize mechanical damage to cells caused by shear forces generated by the stirrer impeller, forced aeration, or perfusion. This compound is more soluble than methylcellulose, and is one of those chemicals that has a higher solubility at 4°C than at 37°C.

ii.

Pluronic F-68 (polyglycol) (BASF, Wyandot) is often added to media (at 0.1%) to reduce the amount of foaming that occurs in stirred and/or aerated cultures, especially when serum is present. It is also helpful in reducing cell attachment to glass by suppressing the action of serum in the attachment process. However, its most beneficial action is to protect cells from shear stress and bubble damage in stirred and sparged cultures, and it is especially effective in low-serum or serum-free media.

2.3 Practical considerations

i. Temperature of medium
Always pre-warm the medium to the operating temperature (usually 37 °C) and stabilize the pH before adding the cells. Shifts in pH during the initial stages of a culture create many problems, including a long lag phase and reduced yield.

ii. Growth phase of cells
Avoid using stationary-phase cells as an inoculum since this will mean a long lag phase, or no growth at all. Ideally, cells in the late logarithmic phase should be used.

iii. Inoculation density
Always inoculate at a high enough cell density. There is no set rule as to the minimum inoculum level below which cells will not grow, as this varies between cell lines and depends on the complexity of the medium being used. As a guide, it may be between 5×10^4 and 2×10^5 cells/ml, or 5×10^3 and 2×10^4 cells/cm^2.

iv. Stirring rate
Find empirically the optimum stirring rate for a given culture vessel and cell line. This could vary between 100 and 500 r.p.m. for suspension cells, but is usually in the range 200–350 r.p.m., and between 20 and 100 r.p.m. for microcarrier cultures.

v. Medium and surface area
The productivity of the system depends upon the quality and quantity of the medium and, for anchorage-dependent cells, the surface area for cell growth. A unit volume of medium is only capable of giving a finite yield of cells. Factors which affect the yield are: pH, oxygen limitation, accumulation of toxic products (e.g. NH_4), nutrient limitation (e.g. glutamine), spatial restrictions, and mechanical/shear stress. As soon as one of these factors comes into effect, the culture is finished and the remaining resources of the system are wasted. The aim is, therefore, to delay the onset of any one factor until the accumulated effect causes cessation of growth: at which point the system has been maximally utilized. Simple ways of achieving this are: a better buffering system (e.g. Hepes instead of bicarbonate), continuous gassing, generous headspace volume, enriched rather than basal media, with nutrient-sparing supplements such as lactalbumin hydrolysate or peptone, perfusion loops through ultrafiltration membranes or dialysis tubing for detoxification and oxygenation, and attention to culture and process design.

2.4 Growth kinetics

The standard format of a culture cycle beginning with a lag phase, proceeding through the logarithmic phase to a stationary phase, and finally to the decline and death of cells, is well documented (see Chapter 1). Although cell growth usually implies increase in cell numbers, increase in cell mass can occur without any replication. The difference in mean cell mass between cell populations is considerable, as would be expected, but so is the variation within the same population.

Growth (increase in cell numbers or mass) can be defined in the following terms:

- *specific growth rate*, μ (i.e. the rate of growth per unit amount (weight/ numbers) of biomass):

$$\mu = \left(\frac{1}{x}\right)\left(\frac{dx}{dt}\right)h^{-1}$$

Equation 3

where dx is the increase in cell mass, dt is the time interval, and x is the cell mass. If the growth rate is constant (e.g. during logarithmic growth), then

$$\ln x = \ln x_0 + \mu t$$

Equation 4

where x_0 is the biomass at time t_0

- *doubling time*, t_d (i.e. the time for a population to double in number/mass):

$$t_d = \frac{\ln 2}{\mu} = \frac{0.693}{\mu}$$

Equation 5

- *degree of multiplication*, n or *number of doublings* (i.e. the number of times the inoculum has replicated):

$$n = 3.32 \log\left(\frac{x}{x_0}\right)$$

Equation 6

2.5 Medium and nutrients

A given concentration of nutrients can only support a certain number of cells. Alternative nutrients can often be found by a cell when one becomes exhausted, but this is bad practice because the growth rate is always reduced (e.g. while alternative enzymes are being induced). If a minimum medium, such as Eagle's basal medium (BME) or minimum essential medium (MEM) is used with serum as the only supplement, then this problem is going to be met sooner than in cultures using complete media (e.g. 199), or in media supplemented with lactalbumin hydrolysate, peptone, or BSA (which provides many of the fatty acids, see Chapter 2). Nutrients likely to be exhausted first are glutamine, partly because it spontaneously cyclizes to pyrrolidone carboxylic acid and is enzymically converted (by serum and cellular enzymes) to glutamic acid, leucine, and isoleucine. Human diploid cells are almost unique in utilizing cystine heavily. The many reports of arginine being a limiting factor are misleading, as often these cells were contaminated with mycoplasma. A point to remember is that nutrients become growth-limiting before they become exhausted. As the concentration of amino acids falls, the cell finds it increasingly difficult to maintain sufficient intracellular pool levels. This is exaggerated in monolayer cultures because as cells become more tightly packed together, the surface area available for nutrient uptake becomes smaller (3).

Glucose is often another limiting factor as it is destructively utilized by cells and, rather than adding high concentrations at the beginning, it is more beneficial to supplement after 2–3 days. In order to maintain a culture some additional feeding often has to be carried out either by complete, or partial, media changes or by perfusion. The data in *Figure 1* compare the growth stimulation to MRC-5 cells as a result of adding equal quantities of media by continuous perfusion (over 24 h) and medium changes. The efficiency of medium changes is probably due to the high extracellular concentration of nutrients this provides, thus stimulating a further replication cycle. These data make it clear that if perfusion is used, the perfusion rate must be high enough to optimize the growth conditions.

Many cell types are either totally dependent upon certain additives or can only perform optimally when they are present. For many purposes it is highly desirable, or even essential, to reduce the serum level to 1% or below. In order to achieve this without a significant reduction in cell yields, various growth factors/hormones are added to the basal medium. The most common additives are insulin (5 mg/litre), transferrin (5–35 mg/litre), ethanolamine (20 μM), and selenium (5 μg/litre) (see Chapter 2).

Cell aggregation is often a problem in suspension cultures. Media lacking calcium and magnesium ions have been designed specifically for suspension cells because of the role of these ions in attachment (see Section 3.1). This

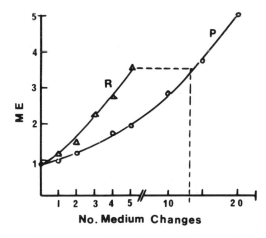

Figure 1. Comparison of MRC-5 growth using continuous perfusion (P) and complete medium changes (R). The same total volumes of medium are used per day for both systems. Growth is given as monolayer equivalents (ME) to demonstrate multilayering.

problem has also been overcome by including very low levels of trypsin in the medium (2 μg/ml).

2.6 pH

Ideally pH should be near 7.4 at the initiation of a culture and not fall below a value of 7.0 during the culture, although many hybridoma lines appear to prefer a pH of 7.0 or lower. A pH below 6.8 is usually inhibitory to cell growth. Factors affecting the pH stability of the medium are buffer capacity and type, headspace, and glucose concentration.

The normal buffer system in tissue culture media is the carbon dioxide bicarbonate system analogous to that in blood. This is a weak buffer system, in that it has a pK_a well below the physiological optimum. It also requires the addition of carbon dioxide to the headspace above the medium to prevent the loss of carbon dioxide and an increase in hydroxyl ions. The buffering capacity of the medium is increased by the phosphates present in the balanced salt solution (BSS). Media intended to equilibrate with 5% carbon dioxide usually contain Earle's BSS (25 mM $NaHCO_3$) but an alternative is Hanks' BSS (4 mM $NaHCO_3$) for equilibration with air. Improved buffering and pH stability in media is possible by using a zwitterionic buffer (4), such as HEPES (10–20 mM), either in addition to, or instead of, bicarbonate (include bicarbonate at 0.5 mM). Alternative buffer systems are provided by using specialist media such as Leibovitz's L-15 (5). This medium utilizes the buffering capacity of free amino acids, substitutes galactose and pyruvate for glucose, and omits sodium bicarbonate. It is suitable for open cultures.

The headspace volume in a closed culture is important because in the initial stages of the culture 5% carbon dioxide is needed to maintain a stable pH in the medium but, as the cells grow and generate carbon dioxide, it builds up in the headspace and this prevents it diffusing out of the medium. The result is an increase in weakly dissociated bicarbonate producing an excess of hydrogen ions in the medium and a fall in pH. Thus, a large headspace is required in closed cultures, typically 10-fold greater than the medium volume (this volume is also needed to supply adequate oxygen). This generous headspace is not possible as cultures are scaled-up, and an open system with a continuous flow of air, supplied through one filter and extracted through another, is required.

The metabolism of glucose by cells results in the accumulation of pyruvic and lactic acids. Glucose is metabolized at a far greater rate than it is needed. Thus, glucose should ideally never be included in media at concentrations above 2 g/litre, and it is better to supplement during the culture than to increase the initial concentration. An alternative is to substitute glucose by galactose or fructose as this significantly reduces the formation of lactic acid, but it usually results in a slower growth rate. These precautions delay the onset of a non-physiological pH and are sufficient for small cultures. As scale-up increases, headspace volume and culture surface area in relation to the medium volume decrease. Also, many systems are developed in order to increase the surface area for cell attachment and cell density per unit volume. Thus pH problems occur far earlier in the culture cycle because carbon dioxide cannot escape as readily, and more cells means higher production of lactic acid and carbon dioxide. The answer is to carry out frequent medium changes, or have a pH control system.

The basis of a pH control system is an autoclavable pH probe (available from Pye Ingold, Russell). This feeds a signal to the pH controller which is converted to give a digital or analogue display of the pH. This is a pH monitor system. Control of pH requires the defining of high and low pH values beyond which the pH should not go. These two set points on the pH scale turn on a relay to activate a pump, or solenoid valve, which allows additions of acid or alkali to be made to the culture to bring it back to within the allowable units. It is rare to have the pH rise above the set point once the medium has equilibrated with the headspace, so the addition of acid can be disregarded. If an alkali is to be added, then sodium bicarbonate (5.5%) is recommended. Sodium hyroxide (0.2 M) can be used only with a fast stirring rate, which dilutes the alkali before localized concentrations can damage the cells, or if a perfusion loop is installed. Normally, liquid delivery pumps are supplied as part of the pH controller. Gas supply is controlled by a solenoid valve and 95% air directly. Above the set point carbon dioxide will be mixed with the air, but below this point only air will be delivered to the culture. This in itself is a controlling factor in that it helps remove carbon dioxide as well as meeting the oxygen requirements of the cultures. pH regulation is very readily adapted to computer control systems.

2.7 Oxygen

The scale-up of animal cell cultures is very dependent upon the ability to supply sufficient oxygen without causing cellular damage. Oxygen is only sparingly soluble in culture media (7.6 μg/ml) and a survey of reported oxygen utilization rates by cells (6) reveals a mean value of 6 μg/10^6 cells/h. A typical culture of 2×10^6 cells/ml would, therefore, deplete the oxygen content of the medium (7.6 μg/ml) in under 1 h. It is necessary to supply oxygen to the medium throughout the life of the culture and the ability to do this adequately depends upon the oxygen transfer rate (OTR) of the system:

$$\text{OTR} = Kla \, (C^* - C) \qquad \text{Equation 7}$$

where OTR is the amount of oxygen transferred per unit volume in unit time, Kl is the oxygen transfer coefficient, and a is the area of the interface across which oxygen transfer occurs (as this can only be measured in stationary and surface-aerated cultures, the value Kla is used; this is the mass transfer coefficient (vol/h). C^* is the concentration of dissolved oxygen when medium is saturated, and c is the actual concentration of oxygen at any given time.

The Kla (OTR/C when $C = 0$) is in units of h^{-1}, and is thus a measure of the time taken to oxygenate a given culture vessel completely under a particular set of conditions.

A culture can be aerated by one, or a combination, of the following methods: surface aeration, sparging, membrane diffusion, medium perfusion, increasing the partial pressure of oxygen, and increasing the atmospheric pressure.

2.7.1 Surface aeration: static cultures

In a closed system, such as a sealed flask, the important factors are the amount of oxygen in the system and the availability of this oxygen to the cells growing under 3–6 mm of medium. Normally a headspace/medium volume ratio of 10:1 is used in order to provide sufficient oxygen. Thus a 1 litre flask (e.g. a Roux bottle) with 900 ml of air and 100 ml of medium will initially contain 0.27 g oxygen (*Table 1*). This amount will support 10^8 cells for 450 h and is thus clearly adequate. The second factor is whether this oxygen can be made available to the cells. The transfer rate of oxygen from the gas phase into a liquid phase has been calculated at about 17 μg/cm^2/h (6). Again, this is well in excess of that required by cells in a 1 litre flask. However, if the surface is assumed to be saturated with dissolved oxygen and the concentration at the cell sheet is almost zero then, applying Fick's law of diffusion, the rate at which oxygen can diffuse to the cells is about 1.5 μg/cm^2/h. At this rate there is only sufficient oxygen to support about 50×10^6 cells in a 1 litre flask, a cell density which in practice many tissue culturists take as the norm. These calculations show the importance of maintaining a large headspace volume; otherwise, oxygen limitation could become one of the growth-limiting factors in static closed cultures. A fuller explanation of these calculations, including Fick's law of diffusion, can be found in a review by Spier and Griffiths (6).

2.7.2 Sparging

This is the bubbling of gas through the culture, and is a very efficient means of effecting oxygen transfer (as proven in bacterial fermentation). However, it may be damaging to animal cells due to the effect of the high surface energy

---

Table 1. Oxygen concentrations on the gas and liquid phases of a Roux bottle culture

Oxygen in 900 ml air:
$900 \times 0.21 \times 32/22400 = 0.27$ g
Oxygen in 100 ml medium:
$100 \times 7.6 \times 10^{-4} = 0.0076$ g

Notes:
0.21	= proportion of oxygen in air
32	= molecular weight of oxygen
22400	= gram-molecular-volume
7.6×10^{-4}	= solubility of oxygen in water at 37°C when equilibrated with air

---

of the bubble on the cell membrane if not properly controlled. This damaging effect can be minimized by using large air bubbles (which have lower surface energies than small bubbles), by using a very low gassing rate (e.g. 5 ml l^{-1} min^{-1}) and by adding Pluronic F-68. A specialized form of sparging, the airlift fermenter (see Section 4.7) has also been used successfully in large unit process monolayer cultures (e.g. multiple plate propagators). When sparging is used, efficiency of oxygenation is increased by using a culture vessel with a large height/diameter ratio. This creates a higher pressure at the base of the reactor, which increases oxygen solubility.

2.7.3 Membrane diffusion

Silicone tubing is very pervious to gases, and if long lengths of thin-walled tubing can be arranged in the culture vessel then sufficient diffusion of oxygen into the culture can be obtained. However, a lot of tubing is required (e.g. 30 m of 2.5 cm tubing for a 1000 litre culture). This method is expensive and inconvenient to use, and has the inherent problem that scale-up of the tubing required is mainly two-dimensional while that of the culture is three-dimensional; however, several commercial systems are available (Braun, Diessel).

2.7.4 Medium perfusion

A closed-loop perfusion system continuously (or on demand) takes medium from the culture and passes it through an oxygenation chamber before it is returned to the culture. This method has many advantages if the medium can be conveniently separated from the cells for perfusion through the loop. The medium in the chamber can be vigorously sparged to ensure oxygen saturation and other additions, such as sodium hydroxide for pH control which would damage the cells if put directly into the culture, can be made. This method is used in glass bead systems (Section 3.3.4) and has proved particularly effective in microcarrier systems (Section 3.3.4), where specially modified spin filters can be used.

2.7.5 Environmental supply

The dissolved oxygen concentration can be increased by increasing the head-space pO$_2$ (from atmospheric 21% to any value, using oxygen and nitrogen mixtures) and by raising the pressure of the culture by 100 kPa (about 1 atm) (which increases the solubility of oxygen and its diffusion rate). These methods should be employed only when the culture is well advanced, otherwise oxygen toxicity effects could occur. Finally, the geometry of the stirrer blade also affects the oxygen transfer rate.

2.7.6 Scale-up

Oxygen limitation is usually the first factor to be overcome in culture scale-up. This becomes a problem in conventional stirred cultures at volumes above

Table 2. Methods of oxygenating a 40 litre Bioreactor (30 litre working volume with a 1.5:1 aspect ratio)

Oxygenating method	Oxygen delivery (mg/l/h)	No. cells ×10⁶/ml supported
AIR (10 ml/l/min at 40 r.p.m.)		
Surface aeration	0.5	0.08
Direct sparging	4.6	0.76
Spin filter sparging	3.0	0.40
Perfusion (1 vol/h)	12.6	2.10
Perfusion (1 vol/h)		
+ Spin filter sparging	15.9	2.65
OXYGEN (10 ml/min at 80 r.p.m.)		
Spring filter sparging	51.0	8.50
+ Perfusion (1 vol/h)	92.0	15.00

(assuming oxygen utilization rate of 2–6 μg/10⁶ cells/h)

10 litres. However, with the current use of high-density cultures maintained by perfusion, oxygen limitation can occur in a 2 litre culture. The relative effectiveness of some of the alternative oxygenation systems in large-scale bioreactors is shown on *Table 2*, and the range of oxygenation procedures in various types of culture vessels has been reviewed (7, 31).

2.7.7 Redox potential

The oxidation–reduction potential (ORP), or redox potential, is a measure of the charge of the medium and thus affected by the proportion of oxidative and reducing chemicals, the oxygen concentration, and the pH. When fresh medium is prepared and placed in the culture vessel it takes time for the redox potential to equilibrate, a phenomenon known as poising. The optimum level for growth of many cell lines is +75 mV, which corresponds to a dissolved pO_2 of 8–10%. Some investigators find it beneficial to control the oxygen supply to the culture by means of redox, rather than an oxygen, electrode. Alternatively, if the redox potential is monitored by means of a redox electrode and pH meter (with mV display), then an indication of how cell growth is progressing can be obtained (8). This is because the redox value falls during logarithmic growth and reaches a minimum value approximately 24 h before the onset of the stationary phase (*Figure 2*). This provides a useful guide to cell growth in cultures where cell sampling is not possible. It is also useful to be able to predict the end of the logarithmic growth phase so that medium changes, addition of virus, or product promoters can be given at the optimum time. The effect of redox potential on cell cultures has been reviewed by Griffiths (9).

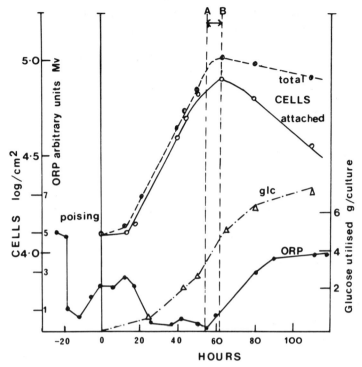

Figure 2. Changes in ORP correlated to cell growth and glucose utilization (8). The minimum ORP value is reached 24 h before the end of logarithmic growth (A).

2.8 Types of culture process

2.8.1 Batch and continuous culture

In standard culture, known as *batch culture*, cells are inoculated into a fixed volume of medium and, as they grow, nutrients are consumed and metabolites accumulate. The environment is therefore continually changing, and this in turn enforces changes to cell metabolism, often referred to as physiological differentiation. Eventually cell multiplication ceases because of exhaustion of nutrient(s), accumulation of toxic waste products, or density-dependent limitation of growth in monolayer cultures. There are means of prolonging the life of a batch culture, and thus increasing the yield, by various substrate feed methods.

(a) Gradual addition of fresh medium, so increasing the volume of the culture (*fed batch*).

(b) Intermittently, by replacing a constant fraction of the culture with a equal volume of fresh medium (*semi-continuous batch*). All batch culture systems retain the accumulating waste products, to some degree, and have a fluctuating environment. All are suitable for both monolayer and suspension cells.

(c) *Perfusion*, by the continuous addition of medium to the culture and the withdrawal of an equal volume of used (cell-free) medium. Perfusion can be open, with complete removal of medium from the system, or closed, with recirculation of the medium, usually via a secondary vessel, back into the culture vessel. The secondary vessel is used to 'regenerate' the medium by gassing and pH correction.

(d) *Continuous-flow culture*, which gives true homeostatic conditions with no fluctuations of nutrients, metabolites, or cell numbers. It depends upon medium entering the culture with a corresponding withdrawal of medium plus cells. It is thus only suitable for suspension culture cells, or monolayer cells growing on microcarriers. Continuous-flow culture is described more fully in Section 4.6.

2.8.2 Comparison of batch, perfusion, and continuous-flow culture (*Figure 3*)

Continuous-flow culture is the only system in which the cellular content is homogeneous, and can be kept homogeneous for long periods of time

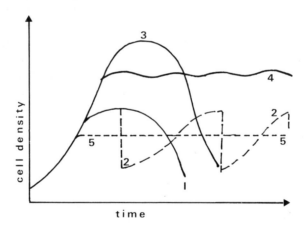

Figure 3. Comparison of culture processes. The numbers 1–5 are explained in the table below.

Culture type	Cell no (millions)	Product yield per litre	
		mg/week*	mg/month*
1 Batch	3	6	12
2 Semi-continuous batch	3	15	60
3 Fed batch	30	60	120
4 Continuous perfusion	10	70	280
5 Continuous flow	2	15	60

* Values allow for turn-around time of non-continuous cultures.

(months). This can be vital for physiological studies, but may not be the most economical method for product generation. Production economics are calculated in terms of staff time, medium, equipment, and downstream processing costs. Also taken into account are the complexity and sophistication of the equipment and process, as this effects the calibre of the staff required and the reliability of the production process. Batch culture is more expensive on staff time and culture ingredients, because for every single harvest a sequence of inoculum build-up steps and then growth in the final vessel has to be carried out, and there is also downtime whilst the culture is readied for its next run. Feeding routines for batch cultures can give repeated but smaller harvests, and the longer a culture can be maintained in a productive state then the more economical the whole process becomes.

Continuous-flow culture in the chemostat implies that cell yields are never maximal because a limiting growth factor is used to control the growth rate. If maximum yields are desired in this type of culture then the turbidostat option has to be used (see Section 4.6). Some applications, such as the production of a cytopathic virus, leave no choice other than batch culture. Maintenance of high yields, and therefore high product concentration, may be necessary to reduce downstream processing costs and these could outweigh medium expenses. For this purpose perfusion has to be used; although for many processes this is more economical than batch culture, it does add to the complexity of the equipment and process, and increases the risk of a mechanical/electronic failure or microbial contamination prematurely ending the production run. There is no clear-cut answer to which type of culture process should be used—it depends upon the cell and product, the quantity of product, downstream processing problems, and product licensing regulations (batch definition of product, cell stability and generation number, etc.). However, a relative ratio of unit costs for perfusion, continuous-flow, and batch culture in the production of monoclonal antibody is 1:2:3.5.

2.9 Summary of factors limiting scale-up

In the following sections a wide variety of alternative culture systems are described. The reason for this large number, apart from scale and different growth characteristics of cells, is the range of solutions which have been used to overcome limitations to scale-up. These will be briefly reviewed so that the underlying philosophy and evolution of the culture systems described in Sections 3–5 can be better understood (see refs 7 and 10).

As already explained, oxygen is the first limiting factor encountered in scale-up. To overcome this, reactors are available with membrane or hollow fibre devices to give bubble-free aeration, often by means of external medium loops. A related factor is mixing, and low-shear options such as airlift reactors and specially designed impellers have been developed. For anchorage-dependent cells the low unit surface area has limited effective scale-up, so a range of devices using plates, spirals, ceramics and, most effective of all, microcarriers, can be used.

The next problem to be overcome is nutrient limitation and toxic metabolite build-up. The most effective solution is to use perfusion by means of spin filters or hollow fibre loops. To maximize the benefit of perfusion, immobilization of suspension cells was needed to prevent cell wash-out, and this resulted in a collection of novel systems based on hollow fibres, membranes, encapsulation, and specialized matrices. Immobilization has many process advantages (see Section 5), and currently the effort is on scaling-up suitable entrapment matrices in fluidized and fixed beds.

3. Monolayer culture

3.1 Introduction

Tissue culture flasks and tubes giving surface areas of 5–200 cm^2 are familiar to all tissue culturists. The largest stationary flask routinely used in laboratories is the Roux bottle (or disposable plastic equivalent) which gives a surface area for cell attachment of 175–200 cm^2 (depending upon type), needs 100–150 ml medium, and utilizes 750–1000 cm^3 of storage space. This vessel will yield 2×10^7 diploid cells and up to 10^8 heteroploid cells. If one has to produce, for example, 10^{10} cells, then over 100 replicate cultures are needed (i.e. manipulations have to be repeated 100 times). In addition the cubic capacity of incubator space needed is over 100 litres. Clearly there comes a time in the scale-up of cell production when one has to use a more efficient culture system. Scale-up of anchorage-dependent cells requires the reduction of the multiplicity of cultures, for more efficient use of staff resources, and to increase significantly the surface area/volume ratio. In order to do this a very wide and versatile range of tissue culture vessels and systems have been developed. Many of these are shown diagrammatically in *Figure 4*, and are described on the following pages. The methods with the most potential are those based on modifications to suspension culture systems because they allow a truly homogeneous, unit process, system with enormous scale-up potential, to be used. However, these systems should be attempted only if time and resources allow a lengthy development period.

Although suspension culture is the preferred method for increasing capacity, monolayer culture has the following advantages:

(a) It is very easy to change the medium completely and to wash the cell sheet before adding fresh medium. This is important in many applications when the growth is carried out in one set of conditions and product generation in another. A common requirement of a medium change is the transfer of cells from serum to serum-free conditions. The efficiency of medium changing in monolayer cultures is such that a total removal of the unwanted compound can be achieved.

(b) If artificially high cell densities are needed then these can be supported by using perfusion techniques. It is much easier to perfuse monolayer cultures because they are immobilized and a fine filter system (to withhold cells) is not required.

(c) Many cells will express a required product more efficiently when attached to a substrate.

(d) The same apparatus can be used with different media/cell ratios which, of course, can be easily changed during the course of an experiment.

(e) Monolayer cultures are more flexible because they can be used for all cell types. If a variety of cell types are to be used then a monolayer system might be a better investment.

It should be noted that the microcarrier system confers some of these advantages to a suspension culture system.

There are four main disadvantages of monolayer compared to suspension systems:

(a) They are difficult and expensive to scale up.

(b) They require far more space.

(c) Cell growth cannot be monitored as effectively, because of the difficulty of sampling and counting an aliquot of cells.

(d) It is more difficult to measure and control parameters such as pH and oxygen, and to have homogeneity throughout.

Microcarrier culture eliminates or, at least, reduces, many of these disadvantages inherent in a monolayer system.

3.2 Cell attachment

Animal cell surfaces and the traditional glass and plastic culture surfaces are negatively charged, so for cell attachment to occur, cross-linking with glycoproteins and/or divalent cations (Ca^{2+}, Mg^{2+}) is required. The glycoprotein most studied in this respect is fibronectin, a compound of high molecular mass (220 000) synthesized by many cells and present in serum and other physiological fluids. Although cells can presumably attach by electrostatic forces alone, it has been found that the mechanism of attachment is similar, whatever the substrate charge (11). The important factor is the net negative charge, and surfaces such as glass and metal which have high surface energies are very suitable for cell attachment. Organic surfaces need to be wettable and negative, and this can be achieved by chemical treatment (e.g. oxidizing agents, strong acids) or physical treatment (e.g. high-voltage discharge, u.v. light, high-energy electron bombardment). One or more of these methods are used by manufacturers of tissue-culture-grade plastics. The result is to increase the net negative charge of the surface (by forming negative carboxyl groups for example) for electrostatic attachment.

Surfaces may also be coated to make them suitable for cell attachment. A tissue-culture grade of collagen can be purchased which saves a tedious preparation procedure using rat tails. The usefulness of collagen as a growth surface is also demonstrated by the availability of collagen-coated micro-carrier beads (Cytodex-3, Pharmacia).

3.2.1 Surfaces for cell attachment

i. Glass

Alum–borosilicate glass (e.g. Pyrex) is the preferred type because soda-lime glass releases alkali into the medium and requires to be detoxified (by boiling in weak acid) before use. After repeated use glassware can become less efficient for cell attachment, but efficiency can be regained by treatment with 1 mM magnesium acetate. After several hours soaking at room temperature the acetate is poured away and the glassware is rinsed with distilled water and autoclaved.

ii. Plastics

Polystyrene is the most used plastic for cell culture, but polyethylene, polycarbonate, Perspex, PVC, Teflon, cellophane, and cellulose acetate are all suitable when pretreated correctly.

iii. Metals

Stainless steel and titanium are both suitable for cell growth because they are relatively chemically inert, but have a suitable high negative energy. There are many grades of stainless steel, and care has to be taken in choosing those which do not leak toxic metallic ions. The most common grade to use for culture applications is 316, but 321 and 304 may also be suitable. Stainless steel should be acid-washed (10% nitric acid, 3.5% hydrofluoric acid, 86.5% water) to remove surface impurities and inclusions acquired during cutting.

3.3 Scaling-up

3.3.1 Step 1: roller bottle

The aims of scaling-up are to maximize the available surface area for growth and to minimize the volume of medium and headspace, while optimizing cell numbers and productivity. Stationary cultures have only one surface available for attachment and growth, and consequently they need a large medium volume. The medium volume can be reduced by rocking the culture or, more usually, by rolling a cylindrical vessel. The roller bottle (see *Protocol 1*) has nearly all its internal surface available for cell growth, although only 15–20% is covered by medium at any one time. Rotation of the bottle subjects the cells to medium and air alternately, as compared with the near anaerobic conditions in a stationary culture. This method reduces the volume of medium required, but still requires a considerable headspace volume to maintain adequate oxygen and pH levels. The scale-up of a roller bottle requires that the diameter is kept as small as possible. The surface area can be doubled by doubling the diameter or the length. The first option increases the volume (medium and headspace) fourfold, the second option only twofold.

The only means of increasing the productivity of a roller bottle and decreasing its volume is by using a perfusion system, originally developed by Kruse (12), and marketed by New Brunswick and Bellco (Autoharvester). This is an expensive option, as an intricate revolving connection has to be made for the supply lines to pass into the bottle. However, cell yields are considerably increased and extensive multilayering takes place.

Protocol 1. Use of standard disposable roller cultures

The procedure is based on the use of a 1400 cm^2 (23 × 12 cm) plastic disposable bottle (Corning or Becton Dickinson).

1. Add 300 ml of growth medium.
2. Add 1.5×10^7 cells.
3. Roll at 12 r.p.h. at 37°C for 2 h, to allow an even distribution of cells during the attachment phase.
4. Decrease the revolution rate to 5 r.p.h., and continue incubation.
5. Examine cells under an inverted microscope using an objective with a long working distance.
6. Harvest cells when visibly confluent (5–6 days) by removing the medium, adding trypsin (0.25%) and rolling. Yields will be very similar to those obtained in flasks, assuming enough medium was added. This method has the advantage of allowing the medium volume/surface area ratio to be altered easily. Thus after a growth phase the medium volume can be reduced to get a higher product concentration.

3.3.2 Step 2: roller bottle modifications

The roller bottle system is still a multiple process, and thus inefficient in terms of staff resources and materials. To increase the surface area within the volume of a roller bottle, the following vessels have been developed (see also *Figure 4*).

i. Spira-Cel

Sterilin have recently replaced their bulk cell culture vessel with a new Spira-Cel roller bottle. This is available with a spiral polystyrene cartridge in three sizes, 3000, 4500, and 6000 cm^2. It is crucial to get an even distribution of the cell inoculum throughout the spiral, otherwise very uneven growth and low yields are achieved. Cell growth can be visualized only on the outside of the spiral, and this can be misleading if the cell distribution is uneven.

ii. Glass tubes

A small-scale example is the Bellco-Corbeil Culture System (Bellco). A roller bottle is packed with a parallel cluster of small glass tubes (separated by silicone spacer rings). Three versions are available giving surface areas of 5×10^3, 1×10^4, and 1.5×10^4 cm^2. Medium is perfused through the vessel from a reservoir. The method is ingenious in that it alternately rotates the bottle 360° clockwise, and then 360° anticlockwise. This avoids the use of special caps for the supply of perfused medium.

An example of its use is the production of 3.2×10^9 Vero cells (2.3×10^5/cm^2) over 6 days using 6.5 litres of medium (perfused at 50 ml/min) in the 10 000 cm^2 version.

iii. Extended surface area roller bottle (ESRB)

In place of the smooth surface in standard roller bottles the surface is 'corrugated', thus doubling the surface area within the same bottle dimensions (available from J. Bibby & Becton Dickinson).

3.3.3 Step 3: large-capacity stationary cultures

i. Multitray unit

The standard multitray unit (Nunc) comprises 10 chambers, each having a surface area of 600 cm^2, fixed together vertically and supplied with interconnecting channels. This enables all operations to be carried out once only for all chambers. It can thus be thought of as a flask with a 6000 cm^2 surface area using 2 litres of medium and taking up a total volume of 12 500 ml. In practice this unit is convenient to use and produces good results, similar to plastic flasks. It is made of tissue-culture-grade polystyrene and is disposable. One of the disadvantages of the system can be turned to good use. In practice it is difficult to wash out all the cells after harvesting with trypsin, etc. However, enough cells remain to inoculate a new culture when fresh medium is added. Given good aseptic technique, this disposable unit can be used 3–4 times. The system is used commercially for interferon production (by linking together multiples of these units) (13). In addition units are also available giving 1200 and 24 000 cm^2.

ii. Hollow fibre culture

Bundles of synthetic hollow fibres offer a matrix analogous to the vascular system, and allow cells to grow in tissue-like densities. Hollow fibres are usually used in ultrafiltration, selectively allowing passage of macromolecules through the spongy fibre wall while allowing a continuous flow of liquid through the lumen. When these fibres are enclosed in a cartridge and encapsulated at both ends, medium can be pumped in and will then perfuse through the fibre walls, which provide a large surface area for cell attachment and growth.

Culture chambers based on this principle are available from Amicon

(Vitafibre). The capillary fibres, which are made of acrylic polymer, are 350 μm in diameter with 75 μm walls. The pores through the internal lumen lining are available with molecular mass cut-offs between 10 000 and 100 000. It is difficult to calculate the total surface area available for growth but units are available in various sizes and these give a very high ratio of surface area to culture volume (in the region of 30 cm^2/ml). Up to 10^8 cells/ml have been maintained in this system. These cultures are mainly used for suspension cells (see Section 5.1.1) but are suitable for attached cells if the polysulphone type is used.

iii. Opticell culture system

This system (Charles River) consists of a cylindrical ceramic cartridge (available in surface area between 0.4 and 12.5 m^2) with 1 mm^2 square channels running lengthwise through the unit. A medium perfusion loop to a reservoir, in which environmental control is carried out, completes the system. It provides a large surface area/volume ratio (40:1) and its suitability for virus, cell surface antigen, and monoclonal antibody production is well documented (14). Scale-up to 210 m^2 is possible with multiple cartridges arranged in parallel in a single controlled unit. Cartridges are available for both attached and suspension cells, which become entrapped in the rough porous ceramic texture.

iv. Plastic film

Bags made of fluoroethylenepropylene copolymer (FEP-Teflon, Du Pont) are biologically inert but are very gas-permeable. Thus a bag of 5 × 30 cm filled with cells and medium to a depth of 2–10 mm can be placed in an incubator. Oxygen supply is sufficient (the cells grow on the inside of the bag and are thus separated by only 25 μm from air) to allow high cell densities to be obtained. The culture is rocked or rotated to keep the medium homogeneous. Cells can be harvested by conventional trypsinization, or mechanically by folding and stretching the bags (15).

A more complex apparatus is the plastic film propagator (16) which uses the same material (FEP-Teflon), but instead of bags it is presented as a long tube and wrapped with a spacer round a reel. Medium is pumped through the tubing and can be either recirculated after passing through a reservoir, or discarded. Gas exchange occurs between the medium and the incubator air space through the wall of the tubing. Lengths of tubing up to 10 m, giving a surface area for growth of 25 000 cm^2, have been used. A more recent system based on the same principle is the SteriCell (Du Pont) which is available as 0.25 and 1 litre plastic containers.

v. Heli-Cel (Sterilin)

Twisted helical ribbons of polystyrene (3 mm × 5–10 mm × 100 μm) are used as packing material for the cultivation of anchorage-dependent cells. Medium is circulated through the bed by a pump, and the helical shape provides good hydrodynamic flow. The ribbons are transparent and therefore allow cell examination after removal from the bed.

vi. Plate heat exchanger

Plate heat exchangers offer a large surface area of stainless steel with adequate provision for circulating medium on one side and water at 37 °C on the other side of the growth surface. This system is available from APV in units of $2\,m^2$ (17).

3.3.4 Step 4: unit process systems

There are basically three systems which fit into fermentation (suspension culture) apparatus (see *Figure 4*):

- cells stationary, medium moves (e.g. glass bead reactor)
- heterogeneous mixing (e.g. stack plate reactor)
- homogeneous mixing (e.g. microcarrier)

i. Bead bed reactor

The use of a packed bed of 3–5 mm glass beads, through which medium is continually perfused has been reported by a number of investigators since 1962. The potential of the system for scale-up was demonstrated by Whiteside

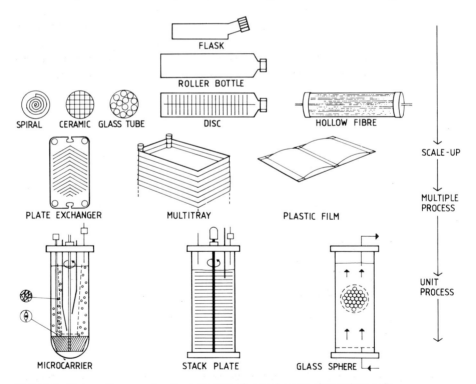

Figure 4. The scaling-up of culture systems for anchorage-dependent cells.

Table 3. Properties of 3 mm glass beads

Weight (g)	0.0375
Surface area (cm^2)	0.283
Volume (cc)	0.015
Packed volume (cc)	0.028
Thus a culture of 10 000 cm^3 =	
35 336	Glass beads
990 ml	Total volume
460 ml	Medium in culture
21 cm^2/ml	Medium
20 × 8 cm	Typical culture size

and Spier (18) who used a 100 litre capacity system for the growth of BHK21 cells (and FMDV).

Spheres of 3 mm diameter pack sufficiently tightly to prevent the packed bed from shifting, but allow sufficient flow of medium through the column so that fast flow rates, which would cause mechanical shear damage, are not needed. The physical properties of glass spheres are given in *Table 3*, and these data allow the necessary culture parameters to be calculated. A system which can be constructed in the laboratory is illustrated in *Figure 5*. Medium is transferred by a peristaltic pump in this example, but an airlift driven system is also suitable and gives better oxygenation. Medium can be passed either up or down the column with no apparent difference in results.

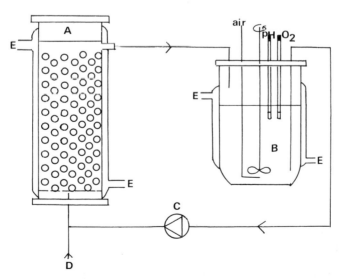

Figure 5. A glass bead bioreactor. A, glass bead bed; B, reservoir; C, pump; D, inoculation and harvest line; E, temperature-controlled water jackets.

Protocol 2. Growing cells in a bead bed reactor

1. Prepare sufficient media at the rate of 1–2 ml/10 cm^2 culture surface area.
2. Circulate media between reservoir and vessel until the system is equilibrated with regard to temperature and pH (allow sufficient time for the glass beads to warm up to 37°C).
3. Add cell inoculum (1 × 10^4/cm^2) to a volume of medium equal to the void volume.
4. Drain the bead column into the reservoir, mix the prepared cell suspension thoroughly, and add to the bead column.
5. Clamp off all tubing and allow the cells to attach (3–8 h depending on cell type).
6. Once cells are attached start perfusing the medium, initially at a slow rate (1 linear cm/10 min). Visually inspect the effluent medium for cloudiness which would indicate that cells have not attached (stop perfusing if this is the case and resume after a further 8 hours).
7. After 24 h this rate can be increased but should be kept below 5 cm/min, otherwise cells may be stripped off (especially mitotic cells). The pH readout at the exit end and a comparison of glucose concentrations in the input and exit sampling points will indicate whether the flow rate is sufficient.
8. Monitor cell growth by glucose determinations. Glucose utilization rates should be found empirically for each cell and medium, but a rough indication is 2–5 × 10^8 cells produced/g glucose.
9. When the culture is assumed to be confluent, drain off the medium, wash the column with buffer, add the void volume of trypsin/versene (or pronase, etc.) and allow to stand for 15–30 min. Cell harvesting can be accelerated by occasionally releasing the trypsin from the column into the harvest vessel and pushing it back into the culture again. Alternatively, feed gas bubbles through the column (only very slowly, otherwise the beads will shift and damage the cells).
10. The culture vessel should be washed immediately after use, otherwise it becomes very difficult to clean. Use a detergent such as Decon, and circulate continuously through the culture system, followed by tap water and then several changes of distilled water. Autoclave moist if control probes (oxygen, pH) are used. Use Pyrex glass beads in preference to the standard laboratory beads which are made of soda glass.

Although 3 mm beads are normally used in order to maximize the available surface area, the use of 5 mm spheres actually gives a higher total cell yield despite the reduced surface area per unit volume (10). A development of this system which significantly increases cell yields is described in Section 5.

ii. Heterogeneous reactor

Circular glass or stainless steel plates are fitted vertically, 5–7 mm apart, on a central shaft. This shaft may be stationary, with an airlift pump for mixing, or revolving around a vertical (6 r.p.h.) or horizontal (50–100 r.p.m.) axis. This multisurface propagator (19) was used at sizes ranging from 7.5 to 200 litres, giving a surface area of up to 2×10^5 cm^2. The author's experience is solely with the horizontal stirred plate type of vessel (*Figure 4*, stack plate), which is easier to use and has been successful for both heteroploid and human diploid cells. The main disadvantage with this type of culture is the high ratio of medium volume to surface area (1 ml to 1–2 cm^2). This cannot be altered with the horizontal types, although it can be halved with the vertically revolving discs.

iii. Homogeneous systems (microcarrier)

When cells are grown on small spherical carriers they can be treated as a suspension culture, and advanced fermentation technology processes and apparatus can be utilized. The method was initiated by Van Wezel (20), who used Dextran beads (Sephadex A-50). These were not entirely satisfactory, because the charge of the beads was unsuitable and also possibly due to toxic effects. However, much developmental work has since resulted in many suitable microcarriers being commercially available (*Table 4*).

The author's experience has largely been with Dextran-based microcarriers (Cytodex and Dormacell) and much of the following discussion is based on these products. The choice of this microcarrier was based on a preference for a dried product which could be accurately weighed and then prepared *in situ*, and the fact that with a density of only 1.03 g/ml this product could be used at concentrations of up to 15 g/litre (90 000 cm^2/litre). However, this preference does not detract from the quality of other microcarriers, most of which have been used with equal success, and many of which offer particular advantages as discussed later.

Culture apparatus

A spinner vessel is not suitable unless the stirring system is modified. Paddles with large surface area are needed. These are commercially available (e.g. Bellco 'μ' and 'Cellon' spinner vessels), but can be easily constructed out of a silicone rubber sheet and attached to the magnet with plastic ties. The advantage of constructing one within the laboratory is that the blades can be made to incline 20–30° from the vertical, thus giving much greater lift and mixing than vertical blades (*Figure 6*). Microcarrier cultures are stirred very slowly (maximum 75 r.p.m.) and it is essential to have a good-quality magnetic stirrer that is capable of giving a smooth stirring action in the range of 20–100 r.p.m. Never use a stirrer mechanism that has moving surfaces in contact with each other in the medium, otherwise the microcarriers will be crushed. Thus, stirrers which revolve on the bottom of the vessel are unsuitable. As mixing,

71

Table 4. Comparison of microcarriers

Trade Name	Manufacturer	Material	SG	Diam (μm)	Area (cm²/g)
Acrobead	Galil	Polyacrolein	1.04	150	5000
Biosilon	Nunc	Polystyrene	1.05	160–300	255
Bioglas	Solohill Eng.	Glass[a]	1.03	150–210	350
Bioplas	Solohill Eng.	Polystyrene[a]	1.04	150–210	350
(Biospheres		Collagen[a]	1.02	150–210	350)
Biocarrier	Biorad	Polyacrylamide	1.04	120–180	5000
Cellfast	QDM Lab.	Silica/Chitosan			10 000
Cytodex 1	Pharmacia	DEAE Sephadex	1.03	160–230	6000
Cytodex 2	Pharmacia	DEAE Sephadex	1.04	115–200	5500
Cytodex 3	Pharmacia	Collagen	1.04	130–210	4600
Cytosphere	Lux	Polystyrene	1.04	160–230	250
Dormacell	Pfeifer & Langen	Dextran	1.05	140–240	7000
DE-53	Whatman	Cellulose	1.03	Fibres	4000
Gelibead	Hazelton Lab.	Gelatin	1.04	115–235	3800
Mica	Muller-Lieheim	Polyacylamide	1.04		350
Micarcel G	Reactifs IBF	Polyacrylamide/Collagen/glucoglycan	1.03		5000
Microdex	Dextran Prod.	DEAE Dextran	1.03	150	250
Superbeads	Flow Lab.	DEAE Sephadex	1.03	150–200	6000
Ventreglas	Ventrex	Glass	1.03	90–210	300
Ventregel	Ventrex	Gelatin	1.03	150–250	4300

[a] Biospheres (glass, plastic, collagen) available at SG of 1.02 or 1.04 and diameters of 150–210 or 90–150 μm (manufactured by Solohill Eng. and distributed by Whatman and Cellon).

and thus mass transfer, is so poor in these cultures, the depth of medium should not exceed the diameter by more than a factor of 2 unless oxygenation systems or regular medium changes are performed. In the basic culture systems, medium changes have to be carried out at frequent intervals. It is worthwhile, therefore, to make suitable connections to the culture vessel to enable this to be done conveniently *in situ*. This will pay dividends in reducing the chances of contaminating the culture. A simplified culture set-up is shown in *Figure 7*.

Initiation of the culture

Many of the factors discussed in Section 2 are critical when initiating a microcarrier culture. Microcarriers are spherical, and cells will always attach to an area of minimum curvature. Therefore a microcarrier surface can never be ideal, however suitable its chemical and physical properties.

Ensure the media and beads are at a stable pH and temperature, and inoculate the cells (from a logarithmic, not a stationary culture) into a third of the final medium volume. This increases the chances of cells coming into contact with the microcarriers. Microcarrier concentrations of 2–3 g/litre should be used. Higher concentrations need environmental control or very

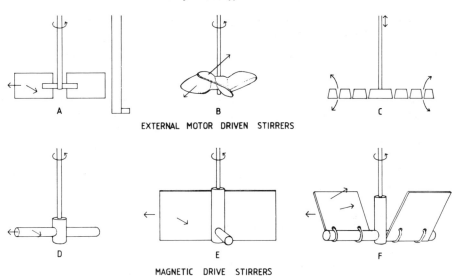

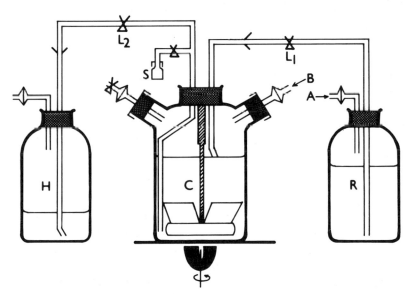

Figure 6. Types of impellers for growing suspension and microcarrier cells. (A) flat disc turbine, $d2:d1=0.33$, radial flow, turbulent mixing; (B) marine impeller, $d2:d1=0.33$, axial flow, turbulent mixing, angle 25°; (C) vibro-mixer; (D) stirrer bar, $d2:d1=0.6$, radial flow, laminar mixing; (E) vertical and (F) angled (25°) paddle, $d1:d2=0.6-0.9$, axial and radial flow, laminar and turbulent mixing. ($d1$ is the diameter of the vessel and $d2$ is the diameter of the impeller.)

Figure 7. A simple microcarrier culture system allowing easy medium changing. To fill the culture (C) open clamp (L1) and push the medium from the reservoir (R) using air pressure (at A). To harvest, stop the stirring for 5 min, open L2 and push the medium from C to H using air pressure at B. S is a sampling point.

frequent medium changes. Historically it was considered best to allow attachment in a stationary culture and to stir for 30 s every 30 min. However, this always gave rise to uneven numbers of cells per bead and now, with the vast improvements in microcarrier quality, stirring can start immediately at the lowest possible setting that gives complete mixing (10–25 r.p.m.).

After the attachment period (3–8 h), slowly top up the culture to the working volume and increase the stirring rate to maintain completely homogeneous mixing. If these conditions are adhered to, and there are no changes in temperature and pH, then all cells which grow on plastic surfaces should readily initiate a microcarrier culture.

Maintenance of the culture

It is very easy to monitor the progress of cell growth in a microcarrier culture. Samples can be easily removed, cell counts (by nuclei counting) and glucose determinations carried out, and the cell morphology examined. As the cells grow, so the beads become heavier and need an increased stirring rate. After 3 days or so the culture will become acidic and need a medium change. Again, this is an extremely easy routine: turn off the stirrer, allow the beads to settle for 5 min, and decant off as much medium as desired. Top up gently with fresh medium (pre-warmed to 37°C) and restart the stirring.

Harvesting

It is very difficult to harvest many cell types from microcarriers unless the cell density on the bead is very high. (Incidentally, do not expect cells to multilayer on microcarriers to the same degree as in stationary cultures.) Harvesting can be attempted by draining off the medium, washing the beads at least once in buffer, and adding the desired enzyme. Stir the culture fairly rapidly (75–125 r.p.m.) for 20–30 min. If the cells detach, a high proportion can be collected by allowing the beads to settle out for 2 min and then decanting off the supernatant. For a total harvest pour the mixture into a sterilized sintered glass funnel (porosity grade 1). The cells will pass through the filter, but the microcarriers will not. Alternatively, insert a similar filter (usually stainless steel) into the culture vessel.

If the microcarrier is dissolved then the cells can be released into suspension completely undamaged, and therefore of better quality than when the cells are trypsinized. Usually this is a far quicker method than trypsinization of cells. Gelatin beads are solubilized with trypsin and/or EDTA, collagenase acts on the collagen-coated beads, and dextranase can be used on the dextran microcarriers.

Scaling-up of microcarrier culture

Scaling-up can be achieved by increasing the microcarrier concentration, or by increasing the culture size. In the first case nutrients and oxygen are very rapidly depleted, and the pH falls to non-physiological levels. Medium changes are not only tedious but provide rapidly changing environmental conditions. Perfusion, either to waste or by a closed loop, must be used to achieve cultures with high microcarrier concentration. This can be brought about only by an efficient filtration system so that medium without cells and microcarriers can be withdrawn at a rapid rate. The only satisfactory means of doing this is with the type of filtration system illustrated in *Figure 8*. This is constructed of a stainless steel mesh with an absolute pore size in the 60–120 μm range. Attachment to the stirrer shaft means that a large surface area filter can be used and the revolving action discourages cell attachment and clogging.

However scaling-up is achieved, oxygen limitation is the chief factor to overcome. This is an especially difficult problem in microcarrier culture because stirring speeds are low (it was pointed out in Section 2.7 that the stirring speed has to be above 100 before it significantly affects the aeration rate). Sparging cannot be used as microcarriers get left literally high and dry above the medium level due to the foaming it causes. The perfusion filter previously described, however, does allow sparging into that part of the culture in which no beads are present. Unfortunately, this means that the most oxygenated medium in the culture is removed by perfusion but at a sparging rate of 10 cm^3 of oxygen/l considerable diffusion of oxygenated

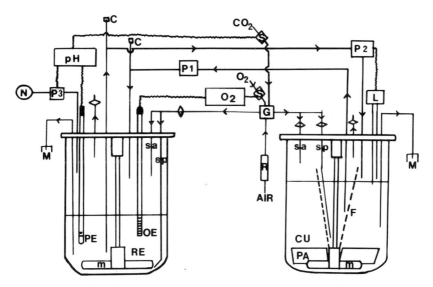

Figure 8. A closed-loop perfusion system with environmental control to allow high cell density microcarrier culture. CU, culture vessel; RE, reservoir; C, connector for medium changes and harvesting; F, filter; G, gas blender; L, level controller; M, sampling device; m, magnet; N, alkali (NaOH) reservoir; OE, oxygen electrode; PE, pH electrode; P, pumps: 1, medium to reservoir (continuous), 2, medium to culture (controlled by L) 3, alkali to reservoir (controlled by PE); R, rotameter; sa, gas supply for surface aeration; sp, gas supply for sparging; S, solenoid valve.

medium occurs into the main culture. However, modified spin filters are available which have a separate compartment for perfusion and sparging (21). Thus oxygen delivery to a microcarrier culture can utilize the following systems (see also *Table 2*):

(a) surface aeration

(b) increasing the perfusion rate of fully oxygenated medium from the reservoir

(c) sparging into the filter compartment (21)

These three systems are used by the author in the system illustrated in *Figure 8* to run cultures at up to 15 g/l Cytodex-3 in culture vessels between 2 and 20 l working volume. A typical experiment is shown in *Figure 9*.

Summary of microcarrier culture

Microcarrier cultures are used commercially for vaccine and interferon production in fermenters up to 4000 l. These processes use heteroploid or primary cells. One of the problems of using very large scale cultures is that the required seed inoculum gets progressively larger. Harvesting of cells from large unit scale cultures is not always very successful, although the availability

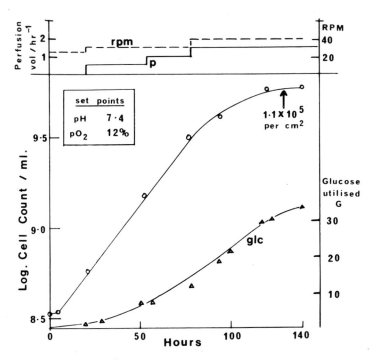

Figure 9. Growth of GPK epithelial cells on Cytodex 3 (10 g/l) in the 10 litre culture system described in *Figure 8*. ○——○, Cell concentration; △——△, glucose utilized.

of collagen and gelatin microcarriers eases this problem considerably. *In situ* trypsinization of cells is the usual method of choice, but it is difficult to remove a high percentage of the cells in a viable condition. Filters incorporated into the culture vessel provide the best solution and are most efficient with the 'hard' (glass, plastic) microcarriers which do not block the filter pores like the Dextran and gelatin microcarriers. Microcarriers cost in the region of £1200–1500/kg. Some consideration must be given to washing and re-using the beads but this, of course, is not possible with the gelatin- or collagenase-treated beads. A development of microcarriers is the porous microcarrier system described in Section 5.

3.3.5 Summary: choice of equipment

A wide range of commercially available and laboratory-made equipment has been reviewed. This should enable a choice to be made, depending upon the amount and type of cells or product needed and the financial and staff resources available. To review the choices available, *Figure 4* and *Table 5* should be consulted.

Table 5. Comparison of monolayer culture systems—relative productivity and scale-up potential

Culture	Unit surface area (cm^2)	Surface area: medium ratio [a]	Max. vol. (l)
Multiple processes:			
Roux bottle	175–200	1–2	
Roller bottle	850–175	3	
—Corrugated	1700	6	
—Spiracell	6000	6	
—Corbeil	15 000	2.5	
Plastic bag	1000	5	
Multitray	24 000	2.5	
Hollow fibres	18 000	72	
Opticell	120 000	30	
Membroferm	200 000	50	
Unit processes:			
Heli-Cel	16 000/l	16	(1) [b]
Glass spheres	7500/l	18	200
Porous glass spheres	80 000/l	150	(10) [b]
Stack plates	1250/l	1.25	200
Plate exchanger	20 000	7	200
Microcarrier (5 g/l)	35 000/l	35	4000
Porous microcarrier	300 000/l	300	(24) [b]

[a] Volume in bioreactor—does not allow for media changes or reservoir.
[b] Current experimental scales—scale-up potential.

4. Suspension culture

As indicated previously in this chapter, suspension culture is the preferred method for scaling-up cell cultures. Some cells, especially those of haemo-poietic derivation, grow best in suspension culture, others, particularly trans-formed cells, can be adapted or selected, while some, for example human diploid cell strains (WI-38, MRC-5), will not survive in suspension at all. A further factor that dictates whether suspension systems can be used is that some cellular products are expressed only when the cell exists in a monolayer or if cell-to-cell contact is established (e.g. for the spread of intracellular viruses through a cell population).

4.1 Adaptation to suspension culture

Different cell lines vary in the ease with which they can be persuaded to grow in suspension. For those that have the potential, there are two basic pro-cedures that can be used to generate a suspension cell line from an anchorage-dependent one.

4.1.1 Selection

This method, as demonstrated by the derivation of the LS cell line from L-929 cells by Paul and Struthers (22), and the HeLa-S3 clone from HeLa cells, depends on the persistence of loosely attached variants within the population. A confluent monolayer culture is lightly tapped or gently swirled, the medium decanted from the culture, and cells in suspension recovered by centrifuga-tion. Cells have to be collected from many cultures to provide a sufficient number to start a new culture at the required inoculum density (at least 2×10^5/ml). This procedure has to be repeated many times over a long period, because many of the cells that are collected are only in the mitotic phase, rather than potential suspension cells (cells round up and become very loosely attached to the substrate during mitosis). Eventually it is possible with some cell lines to derive a viable cell population that divides and grows in static suspension, just resting on the substrate rather than attaching and spreading out.

4.1.2 Adaptation

This method probably works the same way as the selection procedure, except that a selection pressure is exerted on the culture while it is maintained in suspension mechanically. However, with many cell lines the relatively large number that become anchorage-independent suggests it is more than just selection of a few variant cells. The method given in *Protocol 3*, used success-fully by the author for many cell lines, is based on the one published by Capstick et al. (23) for BHK21 cells.

Protocol 3. Adaptation to suspension culture

1. Prepare the cell suspension by detaching monolayer cells with trypsin (0.1%) and versene (0.01%).

2. Make ready at least two, but preferably three or more, spinner vessels by adding the growth medium (calcium- and magnesium-free modification of any recognized culture medium).

3. Add cells at a concentration of at least 5×10^5/ml and commence stirring (the lowest rate which keeps the cells homogeneously mixed, e.g. 250 r.p.m.).

4. Sample and carry out viable cell counts every 24 h.

5. Every 3 days remove the medium from the culture, centrifuge (1000 g.) and resuspend the cells in fresh medium at a density of at least 2.5×10^5/ ml. Depending upon the degree of cell death, cultures will almost certainly have to be amalgamated to keep the cell density at the required level.

6. If, at a medium change, there is significant attachment of cells to the culture vessel, especially at the air–medium interface, then add trypsin (0.01%)/versene (0.01%) mixture to the empty vessel. Stir at 37°C for 30 min and recover the detached cells by centrifugation. If the cells in suspension are badly clumped, they too can be added to the trypsin/versene solution. This treatment is usually necessary at the first and second medium changes, but very rarely after this point. If cell clumping persists, then add 50 µg/ml trypsin or Dispase (Boehringer) to the growth medium.

7. Successful adaptation is recognized initially by an increase in cell numbers after a media change, and subsequently by getting consistent cell yields in a given period of time (indicating a constant growth rate).

8. It is usually necessary to maintain the newly established cell strain in stirred suspension culture, because reversion to anchorage dependence can occur in static cultures. Sometimes cells will adhere to the substrate without becoming completely spread out and continue to grow and divide with a near-spherical morphology.

4.2 Static suspension culture

Many cell lines will grow as a suspension in a culture system used for monolayer cells (i.e. with no agitation or stirring). Cell lines that are capable of this form of growth include the many lymphoblast lines (e.g. MOLT, RAJI), hybridomas, and some non-haemopoietic lines, such as the LS cells described in the previous section. However, with the latter type of cells there is always the danger of reversion to a monolayer (e.g. a small proportion of the LS cell line always attaches and is discarded at each sub-cultivation). Static suspension culture is unsuitable for scale-up, for reasons already stated for monolayer culture.

4.3 Small-scale suspension culture

For the purpose of this chapter, small-scale means under 2 litres. This may seem an entirely arbitrary definition, but it is made on the basis that above this volume additional factors apply. The conventional laboratory suspension culture is the spinner flask, so called because it contains a magnetic bar as the stirrer, and this is driven from below the vessel by a revolving magnet. Details of some of the readily available stirrer vessels, together with suppliers and available size range, are given in *Figure 10*. It is important to get a good-quality magnetic stirrer so that the magnetic field is strong enough to turn the magnet smoothly over the prescribed range of stirring speeds (50–500 r.p.m.) and is reliable over long periods of use. Additional requirements are that it will not get too hot (as the culture vessel sits directly on top) and that

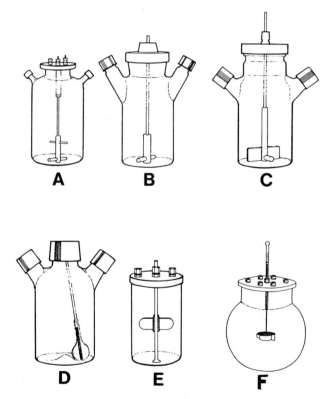

Figure 10. Commercially available spinner cultures. (A) LH Fermentation Biocul (1–20L); (B) Bellco and Wheaton Spinner Flasks (25 ml–2 litres); (C) Bellco and Cellon u spinner (25 ml–2 litres); (E) Techne (25 ml–5 litres); (E) Techne Cytostat (1 litre); (F) Techne BR-06 Bioreactor (3 litres).

the tachometer gives a true reading of stirring speed. Always check that the stirrer will re-start from all required settings after an interruption in power.

4.4 Scaling-up factors

In scale-up, both the physical and chemical requirements of cells have to be satisfied. The chemical factors require environmental monitoring and control to keep the cells in the proper physiological environment. These factors, which include oxygen, pH, medium components, and removal of waste products, are described in Section 2.

Physical parameters include the configuration of the bioreactor and the power supplied to it. The function of the stirrer impeller is to convert energy (measured as kW/m^3) into hydrodynamic motion in three dimensions (axial, radial, and tangential). The impeller has to circulate the whole liquid volume and to generate turbulence (i.e. it has to pump and to mix) in order to create a homogeneous blend, to keep cells in suspension, to optimize mass transfer rates between the different phases of the system (biological, liquid, and gaseous), and to facilitate heat transfer.

Good mixing becomes increasingly difficult with scaling-up, and the power needed to attain homogeneity can cause problems. The energy generated at the tip of the stirrer blade is a limiting factor as it gives rise to a damaging shear force. Shear forces are created by fluctuating liquid velocities in turbulent areas. The factors which affect this are, impeller shape (this dictates the primary induced flow direction) (*Figure 6*), the ratio of impeller to vessel diameter, and the impeller tip speed (a function of rotation rate and diameter). The greater the turbulence the more efficient the mixing, but a compromise has to be reached so that cells are not damaged. Large impellers running at low speeds give a low shear force and high pumping capacity, whereas smaller impellers need high stirring speeds and have high shear effects.

Magnetic bar stirring gives only radial mixing and no lift or turbulence. The marine impeller is more effective for cells than the flat-blade turbine impellers found in many bacterial systems, as it gives better mixing at low stirring speeds. If the cells are too fragile for stirring, or if sufficient mixing cannot be obtained without causing unacceptable shear rates, then an alternative system may have to be used. Pneumatic energy, for example mixing air bubbles (e.g. airlift fermenter) or hydraulic energy (e.g. medium perfusion), can be scaled-up without proportionally increasing the power. To improve the efficiency of mechanical stirring, the design of the stirrer paddle can be altered (e.g. as described for microcarrier culture), or multiple impellers can be used. Some examples are given in *Figure 6*.

A totally different stirring concept is the Vibro-mixer. This is a non-rotating agitator which produces a stirring effect by a vertical reciprocating motion with a path of 0.1–3 mm at a frequency of 50 Hz. The mixing disc is fixed horizontally to the agitator shaft, and conical-shaped holes in the disc caused a pumping action to occur as the shaft vibrates up and down. The shaft is driven by a motor which operates through an elastic diaphragm; this also provides a seal at the top of the culture. A fermentation system using this principle is available commercially (Vibro-Fermenter, Chemap). The advantages of this system are the greatly reduced shear forces, random mixing, reduced foaming, and reduced energy requirement, especially for scaling-up.

The significant effect on vessel design of moving from a magnetic stirrer to a direct drive system is the fact that the drive has to pass through the culture vessel. This means some complexity of design to ensure a perfect aseptic seal while transferring the drive, complete with lubricated bearings, through the bottom or top plate of the vessel. Culture vessels with magnetic coupling are becoming available at increasingly higher volumes, and overcome this problem of aseptic seals.

Scaling-up cannot be proportional; one cannot convert a 1 litre reactor into a 1000 litre reactor simply by increasing all dimensions by the same amount. The reasons for this are mathematical: doubling the diameter increases the volume threefold, and this affects the different physical parameters in different ways. One factor to be taken into consideration is the height/diameter ratio, as this is one of the most important fermenter design parameters. In sparged systems the taller the fermenter in relation to its diameter the better, as the air pressure will be higher at the bottom (increasing the oxygen solubility) and the residence time of bubbles longer. However, in non-sparged systems which are often used for animal cells and rely on surface aeration, the surface area:height ratio is more important and a 1:1 ratio should be maintained.

Mass transfer between the culture phases has been discussed in relation to oxygen (Section 2.7). it is this characteristic of scaling-up which demands the extra sophistication in culture design to maintain a physiologically correct environment. This sophistication includes impeller design, oxygen delivery systems, vessel geometry, perfusion loops, or a completely different concept in culture design to the stirred bioreactor.

4.5 Stirred bioreactors

The move from externally driven magnetic spinner vessels to fermenters capable of scale-up from 1 litre to 1000 litres and beyond, has the following consequences at some stage:

(a) change from glass to stainless steel vessels

(b) change from a mobile to a static system: connection to steam for *in situ* sterilization; requirement for water-jacket or internal temperature control; need for a seed vessel, a medium holding vessel and downstream processing capablity

(c) greater sophistication in environmental control systems to meet the increasing mass transfer equipment

In practice, the maximum size for a spinner vessel is 20 litres. Above this size there are difficulties in handling and autoclaving, as well as the difficulty of being able to agitate the culture adequately. Fermenters with motor driven stirrers are available from 500 ml, but these are chiefly for bacterial growth. It is a significant step to move above the 10–20 litre scale as the cost of the equipment is significant (e.g. £20 000–25 000 for a complete 35 litre system) and suitable laboratory facilities are required (steam, drainage, etc.). There is a wide range of vessels available from the various fermenter suppliers which include some of the following modifications:

- suitable impeller (e.g. marine)
- no baffles
- curved bottom for better mixing at low speeds
- water-jacket rather than immersion heater type temperature control (to avoid localized heating at low stirring speeds)
- top-driven stirrer so that cells cannot become entangled between moving parts
- mirror internal finishes to reduce mechanical damage and cell attachment

As long as adequate mixing, and thus mass transfer of oxygen into the vessel, can be maintained without damaging the cells, there is no maximum to the scale-up potential. Namalva cells have been grown in 8000 litre vessels for the production of interferon. There are many heteroploid cell lines, such as Vero, HeLa, and hybridomas, which would grow in such systems. As long as regulatory agencies required pharmaceutical products to be manufactured predominantly from normal diploid cells (which will not grow in suspension), the only motive for using this scale of culture was for veterinary products. However, the recent licencing of products such as tPA, interferon, and therapeutic monoclonal antibodies from heteroploid lines means that large-scale bioreactors are in widespread use (but see Section 5). There are many applications for research products from various types of cells, and this is served by the 2–50 litre range of vessels. At present, the greatest incentive for large-scale systems is to grow hybridoma cells (producing monoclonal antibody). In tissue culture the antibody yield is 50- to 100-fold lower than when the cells are passaged through the peritoneal cavity of mice, although the purity is greater, particularly if serum-free medium is used. At the moment, the need is to supply antibodies to meet the requirements for diagnostic and affinity chromatography purposes. The future need will be to supply them for therapeutic and prophylactic drug treatments. Many of these cells are fragile and low-yielding in culture, and the specialized techniques and apparatus described in Sections 4.7 and 5 are partly aimed towards this type of cell.

4.6 Continuous-flow culture

4.6.1 Introduction

At submaximal growth rates, the growth of a cell is determined by the concentration of a single growth-limiting nutrient. This is the basis of the chemostat, a fixed-volume culture, in which medium is fed in at a constant rate, mixed with cells, and then leaves at the same rate. The culture begins as a batch culture while the inoculum grows to the maximum value that can be supported by the growth-limiting nutrient (assuming the dilution rate is less than the maximum growth rate). As the growth-limiting nutrient decreases in concentration, so the growth rate declines until it equals the dilution rate. When this occurs, the culture is defined as being in a 'steady state' as both the cell numbers and nutrient concentrations remain constant.

When the culture is in a steady state, the cell growth rate (μ) is equal to the dilution rate (D). The dilution rate is the quotient of the medium flow rate (f) per unit time and the culture volume (V):

$$\mu = D = f/V \text{ day}^{-1} \qquad \text{Equation 8}$$

As the growth rate is dependent on the medium flow rate, the mean generation (doubling) time can be calculated:

$$t_d = \ln 2/D \qquad \text{Equation 9}$$

An alternative system to the chemostat is the turbidostat in which the cell density is held at a fixed value by altering the medium supply. The cell density (turbidity) is usually measured through a photoelectric cell. When the value is below the fixed point, medium supply is stopped to allow the cells to increase in number. Above the fixed point, medium is supplied to wash out the excess cells. This system really works well only when the cell growth rate is near maximum. However, this in fact is its main advantage over the chemostat which is least efficient, or controllable, when operating at the cell's maximum growth rate. Continuous flow culture of animal cells has been well reviewed by Tovey (24).

4.6.2 Equipment

Complete chemostat systems can be purchased from all dealers in fermentation equipment. However, systems can be easily constructed in the laboratory (*Figure 11*) (25). The culture vessel needs a side arm overflow at the required liquid level, which should be approximately half the volume of the vessel. If a suitable 37°C cabinet is not available, then a water-jacketed vessel is needed. Apart from this, all other components are standard laboratory items. Vessel enclosures can be made from silicone (or white rubber) bungs wired onto the culture vessel. A good-quality peristaltic pump, such as the Watson–Marlow range, is recommended.

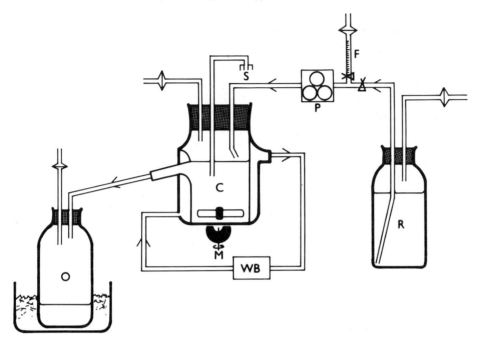

Figure 11. Continuous-flow culture system (25). C, water-jacketed culture vessel; WB, waterbath and circulating system; O, overflow vessel; R, reservoir; P, pump; S, sampling device; F, burette for measuring flow rate; M, magnetic drive.

4.6.3 Experimental procedures

Recommended cells are LS, HeLa-S3, or an established lymphoblastic cell line such as L1210 or a hybridoma. Growth-limiting factors can be chosen from the amino acids (e.g. cystine) or glucose.

Protocol 4. Growth of cells in continuous-flow culture

1. Inoculate the culture vessel at 10^5 cells/ml in preferred medium (e.g. Eagles MEM with 10% calf serum and the growth-limiting factor).

2. The chosen dilution rate is turned on after 24–48 h of growth. The maximum rate must not exceed the maximum growth of the cell line, which is usually within the range 14–27 h doubling time (although some cells can double their number in 9 h). Thus, the dilution rate will be in the range of 0.1 day^{-2} ($t_d = 166$ h) to 1.2 ($t_d = 14$ h).

3. A steady state will become established within 100–200 h, although it may take up to 400 h, especially at the low dilution rates. This will be recognized by the fact that the daily cell counts will not vary by more than the expected counting error.

Protocol 4. *Continued*

4. The culture can be maintained almost indefinitely, assuming it is kept sterile and no breakdown in components occurs. Sometimes the culture has to be terminated because of excessive attachment of cells at the interfaces. A duration of 1000 h is considered satisfactory.

5. Once the steady-state has been demonstrably achieved, the flow rate can be altered and the response of the cell population in establishing new steady-state conditions can be studied.

6. As well as carrying out routine cell counts, measurements can also be made to demonstrate the homogeneity of the cells and the medium. For instance, the size and chemical composition of the cells is remarkably consistent. Also, some of the medium components should be measured (e.g. glucose, an amino acid, lactic acid) to demonstrate again the consistency of the culture over a long period of time.

4.6.4 Uses

Continuous-flow culture provides a readily available continuous source of cells. Also, because optimal conditions or any desired physiological environment can be maintained, the culture is very suitable for product generation (as already shown for viruses and interferon) (24). For many purposes a two-stage chemostat is required so that optimal conditions can be met for cell growth (first stage) and product generation (second stage).

4.7 Airlift fermenter

The airlift fermenter relies on the bubble column principle both to agitate and to aerate a culture. Instead of mechanically stirring the cells, air bubbles are introduced into the bottom of the culture vessel. An inner (draft) tube is placed inside the vessel and mixing occurs because the air bubbles lift the medium (aerated medium has a lower density than non-aerated medium). The medium and cells which spill out from the top of the draft tube then circulate down the outside of the vessel. The amount of energy (compressed air) needed for the system is very low, shear forces are absent, and this method is thus ideal for fragile animal and plant cells. Also as oxygen is continuously supplied to the culture, the large number of bubbles results in a high mass transfer rate. Culture units, as illustrated in *Figure 12*, are commercially available in sizes from 2 to 90 litres (LH Fermentation). The one disadvantage of the system is that scale-up is more or less linear (the 90 litre vessel requires nearly 4 m headroom). Whether it will be possible to use multiple draft tubes, and thus enable units with greatly increased diameters to be used, remains a developmental challenge. However, 2000 litre reactors are in operation for the production of monoclonal antibodies.

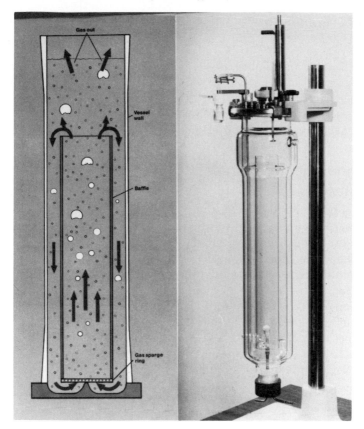

Figure 12. Principle of the airlift fermenter.

5. Immobilized cultures

Immobilized cultures are popular because they allow far higher unit cell densities to be achieved (50–200 × 10^6 cells/ml) and they also confer stability, and therefore longevity, to cultures. Cells *in vivo* are in a three-dimensional tissue matrix, therefore immobilization can mimic this physiological state. Higher cell density is achieved by facilitating perfusion of suspension cells and increasing the unit surface area for attached cells. In addition many immobilization materials protect cells from shear forces created by medium flow dynamics. As cells increase in unit density they become less dependent on the external supply of many growth factors, etc. provided by serum, so real cost saving can be achieved.

The emphasis has been on developing systems for suspension cells because commercial production of monoclonal antibody has been such a dominant factor. Basically two approaches have been used: immurement (confining cells within a medium-permeable barrier), and entrapment (en-meshing cells within an open matrix through which medium can flow un-hindered) (7).

Immobilization techniques that can be scaled-up, such as the porous carriers described in Section 5.3, will become a dominant production tech-nology once the beads and process parameters have been optimized. They are currently excellent laboratory systems for the manufacture of cell products and, with gelatin carriers, production of cells.

5.1 Immurement cultures

5.1.1 Hollow fibres

These have been discussed in Section 3.3.3, and are very effective for suspen-sion cells at scales up to about 1 litre ($1–2 \times 10^8$ cells/ml). Simple systems can be set up in the laboratory by purchasing individual hollow fibre cartridges (e.g. Amicon, Microgon, Minntech). All that is needed is a medium reservoir and a pump to circulate the media through the intracapillary section of the cartridge, and a harvest line from the extracapillary compartment in which the cells and product reside (*Figure 13*). To get better performance there are a number of 'turn-key' systems that can be purchased, from the relatively simple units (e.g. Kinetek, Cellco, Cell-Pharm) to extremely complex and sophisticated units (e.g. Endotronics Acusyst) capable of producing up to 40 g of monoclonal antibody per month.

5.1.2 Membranes

The Dynacell culture system (Millipore) is a small-scale (12.5 ml) multi-membrane module through which medium is pumped from a medium reser-voir via a gas exchange diffuser. In the author's laboratory a unit has been in

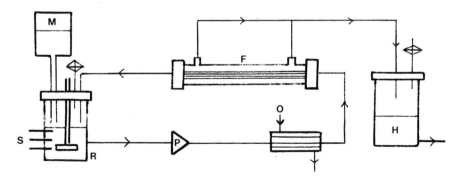

Figure 13. A hollow fibre culture system. F, bioreactor (200 ml); R, reservoir (20 litres); M, medium addition; H, harvest vessel; S, sensors (pH, O_2, temperature, etc.); O, oxygena-tion module; P, pump (200 ml/min).

continuous operation for 85 days producing nearly 1 g of monoclonal antibody in that period. This is a very simple, reliable, and economic (especially in staff time) production system, as long as only small quantities of product are needed.

Large-scale versions of membrane bioreactors are available (e.g. Membroferm, MBR Bioreactor).

5.1.3 Encapsulation

The entrapment of cells in semi-solid matrices, or spheres, has many applications, but the basic function is to stabilize the cell and thus protect it from suboptimal conditions. Cells can be immobilized by adsorption, covalent bonding, cross-linking, or entrapment in a polymeric matrix. Materials that can be used are gelatin, polylysine, alginate, and agarose; the choice largely depends upon the problem being addressed. These techniques have been used to protect cells that are being transported or posted between laboratories; to store cells at 4°C for an extended period of time (e.g. 5 weeks); to avoid immune rejection of transplanted cells; and to protect fragile cells (e.g. hybridomas) from mechanical stress in large-scale culture equipment (26, 27). The latter application not only makes it possible to use such equipment but also allows production of hormones, antibodies, immunochemicals, and enzymes over much longer periods than is possible in homogeneous suspension culture. The matrix allows free diffusion of nutrients and generated product between the enclosed microenvironment and the external medium. Alginate is a polysaccharide and is cross-linked with Ca^{2+} ions. The rate of cross-linking is dependent on the concentration of Ca^{2+} (e.g. 40 min with 10 mM $CaCl_2$). A recommended technique is to suspend the cells in isotonic NaCl buffered with Tris (1 mM) and 4% sodium alginate, and to add this mixture dropwise into a stirred solution of isotonic NaCl, 1 mM Tris, 10 mM $CaCl_2$ at pH 7.4. The resulting spheres are 2–3 mm in diameter. The entrapped cells can be harvested by dissolving the polymer in 0.1 M EDTA or 35 mM sodium citrate. Disadvantages of alginate are that calcium must be present and phosphate absent and that large molecules, such as monoclonal antibodies, cannot diffuse out. For these reasons, agarose in a suspension of paraffin oil provides a more suitable alternative. 5% agarose in PBS free of Ca^{2+} and Mg^{2+} is melted at 70°C, cooled to 40°C, and mixed with cells suspended in their normal growth medium. This mixture is added to an equal volume of paraffin oil and emulsified with a Vibro-mixer. The emulsion is cooled in an ice-bath, growth medium added, and, after centrifugation, the oil is removed. The spheres (80–200 μm) are washed in medium, centrifuged and, after removing the remaining oil, transferred to the culture vessel.

A custom-made unit can be purchased (Bellco Bioreactor). This 3 litre fluidized-bed culture (bubble column) actually encapsulates the cells in hydrogel beads within the culture vessel and comes with a range of modular units to control all process parameters.

5.2 Entrapment cultures

5.2.1 Opticell

This system has already been described (Section 3.3.3). The special ceramic cartridges for suspension cells (S Core) entrap the cells within the porous ceramic walls of the unit. They are available in sizes from 0.42 m^2 to 210 m^2 (multiple cartridges), which will support 5×10^{10} cells with a feed/harvest rate of 500 litres/day and give a yield of about 50 g of monoclonal antibody per day.

5.2.2 Fibres

A simple laboratory method is to enmesh cells in cellulose fibres (DEAE, TLC, QAE, TEAE—all from Sigma). They are autoclaved at 30 mg/ml in PBS, washed twice in sterile PBS, and added to the medium at a final concentration of 3 mg/ml in a spinner/stirred bioreactor. This method has even been found suitable for HDC (28).

5.3 Porous carriers (29)

Microcarriers and glass spheres are restricted to attached cells and, because a sphere has a low surface area/volume ratio, restricted in their cell density potential. A change from a solid to porous sphere of open, interconnecting pores (*Figure 14*), increases their potential enormously (*Table 6*). There are four types of porous (micro)carrier commercially available (*Table 7*). A characteristic of these porous carriers is their equal suitability for suspension cells (by entrapment) and anchorage-dependent cells (huge surface area).

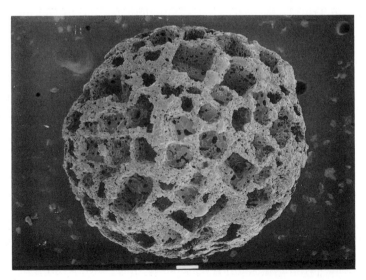

Figure 14. A Siran (Schott Glaswerke) porous glass sphere (120 μm diameter).

Table 6. Porous carriers—advantages compared to solid carriers

(a) Unit cell density 20–50 fold higher
(b) Support both attached and suspension cells
(c) Immobilization in 3D configuration easily derivitized
(d) Short diffusion paths into a sphere
(e) Suitable for stirred, fluidized, or fixed bed reactors
(f) Good scale-up potential by comparison with analogous
    systems (e.g. microcarrier at 4000 l)
(g) Cells protected from shear
(h) Capable of long term continuous culture

Table 7. Commercially available porous (micro) carriers

Trade Name	Supplier	Material	Diam (μm)	Culture mode[a]
Cultispher G	HyClone	Gelatin	170–270 300–500	S
Informatrix	Biomat Corp.	Collagen/ glycosaminoglycan	500	F
Microsphere	Verax Corp.	Collagen	500–600	F
Siran	Schott Glasswerke	Glass	600–1000 4000–6000	F X
Cellsnow	Kirin Co. Ltd	Cellulose	500	S
Microcarrier	Asahi Chem. Ind. Co.	Cellulose	300	S

[a] S, stirred; F, fluidized; X, fixed bed.

The problem with many immobilization materials is that diffusion paths become too long, preventing scale-up. A sphere is ideal in that cells and nutrients have only to penetrate 30% of the diameter to occupy 70% of the total volume. This facilitates scale-up as each sphere, whether in a stirred, fluidized, or fixed-bed culture, can be considered an individual mini-bioreactor.

5.3.1 Fixed bed (Porosphere)

The apparatus and experimental procedures described above (Section 3.3.2.*i*) for solid spheres is equally suited to porous Siran spheres of 4–6 mm diameter (30). The only differences are that the bed should be packed with oven-dried spheres and the void volume of cells plus medium inoculated (2×10^6 /ml) directly to the bed (dry beads permit better penetration of cells), and a larger medium volume is required, and faster perfusion rates (5–20 linear cm/min) should be used (cells are protected from medium shear within the matrix), e.g. 1 litre packed bed needs at least a 15 litre reservoir. After the initial 72 h period, 10 litres of fresh medium is added daily.

5.3.2 Fluidized beds

Fluidized beds are initiated in a similar manner to fixed beds—in fact they are operated as a fixed bed for the first 48 h. The bed is then fluidized with a medium flow velocity of 60 linear cm/min which gives a 20% bed expansion. Medium is withdrawn from the top of the culture above the level of fluidized microcarriers and pumped via a reservoir ($\times$ 20 bed volume) into the base of the bioreactor which is tapered to improve fluidization. A small commercially available fluidized bed bioreactor is now available (Verax Corp.; System One) which has a bed volume of 20 ml and a daily medium throughput/ harvest of 700–1200 ml.

5.3.3 Stirred cultures

The Cultispher-G macroporous gelatin and cellulose (Kirin Cellsnow) micro-carriers are the most suited to stirred bioreactors, and can be used in an identical manner to solid microcarriers, i.e. 2 g/litre in spinner flasks or stirred fermenters. However, about a 40-fold higher concentration of attached cells, and even greater densities of suspension cells, can be achieved. The cells can be released from the microcarriers by collagenase treatment.

References

1. Mosmann, T. (1983). *J. Immunol. Meth.*, **65**, 55.
2. Racher, A., Looby, D., and Griffiths, J. B. (1990). *Cytotechnology*, **3**, 301.
3. Griffiths, J. B. (1972). *J. Cell Sci.*, **10**, 512.
4. Good, N. E. (1963). *Biochemistry*, **5**, 467.
5. Leibovitz, A. (1963). *Am. J. Hyg.*, **78**, 173.
6. Spier, R. E. and Griffiths, J. B. (1984). *Dev. Biol. Stand.*, **55**, 81.
7. Griffiths, J. B. (1988). In *Animal Cell Biotechnology* (ed. R. E. Spier and J. B. Griffiths), Vol. 3, p. 179. Academic Press, London.
8. Toth, G. M. (1977). In *Cell Culture and its Applications* (ed. R. T. Acton and J. D. Lynn), p. 617. Academic Press, New York.
9. Griffiths, J. B. (1984). *Dev. Biol. Stand.*, **55**, 113.
10. Griffiths, J. B. (1990). *Cytotechnology*, **3**, 106.
11. Maroudas, N. G. (1975). *J. Theor. Biol.*, **49**, 417.
12. Kruse, P. J., Keen, L. N., and Whittle, W. L. (1970). *In Vitro*, **6**, 75.
13. Skoda, R., Pakos, V., Hormann, A., Spath, O., and Johansson, A. (1979). *Dev. Biol. Stand.*, **42**, 121.
14. Berg, G. J. and Bodeker, G. D. (1988). In *Animal Cell Biotechnology* (ed. R. E. Spier and J. B. Griffiths), Vol. 3, p. 322. Academic Press, London.
15. Munder, P. G., Modolell, M., and Wallach, D. F. H. (1971). *FEBS Lett.*, **15**, 191.
16. Jensen, M. D. (1981). *Biotech. Bioeng.*, **23**, 2703.
17. Burbidge, C. and Darcey, I. K. (1984). *Dev. Biol. Stand.*, **55**, 255.
18. Whiteside, J. P. and Spier, R. E. (1981). *Biotech. Bioeng.*, **23**, 551.
19. Weiss, R. E. and Schleiter, J. B. (1968). *Biotech. Bioeng.*, **10**, 601.
20. van Wezel, A. L. (1967). *Nature*, **216**, 64.

21. Spier, R. E. and Whiteside, J. P. (1984). *Dev. Biol. Stand.,* **55,** 151.
22. Paul, J. and Struthers, M. G. (1963). *Biochem. Biophys. Res. Comm.,* **11,** 135.
23. Capstick, P. B., Garland, A. J., Masters, R. C., and Chapman, W. G. (1966). *Exp. Cell Res.,* **44,** 119.
24. Tovey, M. G. (1980). *Adv. Cancer Res.,* **33,** 1.
25. Pirt, S. J. and Callow, D. S. (1964). *Exp. Cell Res.,* **33,** 413.
26. Nilsson, K. and Moosbach, K. (1980). *FEBS Lett.,* **118,** 145.
27. Nilsson, K., Scheirer, M., Merten, O. W., Ostberg, L., Liehl, E., and Katinger, H. W. D. (1983). *Nature,* **302,** 629.
28. Litwin, J. (1985). *Dev. Biol. Stand.,* **60,** 237.
29. Griffiths, J. B. (1990). In *Animal Cell Biotechnology* (ed. R. E. Spier and J. B. Griffiths), Vol. 4, p. 147. Academic Press, London.
30. Looby, D. and Griffiths, J. B. (1989). *Cytotechnology,* **2,** 339.
31. Prokop, A. and Rosenberg, M. Z. (1989). *Adv. Biochem. Eng.,* **39,** 29.

4

Cell line preservation and characterization

R. J. HAY

1. Introduction

Literally thousands of different cell lines have been derived from human and other metazoan tissues. (Use of the terms cell line and cell strain in this chapter is as recommended by The Tissue Culture Association committee on nomenclature (1).) Many of these originate from normal tissues and exhibit a definable, limited doubling potential. Other cell lines may be propagated continuously, either having gone through a change from the normal primary population or having been developed initially from tumour tissue. Finite lines of sufficient doubling potential and continuous lines can both be expanded to produce a large number of aliquots, frozen, and characterized for widespread use in research.

The advantages of working with a well-defined cell line free from contaminating organisms may appear obvious. Unfortunately, however, the potential pitfalls associated with the use of cell lines casually obtained and processed require repeated emphasis. Numerous occasions where cell lines exchanged among cooperating laboratories have been contaminated with cells of other species have been detailed and documented elsewhere (2, 3). For example, lines supposed to be human have been found to be monkey, mouse, or mongoose, while others thought to be monkey or mink were identified as rat and dog (2). Similarly, the problem of intraspecies cross-contamination among cultured human cell lines has been recognized for over 20 years and detailed reviews are available on the subject (4). The loss of time and research funds as a result of these problems is incalculable.

While bacterial and fungal contaminations represent an added concern, in most instances they are overt and easily detected, and therefore have less serious consequences than the more insidious contaminations by mycoplasma. That the presence of these micro-organisms in cultured cell lines often negates research findings entirely has been stated repeatedly over the years by a number of specialists in mycoplasma biology (5, 6). However, the difficulties of detection and prevalence of contaminated cultures in the research community suggest that the problem cannot be overemphasized.

In this chapter the procedures used to preserve and characterize cell lines and hybridomas at the American Type Culture Collection (ATCC) are outlined. They have evolved over the past 30 years as awareness of the difficulties of microbial and cellular cross-contaminations has become apparent, and as the need for well-characterized cell lines has increased.

2. Cell line banking

Cell lines suitable for acquisition are selected by ATCC scientists and advisers during regular reviews of the literature. The originators themselves also frequently offer the lines directly for consideration. Detailed information about specific groups of cell lines acquired by the ATCC is available elsewhere (7–9). Generally, starter cultures or ampoules are obtained from the donor, and progeny are propagated according to the donor's instructions to yield the first 'token' freeze. Cultures derived from such token material are then subjected to critical characterizations as described below. If these tests suggest that further efforts are warranted, the material is expanded to produce seed and distribution stocks. Note especially that the major characterization efforts are applied to cell populations in the initial seed stock of ampoules. The distribution stock consists of ampoules that are distributed on request to investigators. The reference seed stock, on the other hand, is retained to generate further distribution stocks as the initial stock becomes depleted.

Although this procedure has been developed to suit the needs of a large central repository, it is also applicable in smaller laboratories. Even where the number of cell lines and users may be limited, it is important to separate 'seed stock' from 'working or distribution stock'. Otherwise the frequent replacement of cultured material, recommended to prevent phenotypic drift or senescence, may deplete valuable seed stock which may be difficult and expensive to replace.

These various steps in the overall accession scheme are summarized in *Figure 1*. It is important to recognize that the characterized seed stock serves as a frozen 'reservoir' for production of distribution stocks over the years. Because seed stock ampoules are used to generate new distribution material, one can ensure recipients that all the cultures obtained closely resemble those received 2, 5, 10 or more years previously. This is a most critical consideration for design of cell banking procedures. Problems associated with genetic instability, cell line selection, senescence, or transformation may be minimized or avoided entirely by strict adherence to this principle.

It is prudent to handle as biohazards all primary tissues or cell lines not specifically shown to be free of adventitious agents, using a Class II vertical laminar flow hood (BS5726; NIH Spec. 03–112). This precaution protects both the cell culture technician and the laboratory from infection or contamination. Some recommend, furthermore, that all human tumour lines be

Accessioning Scheme

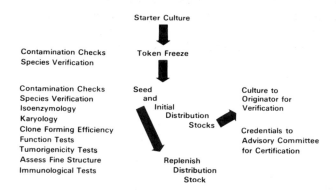

Figure 1. Scheme illustrating a recommended plan for the accession of cell lines to be banked for general distribution. At least the first two of the various characterizations indicated on the left-hand side should be performed before release of any cell line.

treated with similar caution due to the known presence of oncogenes and the recent demonstration of transfection at least within cell culture systems.

3. Cell freezing, and quantitation of recovery

Cellular damage induced by freezing and thawing is generally believed to be caused by intracellular ice crystals and osmotic effects. The addition of a cryoprotective agent, such as DMSO or glycerol, and the selection of suitable freezing and thawing rates, minimizes cellular injury.

While short-term preservation of cell lines using mechanical freezers (−75°C) is possible, storage in liquid nitrogen (−196°C) or its vapour (to −135°C) is much preferred. The use of liquid nitrogen refrigerators is advantageous, not only because of the lower temperatures and, consequently, almost infinite storage times possible, but also due to the total absence of risk of mechnical failures and the prolonged holding times now available. Certainly for all but the smallest cell-line banking activity, storage in a liquid nitrogen refrigerator is essential.

Five safety considerations in processing cell lines require special emphasis:

(a) The cell culture technician may be endangered because of the possibility that liquid nitrogen can penetrate ampoules via hairline leaks during storage. On warming, rapid evaporation of the nitrogen can cause the glass ampoule to explode. Fortunately the frequency of such traumatic accidents declines dramatically as the operator gains experience in ampoule sealing and testing (Sections 3.1.1 and 3.3). However, even with highly accomplished laboratory workers, a remote possibility of explosion still exists. Thus, a protective face mask should be worn whenever ampoules are removed from liquid nitrogen storage and until they have been safely opened in the laminar flow hood.

(b) DMSO can solubilize organic substances and, by virtue of its penetrability through rubber and skin, carry these to the circulation. Thus, special precautions should be exercised when using DMSO to avoid contamination with hazardous chemicals and minimize skin contact.

(c) Evaporation of liquid nitrogen, particularly when inserting or removing ampoules, or when topping up, can rapidly replace the air in the room, so adequate cross-draught ventilation must be provided.

(d) Because of the extremely low temperature there is constant risk of frostbite from the contents of the freezer, and from splashes or spillage. Gloves, a mask, and a lab coat must be worn.

(e) Glass ampoules should not be used with cell lines known to be shedding pathogens (e.g. HIV-1 or other viruses).

3.1 Equipment

3.1.1 Ampoules, marking and sealing devices

The decision on whether to use glass or plastic ampoules will depend on the scale of operation and the extent of anticipated distribution. Glass ampoules can be sterilized, loaded with the appropriate cell suspension, and permanently sealed in comparatively large quantities. They are recommended for large lots of cells (20 ampoules or more) being prepared for long-term use or general distribution. Smaller numbers (1–20) of plastic ampoules are easier than glass to handle, mark, and seal. Problems with the seal may occur in some cases, however, especially if frequent handling or manipulations for shipment are necessary.

The marking of ampoules, especially glass ampoules, requires special consideration in that legibility can easily be obscured as the ampoules are frozen, snapped on and off storage canes, transferred between freezers or shipping containers, and so forth. Glass ampoules should be labelled in advance with ceramic marking ink that can be heat-annealed to the glass surface. The markings can be applied by hand with a straight pen or, if large lots are being processed, by a mechanical labeller (from Markem, for example).

Glass ampoules can best be sealed by pulling on the neck of the ampoule as it is rotated in the highest heat zone of the flame from a gas–oxygen torch. This pull-seal technique is preferred since it reduces the risk of pinhole leaks in the sealed tip. Torches for manual sealing can be obtained from scientific supply houses. For large lots, a torch can be attached to a semi-automatic sealing device (*Figure 2*) available as Bench sealer model 161 from Morgan Sheet Metal Co.

3.1.2 Slow freezing apparatus

The optimum freezing rate for cell lines (usually about −1°C/min) can be achieved through use of apparatus varying in complexity from a tailor-made

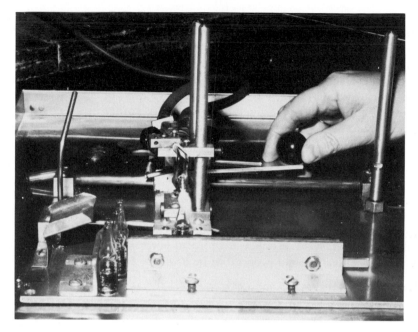

Figure 2. Device for heat-pull sealing of glass ampoules. The ampoules move from right to left and are rotated as each tip is heated in the flame. At maximum temperature the tip is pulled upward as shown, to form a seal of molten glass (left-hand side). The speed at which the handle is moved is critical for production of an appropriate seal.

styrofoam box to a completely programmable freezing unit. The former should have a wall thickness of about 2 cm to approximate the −1 °C/min cooling rate when placed in a mechanical freezer at −70 °C. Alternatively, manufacturers of liquid nitrogen refrigerators supply adapted refrigerator neck plugs, at modest cost, which can be adjusted for slow freezing of small quantities (1–9) of ampoules. For those who produce larger quantities of ampoules and require more precise control of the freezing rate, a controlled-rate freezer, a freezing unit which has a constant rate or programmable, stepwise freezing rate (Cryo-Med; Planer; Union Carbide) should be considered.

3.1.3 Liquid nitrogen refrigerators

Choice of an appropriate refrigerator involves considerations of economy, both in terms of liquid nitrogen consumption and initial outlay, storage capacity required, and desired ease of entry and retrieval. Freezers with narrower neck openings are generally more economical. Some larger models are equipped with lazy susan trays for convenient operation.

Ampoules of cells may be stored immersed in liquid nitrogen or in the vapour phase. The latter has the advantage that ampoules with pinhole leaks will not be exposed to liquid, so the danger of explosion is eliminated. The slightly higher temperature probably offers no disadvantage except perhaps with seed stocks retained for extremely long periods. However, the decreased volume of liquid nitrogen will reduce the holding time.

Useful accessories for refrigerators include roller bases for ease of movement, alarm systems to warn of dangerously-low levels of liquid nitrogen, and racking systems (in larger refrigerators) for ready storage and recovery. Even when an automatic alarm system is used this should be backed up with a regular manual check using a dipstick. Electronic systems can fail, and it has been known for both the automatic fill system and its back-up alarm to fail.

3.2 Preparation and freezing

Cultures in the late logarithmic or just pre-confluent phase of growth should be selected to give the highest possible initial viability. Treat the cultures with trypsin if necessary to produce a uniform single-cell suspension as if for routine subcultivation and proceed as in *Protocol 1*.

Protocol 1. Culture preparation and freezing procedure

1. Just before use, prepare the freeze medium by simple admixture of fresh growth medium and the required cryoprotective agent, generally DMSO or glycerol. The concentrations of choice vary slightly with the cell line and cryoprotective agent used, ranging from 5 to 10% (v/v).

2. Collect the cells in pellet form by centrifugation at about 200 *g* for 10 min and resuspend in an appropriate volume of freeze medium at room temperature. Concentrations of 10^6-10^7 cells/ml are generally satisfactory and practical. For some applications it may be desirable to increase or decrease this by as much as a factor of 10.

3. If the total volume is large, as for a production-level freeze, maintain the cell suspension with gentle agitation in an appropriate stirring vessel. Set this up in advance complete with a Cornwall automatic syringe (available from most laboratory suppliers). If the cell suspension is in a small volume (10–20 ml), dispense by means of a syringe fitted with an 18G cannula. Mix repeatedly during this process.

4. Maintain the pH when necessary by gassing with an appropriate mixture of water-saturated air/carbon dioxide.

5. Dispense the cell suspension in 1 ml aliquots to the ampoule, taking care to proceed rapidly and with uniformity. Cells sediment quickly under unit gravity in the Cornwall tubing, and inaccuracies can result if the process is interrupted or arrhythmic.

6. Seal each ampoule, place on an ampoule cane or in a suitable rack, and immerse totally upright in aqueous methylene blue (0.05%) at 4°C. After

30–45 min remove, wash in cold tap water, and discard any ampoule containing the blue dye.

7. Dry the ampoules thoroughly and begin slow freezing. The optimum rate of cooling is usually −1°C/min but varies among different cell lines and can best be determined empirically before any large-scale freeze is attempted.

8. When ampoules have reached −70°C or lower, remove and transfer them immediately and rapidly to the liquid nitrogen refrigerator.

9. Record the location and specifics of the cell line in question. Two separate cross-index files providing both subject and location cards can be used for this purpose. Alternatively, computerized systems, *with the essential back-up*, may be devised.

3.3 Reconstitution and quantitating recovery

Rapid thawing of the cell suspension is essential for optimal recovery.

Protocol 2. Recovery of cultures

1. **With face and neck protected by a full face mask,** retrieve the selected ampoule rapidly and plunge directly into a warm water bath (37°C). If ampoules have been stored submerged in liquid nitrogen, each ampoule must be thawed separately by depositing directly into about 10 cm of water at 37°C in a 1.5–2 litre bucket with a lid, snapping the lid closed immediately after the ampoule is inserted.

2. Agitate the ampoule contents until the suspension has thawed completely (20–60 s).

3. Immerse the ampoule in 70% ethanol at room temperature.

4. Score standard ampoules using a small file that has been dipped previously in ethanol. Pre-scored ampoules require no such treatment.

5. Use a sterile towel to pick up the ampoule; break sharply at the neck at the pre-scored point. **Continue wearing face mask at least through this stage.**

6. Transfer the contents, by means of a sterile Pasteur pipette, to a centrifuge tube or culture vessel. Add 10 ml of complete growth medium. In some cases, slow addition over a 1–2 min interval at this point may be beneficial.

7. Centrifuge at 200 g for 10 min. Remove the supernatant, add a fresh aliquot of medium, and mix the suspension. This centrifugation step may be omitted in some cases since the residual concentration of cryoprotective agent is low and the stress of centrifugation can be harmful.

8. Initiate the culture by standard procedure. If the centrifugation step was omitted, change the medium after 24 h.

A variety of methods may be used to quantitate cell recovery after freezing. Of course this is the first characterization step to be performed after preservation.

3.3.1 Dye exclusion for quantitating cell viability

A very approximate estimate of the viability of cells in a suspension may be obtained by the dye exclusion test. A solution of the dye in saline is added to the suspension and the percentage of cells which do not take up the stain is determined by direct count using a haemocytometer. Trypan blue or erythrocin B are probably the stains most commonly used for this procedure. The former reportedly has a higher affinity for protein in solution than for non-viable cells. This may reduce the accuracy of the estimate if the suspension contains much more than 1% serum. Furthermore, because solutions of erythrocin B are clear, microbial growth or precipitates are immediately apparent. This is not true for stock solutions of trypan blue. *Protocol 3* describes the recommended procedure.

Protocol 3. The dye exclusion test

1. Obtain a uniform suspension of cells in growth medium by any standard procedure. If the estimate is for freeze-characterization, reconstitution of contents from about 5% of the ampoules prepared is recommended.
2. Dilute an accurately measured aliquot of the cell suspension using the stock dye solution. This consists of 100 mg erythrocin B/100 ml of an isotonic PBS adjusted to pH 7.2–7.4 with 1 M NaOH. For ease and accuracy of count a final density of about $0.3–2 \times 10^6$ cells/ml is satisfactory.
3. Score the number of stained versus total number of cells keeping in mind that, in general, a larger sampling (300 cells or more) will give a more accurate quantitative estimate. Ideally the count should be performed within 1–5 min after mixing of the dye and cell suspension.

3.3.2 Clone-forming efficiency

The dye-exclusion test for cell viability generally overestimates recovery. For lines consisting of adherent cells, the clone-forming ability of cells from the reconstituted population represents a more accurate overall estimate of survival. Of course the choice of growth medium used, the substrate on which the clones develop, and the incubation time may all have an effect on the end result. Thus for comparisons among different freezes, conditions for selected lines must be standardized. A representative outline as used at the ATCC is provided in *Protocol 4*.

Protocol 4. Estimation of clone-forming efficiency

1. Pool the reconstituted contents of ampoules from about 5% of a freeze lot and serially dilute the suspension to provide inocula of 100–10^4 cells/

culture depending upon the cell line, plate size, and expected recoveries. For example, cells of a line with high plating efficiency could be added at 10 cells, 100 cells, 1000 cells, and 10 000 cells to four separate plate sets, assuming 9 cm plates were used.

2. Make the serial dilutions by 10-fold reductions in cell number by transferring 1 ml of suspension to 9 ml of the selected, complete growth medium. Half-log dilutions can be made by transferring 1.0 ml of cell suspension to 2.16 ml of diluent medium. Discard the 1 ml pipette and use a fresh 1 ml pipette to mix and make each subsequent, lower-density transfer for dilution.

3. Inoculate 1 ml of each dilution, beginning with the lowest cell concentration, to each of three T-75 flasks (or other suitable culture vessel) containing 8 ml of growth medium and incubate at 37°C.
   Note: If media with high bicarbonate are used with low cell inocula, even sealed plastic flasks will have to be equilibrated *and incubated* with the appropriate air/carbon dioxide mixture to maintain pH.

4. Renew the fluid on test cultures on the fourth or fifth day and thereafter every third or fourth day.

5. After 12–14 days total, remove the fluid, and fix the culture with a 10% solution of formaldehyde or other suitable fixative.

6. Remove the fixative, rinse the flask interior gently with several changes of tap water, and stain with 1% aqueous toluidine blue or other simple stain for 1–5 min.

7. Remove the staining solution, rinse out residual fluid with tap water and count clones consisting of 16 or more cells, using a dissecting microscope or automated colony counter.

8. Calculate the percentage clone-forming efficiency:

$$\frac{\text{clones formed}}{\text{number of cells inoculated}} \times 100$$

3.3.3 Proliferation in mass culture

The vitality of reconstituted cells from either adherent or non-adherent lines can also be documented by simply quantitating proliferation during the initial 1–2 weeks after recovery from liquid nitrogen, as in *Protocol 5*.

Protocol 5. Measuring vitality of reconstituted cells

1. Using the cell suspension pooled from 5% of a freezing lot, inoculate three sets of culture vessels (e.g. T-25 flasks) at different densities. Typically one might choose the inoculum expected to yield a confluent or maximum density culture in 7 days for one set; twice that for a second set, and 3–5 times that (depending on viable count) for the third set.

Protocol 5. *Continued*

2. Renew the culture fluid (or add fresh medium for a suspension culture) after 4 days.

3. Harvest the three sets of cultures by standard trypsinization if an adherent culture or by direct sampling if a suspension culture, and determine the cell yield by either electronic cell count or haemocytometry.

4. Calculate the increase (cells recovered/cells inoculated) and compare this with what one would expect under similar conditions with cells taken during logarithmic or pre-confluent growth phase.

Typical results from the above tests have been recorded for a wide variety of cell lines in the *ATCC Catalogue of Cell Lines and Hybridomas* (8).

4. Cell line characterizations

In addition to recoverability from liquid nitrogen, the absolute minimum recommended characterizations include verification of species and demonstration that the cell line is free of bacterial, fungal, or mycoplasmal contamination.

4.1 Species verification

Species of origin can be determined for cell lines by a variety of immunological tests by isoenzymology and/or by cytogenetics (10). Advantages and problems associated with each of these will be included with the appropriate sections.

4.1.1 Fluorescent antibody staining

The indirect fluorescent antibody staining technique is used at the ATCC as one method for verifying the species of cell lines. The technique involves two general steps. Firstly, species-specific antiserum, produced in rabbits, is used to label test cells plus positive and negative control cell populations (see *Protocol 6*). Secondly, anti-rabbit globulin, produced in goats and coupled to the fluorescent dye, fluorescein isothiocyanate (FITC) is applied. The second reagent binds to the rabbit antibody which has attached to the target cells and, by virtue of the fluorescence, the antibody–antigen complexes can be visualized.

To verify species on a test cell line, follow the procedure in *Protocol 7*.

Protocol 6. Preparation of antiserum to cultured cells

1. Harvest cells of known species by scraping from the culture surface with a rubber policeman.

2. Wash by suspending in HBSS^a with subsequent centrifugation at 200 g for 10 min, and repeat three times.

3. Resuspend in HBSS such that the viable cell count is about 5×10^5/ml (first week), 10^6/ml (second week), and 10^7/ml (third week).

4. Inoculate 1 ml to each marginal ear vein of a healthy rabbit twice weekly for 3 weeks, increasing the dose each week, as in step 3.

5. Administer three or more additional booster injections at 10^7 cells/ml (1 ml/ear) on a weekly basis.

6. After the third booster injection, perform test bleedings, collect, and serially dilute antisera. Mix an equal volume of cell suspension (10^6/ml) and evaluate cytotoxicity using the viable staining technique described earlier.

7. If the titres are satisfactory (1:8 or greater) collect the blood by cardiac puncture, permit it to clot for 1–2 h at room temperature, and centrifuge at 200 g for 15 min.

8. Remove serum, inactivate complement by heating at 56°C for 30 min, dilute and distribute in 0.2 ml aliquots for storage at −70°C.

[a] All solutions are prepared in double glass-distilled water unless otherwise indicated. See ref. 10 for more detail.

Protocol 7. Species verification on a test cell line

1. Harvest by trypsinization if necessary and wash three times by suspending the cells in PBS at pH 7.5 with subsequent centrifugation to form a cell pellet.

2. Resuspend the washed cells in PBS to give a density of $3–4 \times 10^6$/ml.

3. Mix 0.1 ml of cell suspension and 0.1 ml of diluted antiserum and place in a humidified incubation chamber at room temperature for 30 min. The appropriate dilution of antisera will have to be determined empirically for each antiserum preparation with positive control cells through an initial preliminary trial. Non-specific absorption can usually be excluded by further diluting the antiserum.

4. Samples are then washed to remove unabsorbed antiserum using three complete changes of PBS and are incubated for 30 min in the dark with 0.1 ml of FITC-conjugated, goat anti-rabbit antiserum (obtainable commercially).

5. After a final three additional washes with PBS, a drop of the final cell suspension is sealed under a coverslip. This is examined by fluorescence microscopy at 500 × using number 50 barrier (LP 440 nm) and BG12 exciter (330–380 nm) filters on a Zeiss Universal microscope with an epi-illuminator.

Protocol 7. *Continued*

6. Positive reactions are seen as staining of brilliant fluorescence intensity at the cell periphery (*Figure 3*).

7. Controls consisting of cells of the suspected species, one related species, and a distant species are included with each test.

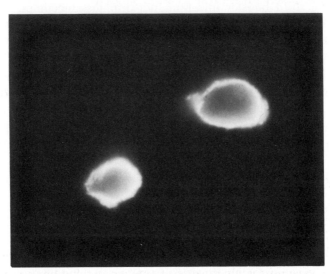

Figure 3. Indirect fluorescent antibody test for cell species. Note the halo of fluorescence around the cells showing a typical positive reaction. (Photograph courtesy of M. L. Macy.)

This method is an adaptation of that described by Stulberg (10). Its advantages include simplicity and the ability to allow identification of even minor cellular contaminants among populations. Stulberg reported that as few as one contaminating cell among 10 000 could be identified under appropriate conditions of resolution.

4.1.2 Isoenzyme profiles

Isozyme analyses performed on homogenates of cell lines from over 25 species have clearly demonstrated the utility of this biochemical characteristic for species verification (11). By determining the mobilities of three isozyme systems (glucose-6-phosphate dehydrogenase (G6PD), lactic dehydrogenase (LDH), and nucleoside phosphorylase (NP)) using vertical starch gel electrophoresis, one can identify species of origin of cell lines with a high degree of certainty. The procedures are relatively straightforward, provide consistent results, and do not require expensive equipment.

A method for preparing cell extracts is summarized in *Protocol 8*. The compositions of stock buffers for starch electrophoresis are given in *Table 1*.

Ingredients for electrophoretic separations and staining for the three enzyme systems are provided in *Table 2*.

Protocol 8. Preparation of cell extract

1. Obtain a cell suspension by standard methods and wash three times in 0.9% NaCl solution (pH 7.1) containing EDTA at 6.6×10^{-4}M by centrifugation with subsequent resuspension as usual.
2. Resuspend the resulting cell pellet at $2–5 \times 10^7$ cells/ml in saline-EDTA.
3. Produce a homogenate by freezing and thawing rapidly three times in liquid nitrogen or a dry ice–alcohol bath, by ultrasonic treatment or by mixing with an equal volume of octylalcohol (4°C overnight).
4. Dispense aliquots of the homogenate to Eppendorf tubes, clarify by centrifugation in a microfuge for 1–2 min and store at −70°C for subsequent assay.

Protocol 9. Preparation of starch gel

1. Suspend 60 g of electrostarch in 500 ml of the appropriate buffer and produce a uniform suspension by stirring and heating to 90°C. This may be accomplished using a boiling water bath, a stirrer–hot plate, or microwave oven. The stirring must be constant to avoid lumpiness and burning.
2. Apply a vacuum for about 60 s to remove gas bubbles, and allow the suspension to cool uniformly to 60°C by swirling gently.
3. For LDH or G6PD systems add NADP at this point (see *Table 2*) and mix gently but thoroughly.
4. Pour the starch into the gel mould according to the manufacturer's instructions and allow it to cool for 2–3 h undisturbed at room temperature. The Buchler vertical gel electrophoresis apparatus (Buchler Instruments) is used in the author's laboratory.

Protocol 10. The electrophoresis procedure

1. Thaw cell extracts and centrifuge for 1–2 min in the microfuge.
2. Carefully remove the 10-slot comb from the mould according to manufacturer's instructions and add 0.03–0.04 ml of the extracts to the slots using a fine-bore Pasteur pipette or syringe with a 26-gauge (26G) needle. Use care to avoid overflow of extracts among adjacent slots.
3. Cover the entire area of the gel, exposed following removal of the slot-former, with heated liquid petroleum jelly.

Protocol 10. *Continued*

4. Add buffer to the electrode chambers and secure the loaded gel mould in place with the retention spring (see manufacturer's instructions).

5. Apply the flannelette wicks, place the unit at 4°C and begin the run using 4–5 V/cm of gel slab.

Protocol 11. Visualizing the isozymes

1. First remove the starch gel slab, cut off and discard 5 cm from each end and cut the gel horizontally into three identical parts using the gel-cutting-device wire (see manufacturer's instructions).

2. Separate the three layers and place in staining boxes.

3. Cover the entire surface of the gel layers with the appropriate noble agar–stain mixture and incubate at 37°C in the dark. Bands will appear in 1–3 h depending on enzyme type and on the activity in the extract.

Table 1. Stock buffers for electrophoresis

Tris-citrate (TC)	1 × stock
Tris	17 g
Citric acid · monohydrate	9.1 g
Distilled water	1000 ml
pH	7.1
Tris-EDTA-borate (TEB)	5 × stock
Tris	109.0 g
EDTA Na$_2$	7.6 g
Boric acid	6.2 g
Distilled water	1000 ml
pH	8.6
Gel buffers: 50 ml of 1 × stock + 450 ml H$_2$O	
Tray buffers: 1 × stock solution	

The zymograms in *Figure 4* illustrate typical mobilities which could be expected in a starch gel for isozymes of commonly used lines from a variety of species.

More recently a kit (AuthentiKit™, Innovative Chemistry) has been developed for cell line identification. Pre-cast 1% agarose gels on a polystyrene film backing, buffers, enzyme substrates with stabilizers, and appropriate control extracts are available. The specially constructed electrophoretic chambers can be connected to power supplies on hand for related applica-

Table 2. Procedures for LDH, G6PD, and NP[a]

Enzyme system	Chamber buffer	Starch buffer	Staining solutions[b]
LDH	TC	TC (0.1 ×)	10 ml water 5 ml 0.5 M Tris pH 7.5 5 ml 1.0 M Na lactate 5 ml NAD[c] (10 mg/ml) 5 mg MTT[c] 2 mg PMS[c]
G6PD	TEB	TEB (0.1 ×)	5 ml water 5 ml 0.5 M Tris pH 7.5 5 ml 0.025 M glucose-6-phosphate 5 ml 0.1 M MgCl$_2$.6H$_2$O 5 ml 0.005 M NADP[c] 5 mg MTT[c] 2 mg PMS[c]
NP	TEB	TEB (0.2 ×)	20 ml water 5 ml 0.1 M NaH$_2$PO$_4$.H$_2$O 50 ml inosine 100 μl xanthine oxidase[d] 5 mg MTT[c] 2 mg PMS[c]

[a] All gels are run for 16–18 h at 4°C at 4–5 V/cm (160–180 V). NADP (5 ml at 0.005 M) is added to the cathode buffer as well as to the staining solution for G6PD.

[b] These stock solutions are mixed just before use with an equal volume (25 ml) of 2% noble agar in water which has been liquefied and cooled to 45–50°C.

[c] MTT, dimethylthiazol diphenyltetrazolium; PMS, phenazine methosulphate.

[d] At 100 mg/ml—add just before mixing with agar.

tions, or those offered optionally with other kit hardware. Over 15 different enzyme systems can be evaluated.

Advantages of the kit include the convenience of ready-made gels and reagents plus the significantly lower times required for electrophoretic separations (15–45 min). After drying, the gels can be retained for years to document cell line characteristics (*Figure 5*). National repositories and other laboratories handling many cell lines from diverse species have used the AuthentiKit™ effectively.

4.1.3 Cytogenetics

Karyologic techniques have long been used informatively to monitor for interspecies contamination among cell lines. In many instances the chromosomal constitutions are so dramatically different that even cursory microscopic

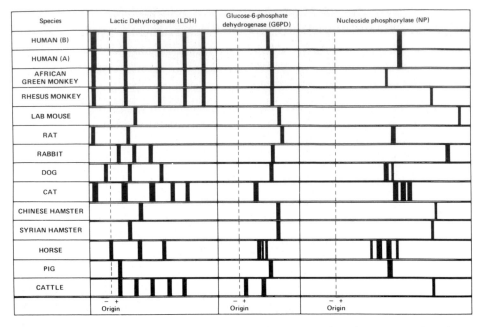

Figure 4. Zymograms for isozymes of lactic dehydrogenase (LDH), glucose-6-phosphate dehydrogenase (G6PD) and nucleoside phosphorylase (NP) showing typical banding patterns of cell lines from some of the more commonly used species.

observations are adequate. In others, as for example comparisons among cell lines from closely related primates, careful evaluation of banded preparations (10) is required (see Section 4.3.3).

The standard method described in *Protocol 12* involves the swelling of metaphase-arrested cells by brief exposure to hypotonic saline (*Table 3*). The cells are then fixed, applied to slides to optimize spreading, stained and mounted for microscopic observation.

Table 3. Stock solutions for routine karyology

Colcemid stock
   1 mg/50 ml double-distilled water and store frozen in small aliquots.
KCl
   0.075 M in double-distilled water
Fixative
   3 parts anhydrous methyl alcohol
   1 part glacial acetic acid
   (combine just before use)
Acetic orcein stain
   2 g natural orcein dissolved in 100 ml of 45% acetic acid.
Giemsa stain
   10% in 0.01 M phosphate buffer at pH 7 (solution available commercially).

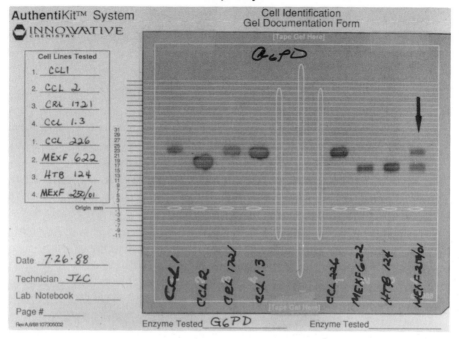

Figure 5. An agarose gel processed for glucose-6-phosphate dehydrogenase using the Authenti Kit™. The gel was loaded with extracts from murine and human cells plus one murine/human mixture. Lanes 2 in the left section (LHS) and 2 and 3 in the right section (RHS) are human. Lanes 1, 3, and 4 (LSH) and 1 (RHS) are from extracts of murine lines. Lane 4 (RHS and arrow) is from the mixed line.

Protocol 12. Karyological procedure

1. To a culture (T-75) in the exponential growth phase add colcemid to give a final concentration of 0.1–0.4 μg/ml.

2. Incubate for 1–6 h, selecting the length of this period roughly by the cycling time of the cell population under study. Diploid human cells with relatively long doubling times would generally require longer incubation than would a rapidly-proliferating Chinese hamster ovary cell line.

3. Gently decant the supernatant and treat the adherent layer with trypsin as for a standard subcultivation. Omit this step if working with a suspension culture. Place the suspension in a 15 ml centrifuge tube. Add serum to a final concentration of approximately 10% of the cell suspension to neutralize further trypsin action on cells.

4. Collect the suspended cells by centrifugation, discard the supernatant, and re-suspend the pellet in the hypotonic KCl solution at 37°C.

Protocol 12. *Continued*

5. After 10–15 min incubation at 37°C, sediment the cells by centrifugation at 100 *g* for 10 min and decant the supernatant. Re-suspend the cells in a small amount of the hypotonic KCl (1–2 times volume of cell pellet).

6. Slowly add 5 ml of freshly-made fixative while agitating the tube manually by pipetting, snapping, or using a vortex mixer.

7. After 20 min or more, repeat the centrifugation step, decant the supernatant, add 5 ml of fresh fixative, mix by agitation, and let stand at room temperature for 10–15 min.

8. Repeat step 7, centrifuge at 100 *g* for 2 min, remove the supernatant, and re-suspend the cells in a small amount of fresh fixative (10–15 times the volume of cell pellet).

9. To prepare slides, add the suspension dropwise onto clean, cold (4°C) slides held at about at 45° angle, blow gently to spread the cell suspension over the slide, and allow to dry completely in air at room temperature.

10. Examine the preparation for general arrangement using phase contrast optics. Metaphases should be spread evenly over the slide surface without overlapping. The densities can be adjusted either by altering the number of drops of suspension applied or by changing the concentration of cells in the suspension.

11. Stain with either acetic acid orcein for 3–5 min or Giemsa for 10–15 min, rinse in tap water and air dry. Mount in Permount, if so desired.

The frequency of introduction of artifacts through this method will vary depending upon the cell line and the operator's experience. Rupturing of cells will occur, for example, and apparent losses or gains in chromosomes will result. However, by counting the chromosomes in 50–100 well-spread metaphases and recording the modal number, the cytogeneticist can obtain a reliable estimate of the true figure for a specific line. Representative data on hundreds of different lines are available in ref. 8.

The karyotype is constructed by cutting chromosomes from a photomicrograph and arranging them according to arm length, position of centromere, presence of secondary constrictions, and so forth. The *Atlas of Mammalian Chromosomes* (12) gives examples of conventionally stained preparations from over 550 species. For a more critical karyotypic analysis, chromosome banding techniques are required, see Section 4.3.3.

4.2 Tests for microbial contamination

The tests included here are suitable for detection of most micro-organisms that would be expected to survive as contaminants in cell lines or culture fluids. Techniques for detecting protozoan contamination are omitted, because these organisms are rarely found in continuous lines and descriptions of the methodology are available elsewhere (13).

Commercial dry powders are entirely satisfactory for preparation of test media for bacteria, fungi, and mycoplasma, provided that positive controls are included at least with the initial trials on each lot obtained.

4.2.1 Detection of bacteria and fungi

To examine cell cultures or suspect media for bacterial or fungal contaminants follow the procedure in *Protocol 13*.

Protocol 13. Microscopic examination for contaminants

1. Using an inverted microscope, equipped with phase contrast optics if possible, examine cell culture vessels individually. Scrutiny should be especially rigorous in cases in which large-scale production is involved. Check each culture first using low power.
2. After moving the cultures to a suitable isolated area, remove aliquots of fluid from cultures that are suspect and retain these for further examination. Alternatively, autoclave and discard all such cultures.
3. Prepare wet mounts using drops of the test fluids, and observe under high power.
4. Prepare smears, heat-fix, and stain by any conventional method (e.g. Wright's stain), and examine under oil immersion.
5. Consult refs 14 and 15 for photomicrographs of representative contaminants and further details.

Microscopic examination will detect only gross contaminations, and even some of these cannot be readily detected by simple observations. Therefore, an extensive series of culture tests is also required to provide reasonable assurance that a cell line stock or medium is free of fungi and bacteria. To perform these on stocks of frozen cells, follow *Protocol 14*.

Protocol 14. Culture tests

1. Pool and mix the contents of about 5% of the ampoules from each freeze lot prepared using a syringe with a 18G cannula. It is generally recommended that antibiotics be omitted from media used to cultivate and preserve stock cell popuations. If antibiotics are used, the pooled suspension should be centrifuged at 2000 *g* for 20 min and the pellet should be resuspended in antibiotic-free medium. A series of three such washes with antibiotic-free medium prior to testing will reduce the concentration of antibiotics, which would obscure contamination in some cases.

Protocol 14. *Continued*

2. From each pool, inoculate each of the test media listed in *Table 4* with a minimum of 0.3 ml of the test cell suspension and incubate under the conditions indicated. Include a suitable range of bacteria and fungi as positive and negative controls. A recommended grouping consists of *Pseudomonas aeruginosa, Micrococcus salivarius, Escherichia coli, Bacteroides distasonis, Penicillium notatum, Aspergillus niger*, and *Candida albicans*.

3. Observe as suggested for 14–21 days before concluding that the test is negative. Contamination is indicated if colonies appear on solid media or if any of the liquid media become turbid.

Table 4. Suggested regimen for detecting bacterial or fungal contamination in cell cultures. For further detail see ref. 15

Test medium	Temperature (°C)	Aerobic state	Observation time (days)
Blood agar with			
fresh defibrinated	37	aerobic	14
rabbit blood (5%)	37	anaerobic	
Thioglycollate broth	37	aerobic	14
	26		
Trypticase soy broth	37	aerobic	14
	26		
Brain heart infusion	37	aerobic	14
broth	26		
Sabouraud broth	37	aerobic	21
	26		
YM broth	37	aerobic	21
	26		
Nutrient broth with	37	aerobic	21
2% yeast extract	26		

4.2.2 Mycoplasma detection

Contamination of cell cultures by mycoplasma can be a much more insidious problem than that created by growth of bacteria or fungi. While the presence of some mycoplasma species may be apparent because of the degenerative effects induced, others metabolize and proliferate actively in the culture without producing any overt morphological change in the contaminated cell line. Thus cell culture studies relating to metabolism, surface receptors, virus–host interactions, and so forth are certainly suspect, if not invalidated entirely, when conducted with cell lines harbouring mycoplasma.

Nine general methods are available for detection of mycoplasma contamination (5, 6). Both the direct culture test and the 'indirect' test employing a bisbenzimidazole fluorochrome stain (Hoechst 33258) for DNA are used routinely at the ATCC to check incoming cell lines and all cell distribution stocks for mycoplasma (16, 17).

The serum, yeast extract, and other ingredients used for mycoplasma isolation and propagation should be pre-tested for absence of toxicity and for growth-promoting properties before use (5). Positive controls consisting of *M. arginini, M. orale*, and *A. laidlawii* are recommended since they are among the most prominent species isolated from cultured cells (6, 16).

Protocol 15. Testing a culture for the presence of mycoplasma

1. Be sure that the growth medium is free of antibiotics, especially gentamycin, tylocine, or other antimycoplasma inhibitors.

2. Remove all but about 3 ml of fluid from a confluent or dense culture which has not been fed for at least 3 days, and scrape off parts of the monolayer with a rubber policeman or disposable scraper. Suspension cultures near saturation may be sampled directly.

3. Inoculate 1.0 ml of the cell suspension to a tube of mycoplasma broth and 0.1 ml on to a plate of mycoplasma agar (*Table 5*).

4. Incubate the broth aerobically at 37°C and the agar at 37°C in humidified 95% nitrogen/5% carbon dioxide. A change in the pH of the broth or development of turbidity warns of mycoplasma contamination.

5. At 7 and 14 days transfer 0.1 ml of the broth to a fresh plate of mycoplasma agar, and incubate at 37°C anaerobically.

6. Using an inverted microscope examine all agar plates at 100 × and 300 × weekly for a minimum of 3 weeks. A photomicrograph of mycoplasma colonies on agar is shown in *Figure 6*.

Figure 6. Microphotograph showing colonies of mycoplasma (arrows) in an agar plate. The smaller material between colonies is from the inoculum of cultured cells and debris.

Table 5. Recipes for media used in culture test for mycoplasma

A. *Stock supplements*

Ingredient	Concentration (g/litre)	Ingredient	Concentration (g/litre)
Dextrose	50	D-calcium pentothenate	0.024
L-arginine. HCl	10	Pyridoxal. HCl	0.020
Thymic DNA	0.02	Folic acid	0.010
Choline chloride	0.922	Cyanocobalamin	0.003
i-Inositol	0.110	D-biotin	0.002
Niacinamide	0.024	Thiamine. HCl	0.010

Dissolve double-distilled water, make to 1 litre, sterilize by filtration, adjust pH to 7.2–7.4 if necessary and store in 100 ml aliquots at −70°C.

B. *Mycoplasma broth*

Ingredient		Instructions
Broth base (Becton Dickinson)	14.7 g	In 600 ml double-distilled water and autoclave
Phenol red	0.02 g	
Selected horse serum	200 ml	Sterile mix with broth base at room temperature
Yeast extract, 25%	100 ml	Store at −20°C in 100 ml aliquots at 25%. Final concentration 2.5%
Stock A	100 ml	

Adjust pH to 7.2–7.4 if necessary, aliquot to sterile test tubes (10 ml/tube), store at 4°C and cap for use within 4 weeks.

C. *Mycoplasma agar medium*

Ingredient		Instructions
Agar base (Becton Dickinson)	23.7 g	In 600 ml double-distilled water and autoclave. Cool to 50°C in water bath
Selected horse serum	200 ml	Warm to 37°C and sterile mix with agar base
Yeast extract (25%)	100 ml	
Stock A	100 ml	

Dispense 10 ml aliquots quickly to 60 mm plates and store at 4°C in a closed container for use within 4 weeks.

Since many strains of mycoplasma, especially *M. hyorhinis*, are difficult or impossible to cultivate in artificial media, the 'indirect' test method in *Protocol 16* should also be included.

116

Protocol 16. Indirect test for mycoplasma

1. Inoculate Petri plates (60 mm) containing 10.5×22 mm glass cover slips (No. 1) with 10^5 indicator cells in 4 ml of EMEM supplemented with 10% FBS. The Vero or 3T6 lines are commonly used. These cells help to amplify low-level infections which may otherwise be undetectable.

2. After incubating these indicator cultures for 16–24 h at 37°C in an air (95%)/carbon dioxide (5%) atmosphere, add 0.5 ml of the test cell suspension obtained as outlined for the direct test.

3. Return the plates to the incubator for an additional 6 days.

4. Remove the culture plates from the incubator, aspirate the fluid and immediately add fixative (*Table 6*). It is important not to let the culture dry before fixation as this may introduce artifacts.

5. After 5 min replace the 'spent' fixative with a fresh volume of the same solution such that the specimen and cover slip are well immersed.

6. Aspirate, allow the specimen to air dry and prepare a fresh working stain solution from the $1000 \times$ stain stock (*Table 6*).

7. Add 5 ml of the staining solution to each dish. Cover and let stand for 30 min at room temperature.

8. Aspirate and wash the culture three times with 3–5 ml of distilled water, removing the final wash completely to allow the cover slip to dry.

9. Remove the cover slip and place with cells up on a drop of mounting fluid (*Table 6*) on a microscope slide.

10. Add another drop of mounting fluid on top of the cover slip and place a larger clean cover slip above this, using standard mounting technique to avoid trapping bubbles.

11. Examine each specimen at $500 \times$ using a fluorescence microscope fitted with number 50 barrier (LP440 nm) and BG12 exciter (330/380 nm) filters.

Note: Confluent cells do not spread sufficiently to stain adequately with Hoechst 33258. Ensure that cultures are subconfluent at the time of staining.

Positive and negative controls should be included with each test series. For this indirect test the Vero or 3T6 cells inoculated with medium serve only as negative controls. *M. hyorhinis, M. arginini, M. orale*, and *A. laidlawii* are suitable controls, in that order of preference. The nucleic acid of the organisms is visible as particulate or fibrillar matter over the cytoplasm, with the cultured cell nuclei more prominent (*Figure 7*). With heavy infections, intercellular spaces also show staining.

Preparation of positive controls, i.e. deliberately infecting cultures, creates a potential hazard to other stocks, so infected cultures should be prepared at a

Table 6. Solutions for the indirect fluorochrome test for mycoplasma

Stock fluorochrome (1000 ×)
Hoechst 33258 stain	5 mg
Thimersol (merthiolate)	10 mg

Mix thoroughly for 30 min at room temperature in 100 ml of HBSS without sodium bicarbonate or phenol red.
Store at 4°C in an amber bottle wrapped completely in aluminium foil. The stain is sensitive to light and heat.

Working fluorochrome stock (1–5 ×)
Make up by adding 0.1–0.5 ml of 1000 × stock to 100 ml HBSS without bicarbonate or phenol red and mix again as for 1000 × stock.

Mounting medium
Citric acid (0.1 M)	22.2 ml
Disodium phosphate (0.2 M)	27.8 ml
Glycerol	50 ml

Store at 4°C and check pH just before use. pH 5.5 is optimal for fluorescence.

Fixative
Glacial acetic acid	1 part
Anhydrous methanol	3 parts

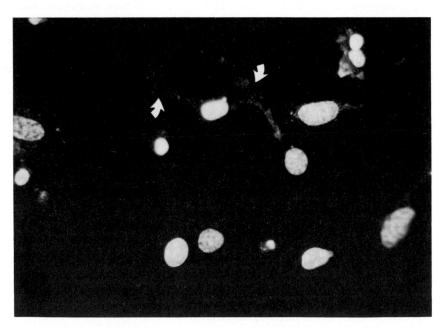

Figure 7. Microphotograph showing mycoplasmal DNA (arrows) demonstrated by Hoechst staining. The larger fluorescing bodies are nuclei of the substrate cell line Vero. (Photograph courtesy of M. L. Macy.)

118

time when, or in a place where, other cultures will not be at risk. Fix and store positive controls at $-4\,°C$ over desiccant until required.

In all quality-control work with mycoplasma or unknown cultures it is prudent to work in a vertical laminar flow hood, preferably isolated from other standard cell culture activity. One should always be aware of the danger of contaminating clean cultures by aerosols from mycoplasma-containing cultures manipulated in the same area, or manually during processing of multiple culture flasks. Appropriate disinfection between uses of a hood or work area is strongly recommended.

Recently a nucleic acid hybridization test for mycoplasma has been developed. The sensitivity and reliability compare favourably with those of the Hoechst stain, and a numerical isotope count is utilized as opposed to the more subjective evaluation of a cytological preparation. The assay uses a tritium-labelled, single-stranded DNA probe complementary to ribosomal RNA of mycoplasma and acholeplasma specifically to detect contaminants. Reagents and detailed methodology for the test are available in kit form from Gen-Probe.

Further detail on these methods and other aspects of mycoplasma infection of cultured cells is available in refs 5, 6, 16, and 17.

4.2.3 Testing for the presence of viruses

Of the various tests applied for detection of adventitious agents associated with cultured cells, those for endogenous and contaminant viruses are the most problematical. Development of an overt and characteristic cytopathogenic effect will certainly provide an early indication of viral contamination. However, the absence of this effect definitely does not indicate that the culture is virus-free. In fact, persistent or latent infections may exist in cell lines and remain undetected until the appropriate immunological, cytological, ultrastructural, and/or biochemical tests are applied. Additional host systems or manipulations, for example treatment with halogenated nucleosides, may be required for virus activation and isolation.

It should be emphasized at the outset that the protocols presented below represent an expedient compromise established to monitor for readily detectable viruses associated with cell lines. Egg inoculations plus selected co-cultivations and haemadsorption tests are included in addition to routine examinations for cytopathogenic effect using phase-contrast microscopy. Similar general tests are recommended by government agencies in cases where cell lines are to be used for biological production work.

Protocol 17. Examining a culture for morphological evidence of viral contamination

1. Hold each flask or bottle so that light is transmitted through the monolayer and look for plaques, foci, or areas that lack uniformity. Cultures from

Protocol 17. *Continued*

frozen stocks should be set up from the pooled contents of about 5% of the ampoules from each lot.

2. Using an inverted microscope, equipped with phase-contrast optics wherever possible, examine cell culture vessels individually, paying special attention to any uneven areas in gross morphology observed previously.

3. Prepare cover-slip cultures if higher power or additional study is required. The cover slips with monolayers can be fixed and stained by standard histological procedure, and the morphology can be compared with that of appropriate positive and negative controls.

The presence of certain viruses which do not produce a CPE in cultured cells can be demonstrated by the haemadsorption test. Infected cells in monolayer adsorb indicator red blood cells after brief exposures in the cold.

Protocol 18. Haemadsorption test

1. Establish test cultures in T-25 flasks using an inoculation density such that the monolayers become confluent in 48–72 h. Use pooled cells from about 5% of the ampoules from any given freeze lot to be tested.

2. Remove the fluid from confluent cultures and rinse the test monolayers with 5 ml HBSS free of divalent cations.

3. Add 0.5 ml each of freshly-prepared suspensions of chick, guinea pig, and human type O erythrocytes (0.5% v/v, washed three times in succession with saline). Then place the flask with monolayer down at 4°C for 20 min.

4. Observe macroscopically and microscopically under low power for clumping and adsorption of RBCs to the monolayer.

5. Repeat on all test cultures not exhibiting haemadsorption before recording a negative result.

A suitable positive control can be established by infecting a flask of RhMK with 0.2 ml of undiluted influenza virus 48–72 h before testing.

The inoculation of test cells or cell homogenates into embryonated chicken eggs provides an additional sensitive test for the presence of viruses.

Protocol 19. Preparation of eggs for inoculation

1. Drill a small hole in the air sac (blunt end) of an 8–9 day embryonated egg using an electric drill and a 1.5 mm ($^1/_{16}$ inch) burr-type bit. In this and subsequent operations, work with sterile instruments. Swab areas of the shell to be drilled with 70% ethanol before and after each manipulation. The drill bits may be placed in 70% ethanol before use.

2. Using an egg candling lamp, locate the area of obvious blood vessel development, and at a central point carefully drill through the shell leaving the shell membrane intact.

3. Place 2 or 3 drops of HBSS on the side hole and carefully pick through the shell membrane with a 26G syringe needle. The saline will seep in and over the chorioallantoic membrane (CAM) to facilitate its separation from the shell membrane.

4. Apply gentle suction to the hole in the air sac using a short piece of rubber tubing with one end to the mouth and the other pressed to the blunt end of the egg. Use the candling lamp to monitor formation of the artificial air sac over the CAM.

5. Seal both holes with squares of adhesive or laboratory tape and incubate the eggs horizontally at 37°C. Standard cell culture incubators and walk-in rooms are entirely adequate for egg incubations. High humidity or air/ carbon dioxide boxes are not satisfactory.

The embryonated eggs with artificial air sacs are ready to receive the test cells or homogenates in 24–48 h. It is important to examine the eggs prior to injection and discard any which show signs of death or degeneration. The presence of intact blood vessels and embryo movements generally indicate that the preparation is healthy. In this event the procedure in *Protocol 20* may be followed.

Protocol 20. Inoculation of embryonated eggs

1. Obtain suspensions of test cells in the appropriate growth medium and adjust the concentration such that 0.2 ml contains $0.5–1 \times 10^7$ cells. If the cells are or may be tumorigenic, freeze in a dry-ice alcohol bath and thaw quickly in a water bath at 37°C. Repeat twice.

2. Remove the seal from the side holes in the embryonated eggs and inject 0.2 ml of the cell suspension or homogenate on to the CAM of each of 5–10 eggs.

3. Using the candling lamp, examine the embryos 1 day after adding the cell suspension; discard any embryos that have died. Repeat the examination periodically for 8–9 days.

4. If embryos appear to be viable at the end of the incubation period, open the eggs over the artificial air sac and examine the CAM carefully for oedema, foci, or pox. Check the embryo itself for any gross abnormalities such as body contortions or stunting.

5. In cases in which viral contamination is indicated, repeat all steps both with a second aliquot of the suspect cells and with fresh fluid samples from eggs in which the embryos have died or appear abnormal.

Positive controls may be established by inoculating eggs with influenza virus, Newcastle disease virus, and/or Rous sarcoma virus. Negative controls receive only an injection of standard cell culture medium. Consult ref. 18 for diagrams and more detail on egg inoculations.

The co-cultivation of test and presumptive-substrate cell lines provides another suitable method for detection of viruses, and this technique is especially useful for suspension cell cultures. The substrate or target line of choice will depend upon the species from which the test cell line originated. For example, for human cell lines, one could co-cultivate with WI-38, MRC-5, or primary human embryonic kidney cells (HEK). A cell line from a second species of choice in this example might be the African green monkey.

After selecting appropriate substrate lines, initiate and maintain the co-cultivation as described in *Protocol 21*.

Protocol 21. Co-cultivation procedure

1. Inoculate a T-75 flask with 10^6 cells from each line in a total of 8 ml of an appropriate growth medium. In some cases the inocula may have to be adjusted in an attempt to maintain both cell populations during the co-cultivation period. For example, if a very rapidly proliferating line is co-cultivated with a test line that multiplies slowly, the initial ratio of the former to the latter could be adjusted to 1:10. Similarly, the population that multiplies slowly might have to be re-introduced to the co-cultivation flasks if it were being overgrown by the more rapidly dividing cells.

2. Change the culture fluid twice weekly and subcultivate the populations as usual soon after they reach confluence. If the test line grows in suspension, that population will have to be recovered by centrifugation and subcultivated by dilution as usual.

3. Examine periodically for CPE and haemadsorption over at least a 2–3 week period, using procedures described above.

Viral isolates may be identified through standard neutralization (haemadsorption inhibition, plaque inhibition, haemagglutination inhibition), or complement fixation tests.

It should be emphasized that, in spite of these screens, latent viruses and viruses which do not produce overt CPE or haemadsorption will escape detection. Some of these could be potentially dangerous for the cell culture technician or for exposed animals. For example, Hantaan virus, the causative agent for Korean haemorrhagic fever, replicates in tumour and other cell lines. Outbreaks of the disease in individuals exposed to infected colonies of laboratory rats have been reported separately in five countries. An incident of transmission during passage of a cell line was confirmed in Belgium. As a result of these observations, rat and mouse–rat hybrid cell lines in the ATCC repository were screened using an indirect immunofluorescent antibody assay (19) and were found to be negative.

Concern over laboratory transmission of the human immunodeficiency viruses should also be mentioned. At least two cases of probable infection during processing have been described in US laboratories; one is presumed to be due to parenteral exposure and the second to work with highly concentrated preparations (20). In the latter circumstance, strict adherence to Biosafety Level 3 (UK Category 3) containment and practices is essential. For more detail on safety precautions for work with cell lines, consult refs 21–23.

A host of other viruses which may be present in cell lines are revealed through specific immunological or biochemical testing. Murine pathogens such as *Ectromelia*, hepatitis, or lymphocytic choriomeningitis viruses (and others) are detected through mouse antibody profile (MAP) tests. Murine products such as ascitic fluid may transmit some of these viruses to cause serious infections in animal colonies. Accordingly, it is prudent to screen murine products and lines.

For additional detail and methodology consult refs 18, 21, and 24.

4.3 Testing for intraspecies cross-contamination

With the dramatic increase in numbers of cell lines being developed, the risk of intraspecies cross-contamination rises proportionately. The problem is especially acute in laboratories where work is in progress with the many different cell lines of human and murine origin (hybridomas) available today. Tests for polymorphic isoenzymes and DNA, surface marker antigens, and unique karyology are all important tools to detect cellular cross-contamination within a given species.

4.3.1 Isoenzymology

A method for the verification of cell line species by determining mobilities of G6PD, LDH, and NP isozymes is outlined in Section 4.1.2. By using similar technology one can also screen for intraspecies cross-contamination.

In cell lines from various individuals of the same species there are often different co-dominant alleles for a given enzyme locus, the products of which are polymorphic and electrophoretically resolvable. In most cases the phenotype for these allelic isozymes (allozymes) is extremely stable. Consequently when allozyme phenotypes are determined over a suitable spectrum of loci they can be used effectively to provide an allozyme genetic signature for each specific line under study (25–30).

With human cell lines, the definition of allozyme phenotypes for seven or more enzymes—for example adenosine deaminase (ADA), G6PD, esterase D (ESD), peptidase-D (PEP-D), phosphoglucomutases 1 and 3 (PGM_1, PGM_3), and 6-phosphogluconate dehydrogenase (PGD)—has been shown to be sufficient to provide identification with a high degree of confidence. The ability to discriminate identities among cell lines using this method can be increased of course, by determining the profiles of additional allozymes. Some of the more useful for this purpose include acid phosphatase (ACP1), glyoxalase 1 (GL01), malate dehydrogenase (ME2), α-fucosidase (FUCA), and adenylate kinase (AK1). Methodology is described in detail elsewhere (24–30).

Polymorphic isozymes have also been reported for murine cell lines and hybrids. Allozymes of potential use in monitoring for intraspecies cross-contamination in this case include the esterases (ES-1 and ES-2), dipeptidase 1 (Dip-1), glutamic-oxaloacetic transaminase 2 (GOT-2), isocitric dehydrogenase 1 (NADP-dependent form), NADP-malic enzyme (Mod-1, supernatant form), glucose-phosphate isomerase 1 (GPI-1), and phosphoglucomutases 1 and 2 (PGM_1 and PGM_2). While methodology for their electrophoretic separation is available (31), these allozymes have not yet been fully exploited in the development of signatures for specific murine cell lines.

4.3.2 Tests for blood group and histocompatibility antigens

The blood group and human leukocyte antigens (HLA) on the plasma membrane of human cells in culture provide additional, highly useful markers for identification. Lack of expression or partial expression of these has been documented in a number of cases.

Blood group antigens are present on normal human epithelia in primary culture (32) and on some continuous epithelial lines. The standard test for these can be applied as described in *Protocol 22*.

Protocol 22. Standard tests for blood group antigens

1. Obtain a cell suspension by dissociation or harvest and wash twice in PBS by centrifugation with subsequent resuspension as usual.

2. Perform a cell count and adjust the final cell concentration to $1-2 \times 10^6$/ml.

3. Place a drop of the suspension on each of four separate locations on a large microscope slide and add a drop of anti-human A, B, or AB typing antiserum to each. Add a drop of PBS to the fourth drop of cell suspension to provide a negative control.

4. Immediately mix each pool separately with glass rods and observe under low power for agglutination. The negative control may also show some cell clumping but this should be minor when compared with the positive test suspension–antiserum mix.

Experience in this laboratory indicates that in many cases the donor's blood group type is expressed even on lines from malignant tumours (8, 9). Not infrequently, however, cells of other human lines from A, B, or AB donors will not react, thus giving a false negative, type O reading. The hypothesis that this could be due to removal of the antigen during dissociation has been tested by repeating the assay on cells maintained in suspension culture for

2–18 h. In spite of the fact that this should allow for replacement synthesis of the surface antigen, inappropriately negative lines remained negative.

This problem with blood group antigen detection on cultured epithelia should be recognized. The simple test, coupled with others for intraspecies contamination among lines, is nevertheless valuable in initial screening. Examples of results for a variety of lines are given in refs 8 and 9.

The major histocompatibility system in humans consists of HLA antigens present on the plasma membrane of most nucleated cells. These numerous antigens, which are coded by co-dominant genes (77 alleles) of five closely linked loci on chromosome 6, provide one of the most polymorphic human group systems known. The antigens are detected routinely by a two-stage, complement-dependent cytotoxity test, and dye-exclusion is used to estimate loss of viability (Section 3.3.1).

Protocol 23. Detection of HLA antigens (simplified and standard procedure)

1. Load wells of plastic histocompatibility typing plates at 1 μl/well with the desired panel of antisera using a Hamilton dispenser. Antisera, which are available commercially, are generally obtained from individuals immunized to HLA antigens by pregnancy or blood transfusions. Typing plates loaded with a spectrum of HLA antisera can also be purchased for routine clinical work. Negative controls should be included with each cell line and run.

2. Obtain a cell suspension, wash as for the blood group antigen assay using medium RPMI 1640 without serum, and adjust the cell concentration such that 1 μl with 10^5 cells can be added to each well using a single-place Hamilton dispenser.

3. Mix by placing the typing plate against a Yankee pipette shaker and incubate at 20°C for 30 min.

4. Add 5 μl of rabbit complement to each well. Incubate at 20°C for 60 min.

5. Dispense 3 μl of 5% aqueous eosin to each well. Wait 2 min for dead cells to stain.

6. Fill wells with buffered formalin (pH 7.2) and add a cover glass (50 × 75 mm) to flatten the droplets.

7. Observe and record the incidence of staining using an inverted microscope. The degree of staining is usually by approximation rather than actual cell count.

While this procedure can be applied successfully for typing some cell lines, it should be emphasized that modifications will be required in many individual cases. The major variables are non-specific antibodies present in the rabbit complement or HLA antisera.

Rabbit serum is the most satisfactory source of complement for this reaction because it contains natural antibodies to human cells. The interaction of these with other cell surface antigens enhances the complement-dependent cytotoxic effect of the anti-HLA antibody–antigen union. The titres and specificities of these natural antibodies differ even among pooled rabbit sera. The problem may be overcome by varying the incubation times, by trying different sources and dilutions of complement, by diluting the rabbit serum with human serum, or by absorption of the rabbit serum with cultured cells (33). Each cell line may have to be examined separately, since the end result depends upon the multiple interactions between antibodies present with the spectrum of antigens on the surface of each cell type.

The presence of given HLA allo-antigens on a particular cell line can be confirmed by absorption-inhibiting typing. In this case the ability of the cells to absorb HLA allo-antibodies from antisera of known specificity is determined by quantitating the loss of cytotoxic effect after absorption. To accomplish this, the pre- and post-absorption antisera are titrated in parallel against a panel of lymphocytic lines of known HLA profile (33).

4.3.3 Giemsa banding

A powerful method for cell identification, which is close to absolute in some cases, involves karyotype analysis after treatment with trypsin and the Giemsa stain (Giemsa or G-banding). The banding patterns made apparent by this technique are characteristic for each chromosome pair and permit recognition by an experienced cytogeneticist even of comparatively minor inversions, deletions, or translocations. Many lines retain multiple marker chromosomes, readily recognizable by this method, which serve to identify the cells specifically and positively.

Many modifications of the original technique (34) are available. *Protocol 24* can be applied to metaphase spreads obtained as in step 10 of *Protocol 12*.

Protocol 24. Cell identification by Giemsa banding

1. Use air-dried slides within 2–7 days.
2. Incubate in 0.025 M phosphate buffer (3.4 g KH_2PO_4/l adjusted to pH 6.8) at 60°C for 10 min. The buffer is also used in subsequent steps.
3. Prepare a staining solution just before use consisting of 6.5 ml phosphate buffer, 0.55 ml of a trypsin solution (1% Difco 1:250 distilled water), 2.5 ml of 100% methanol and 0.22 ml of stock Giemsa solution (commercial). The 1% trypsin stock may be made in bulk and stored in aliquots at −70°C.
4. Flood each slide with about 1 ml of staining solution and leave for 15 min.
5. Rinse briefly with distilled water and air dry completely.

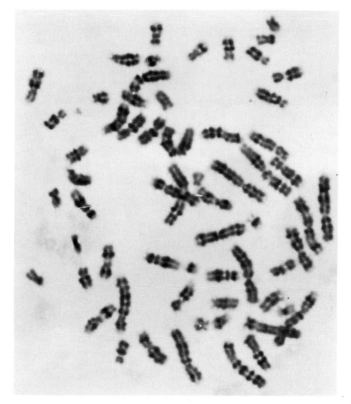

Figure 8. Giemsa banded chromosomes from the cell line ATCC.HTB 16 (U-138 MG) isolated from a human glioblastoma. (Photograph courtesy of T. R. Chen.)

Completely dried slides are used to examine under bright field, oil-immersion planapochromat objectives without cover slip. Oil can be placed directly on the slide. However, care must be taken not to scratch the cell surface. Oil must be removed as completely as possible immediately after the use of slides. A few changes in xylene should generally be satisfactory. A typical banded preparation is shown in *Figure 8*. See refs 35 and 36 for further detail and additional example preparations.

4.3.4 DNA fingerprinting

The application of recombinant DNA technology and cloned DNA probes to identify and quantitate allelic polymorphisms provides a reasonably straight-forward and robust means for cell line identification. Jeffreys and coworkers (37–39) developed DNA probes from tandem repetitive minisatellite regions of the human genome. These hybridize to DNA fragments from multiple, hypervariable loci separated by electrophoresis and visualized by autoradiography after Southern blotting. Such fingerprints are individual-

Table 7. Stock reagents for DNA fingerprinting

Lysis buffer	
NaCl	0.15 M
EDTA	0.1 M
Tris	0.02 M
Sodium dodecyl sulphate (SDS)	1% w/v
pH	8.0
Proteinase K stock in 0.15 M NaCl	20 mg/ml
RNase stock in 10 mM Tris at pH 7.5	10 mg/ml
TE Buffer	
Tris	10 mM
EDTA	1 mM
pH	8.0
Enzyme buffer (10×) (e.g. for Hae III)	
Tris–HCl	500 mM
$MgCl_2$	100 mM
NaCl	500 mM
pH	8.0
TAE Buffer (40×)	
Tris	1.6 M
Sodium acetate	0.8 M
EDTA	40 mM
pH	7.2
Loading buffer (10×)	
Glycerol	50%
40 × TAE	1×
Bromophenol blue, 10%	1%
Xylene cyanol, 10%	1%
SSC Buffer (20×)	
NaCl	3 M
Na_3 citrate.$2H_2O$,	0.3 M
pH	7.0
TBE buffer (1×)	
Tris	0.05 M
Boric acid	0.05 M
EDTA	1 mM
pH	8.2
Polymerase buffer (10×)	
Tris	70 mM
NaCl	500 mM
$MgCl_2$	70 mM
pH	7.4
HS (high salt) buffer	
Tris	50 mM
$MgCl_2$	10 mM
NaCl	100 mM
pH	8.0
Denhardt's solution (100×)	
Polyvinyl pyrrolidone	10 g
BSA	10 g
Ficoll 400	10 g
Double-distilled water to make	500 ml

Filter sterilize, store at 4°C, and mix well before use.

specific and have been used successfully in forensic medicine (40), in paternity suits (39), and more recently for cell line identification (41).

Steps in the process involve extraction of cellular DNA, digestion with restriction endonucleases, electrophoresis in an agarose gel, Southern blotting to transfer DNA fragments to a nylon membrane, hybridization with a radioactively-labelled, single-strand DNA probe and autoradiography to locate hybrids on the membrane. Apparatus and reagents for these procedures may be obtained from Life Technologies, New England Biolabs, Associated Bag, Fotodyne, VWR, and Sigma. The formulae for the reagents required for the various steps are provided in *Tables 7* and *8* and *Protocols 25* and *26*.

Table 8. Pre-hybridization and hybridization solutions

Stock	Prehybridization Final conc.	200 ml	Hybridization Final conc.	200 ml
1 M NaPO$_4$, pH 7.0	50 mM	10.0	50 mM	10.0
ss-DNA (20 mg/ml) (boil 20 mins)	500 µg/ml	5.0	100 µg/ml	1.0
Denhardt's (100×) (*Table 7*)	10×	20.0	1×	2.0
20× SSC (*Table 7*)	5×	50.0	5×	50.0
Dextran sulphate (50%)	5%	20.0	10%	40.0
20% SDS	1%	10.0	1%	10.0
Water (double-distilled)		85.0		87.0

Protocol 25. Preparation of buffered saturated phenol

1. Melt redistilled crystallized phenol at no more than 50°C and saturate with an equal volume of distilled water.

2. Remove aqueous phase and add aqueous Tris base (50 mM) at pH 10.5 and mix.

3. Spin at about 500 *g* at room temperature to separate phases and decant and discard the top aqueous layer.

4. Add an equal volume of aqueous Tris–HCl (50 mM) at pH 8.0, mix and repeat step 3.

5. Store at 4°C.

Protocol 26. Preparation of a single-strand probe from M13 recombinant

This method produces a single-stranded ^{32}P-labelled insert fragment from M13 sequencing substrates.

1. Thaw ss-M13 DNA with insert.

2. Mix:
   - 4 μl M13 DNA with insert (400 ng DNA) at 100 μg/ml
   - 2.6 μl double distilled water
   - 1.0 μl M13 polymerase buffer (*Table 7*) (vortex)
   - 3.3 μl of 1.2 μg/ml 17-mer primer (vortex, spin)

   Anneal at 60°C for 30 min.

3. Add:
   - 10 μl AGT mix (dATP, dGTP, dTTP, 0.5 mM each)
   - 6.0 μl 10 mCi/ml ^{32}P-dCTP
   - 3.0 μl M12 TE buffer (vortex)
   - 0.7 μl Klenow polymerase, 5000 U/ml (vortex, spin)

   Incubate 37°C for 15 min.

4. Add 2.5 μl 0.5 mM CTP (vortex, spin).
   Incubate 37°C for 15 min.

5. Add:
   - 3.0 μl HS buffer (*Table 7*)
   - 1.2 μl of 0.1 M spermidine (vortex)
   - 2.0 μl EcoR1 (10 units) (vortex, spin)

   Incubate 37°C for 20 min.

6. Add:
   - 5.2 μl × 1.5 M NaOH, 0.1 M EDTA, pH 8.0
   - 8.0 μl 10 × loading buffer

7. Load on to 1.2% agarose gel with wide slots.

8. Allow to sit for 5 min.

9. Run at 100 V until loading dye has moved 6 cm.

10. Remove gel and wrap in plastic wrap.

11. Expose gel to X-ray film for about 5 min, develop film, and determine location of insert.

12. With a clean blade, cut out insert and place in a small glass tube, add 1 mg carrier DNA (single-stranded herring DNA). Add 500 μl double-distilled water.

13. Store at −20°C until needed.

14. Place gel piece in dialysis bag with 200 μl of 1 × TBE buffer (*Table 7*) and electroelute. Probe will remain inside the bag and can be removed by low-speed centrifugation in a 15 ml tube. The probe is now ready for use in hybridization.

Protocol 27. Preparation of cellular DNA

1. Suspend a pellet of about 10^7 cells in 10 ml of lysis buffer (*Table 7*), add stock proteinase K solution (*Table 7*) to give a final concentration of 0.1 mg per ml and incubate at 65°C for 3–18 h, mixing occasionally to disperse clumps.

2. Add stock RNase solution (*Table 7*) to give a final concentration of 100 μg/ml, incubate at room temperature for 30–60 min and extract by mixing with an equal volume of buffered-saturated phenol (*Protocol 21*).

3. Separate the phases by centrifugation at 500 × g (15 min), transfer the upper aqueous phase to a new tube and adjust to the original volume with TE buffer (*Table 7*).

4. Extract with an equal volume of buffered-saturated phenol: chloroform: isoamyl alcohol mixed at a 25:24:1 (v/v) ratio and repeat step 3.

5. Extract with an equal volume 24:1 mixture of chloroform and isoamyl alcohol and repeat step 3.

6. Add one tenth the volume of sodium acetate (3M), plus 2 volumes of ice-cold ethanol, and mix gently to precipitate DNA. Transfer the white coiled mass of DNA using a transfer pipette to a 1.5 ml microfuge tube containing 1 ml of 70% ethanol, invert to mix (wash) the DNA mass. Spin briefly in a microfuge, discard ethanol wash, and air dry the DNA pellet.

7. Resuspend the DNA in 1 ml of TE buffer, mix overnight at 4°C, determine the concentration of DNA by measurement of absorbance at 260 nm, and store the sample at −20°C until required for further analysis.

When a convenient number of DNA samples has been accumulated, continue by digesting the DNA as in *Protocol 28*.

Protocol 28. DNA digestion procedure

1. Add a sufficient volume of each DNA suspension to a tube containing the digestion mixture (see below) to provide 10 μg DNA in each tube, mix

Protocol 28. *Continued*

briefly, spin in a microfuge for a few seconds and incubate at 37°C for 3–4 hours. The digestion mixture contains:

10 µl	10× enzyme buffer (*Table 7*)
8 µl	the particular endonuclease
x µl	DNA sample to give 10 µg DNA
82 − *x* µl	sterile distilled water
100 µl	total volume

2. Add 10 µl of sodium acetate (3 M) pH 7.2, 4 µl of EDTA (0.5 M) pH 8.0, mix, add 114 µl of buffered-saturated phenol, vortex to mix and spin for 5 min in the microfuge to separate phases.

3. Transfer the upper aqueous phase to a new tube, add 250 µl ice-cold ethanol, place at −60°C for 30 min, spin in the microfuge for 20 minutes to pellet DNA, and decant the ethanol wash. Add 250 µl ice-cold 70% ethanol, mix briefly, microfuge 5 min, and decant the supernatant.

4. Dry the digested DNA pellet, resuspend in 17 µl of 1 × TAE buffer (*Table 7*) by gentle pipetting, add 3 µl of 10 × loading dye (*Table 7*), vortex, incubate for 5 min at 50–60°C to complete resuspension, microfuge for a few seconds, and load on the agarose gel.

During the incubation and precipitation steps described above, prepare the agarose gel to proceed with electrophoresis as in *Protocol 29*.

Protocol 29. Preparation of agarose gel

1. Add 2.75 g of agarose to 275 ml of 1× TAE buffer and dissolve by heating on a hotplate.

2. Cool to 60°C and pour the solution into a large, 20 × 20 cm gel mould and insert a 20-place, 1 mm tooth comb.

3. After the gel has hardened, place it in the gel box with sufficient 1 × TAE buffer to cover, remove the comb carefully and apply the DNA samples, indicators, and standards from right to left.

4. Run at 60 V (constant) overnight or until the bromophenol blue marker is at the bottom of the gel.

The Southern blotting procedure transfers DNA fragments from the gel to a nylon membrane which can then be manipulated easily in the hybridization and subsequent steps. The nylon membrane is placed on the gel, and sodium hydroxide solution carrying DNA is drawn through to wick, filter, and blotting papers. DNA is bound to the nylon membrane. It is important to include steps to permit exact orientation of the nylon membrane to the gel and markers to identify molecular weights of the DNA fragments isolated.

When electrophoresis is complete, proceed with the Southern blotting as in *Protocol 30*.

Protocol 30. Southern blotting procedure

1. Remove the gel still in the casting mould and submerge in 1 litre of distilled water in a 20×25 cm (8×10 inch) tray, add a few drops of ethidium bromide solution (10 mg/ml), mix and stain for about 20 min.

2. Wash in distilled water to remove excess stain, and observe the gel under ultraviolet light.

3. Place a clear ruler along the markers on the gel with the rule edge lined up to the bottom of the wells to permit ready measurement. Make a Polaroid photograph of the gel to facilitate orientation later.

4. With a razor blade trim away the lanes with the markers (vertical). Likewise, trim just above the lanes leaving wells intact and trim a small strip from the bottom of the gel (both horizontals).

5. Return the gel to the tray and denature by soaking in 1 litre of 0.4 M NaOH for 30 min on a platform with gentle rocking.

6. Wet the nylon membrane (19×23 cm) with distilled water, then soak in 0.4 M NaOH.

7. Wet two Whatman paper wicks (19.5×40 cm) in 0.4 M NaOH, place on the gel and flip the gel over with the inner U-shaped piece from the transfer apparatus (see supplier's directions).

8. Using the wicks as support, remove the U-shaped piece and transfer the gel to the transfer box. Place Parafilm around the edges of the gel.

9. Place the nylon membrane on the gel, then sequentially two pieces of Whatman filter paper (19×23 cm) presoaked in 0.4 M NaOH. As each is added, remove excess liquid and any air bubbles that may be trapped between the layers by gently rolling a pipette over the surface.

10. Add a dry filter paper, then an additional 15 cm of filter papers, place a weight of about 600 g on top and blot for 6 h to overnight.

11. Transfer the nylon membrane with gel facing up to a clean dry surface, mark position of the wells, and outline the gel with a lab marker.

12. Label the membrane, wash twice for 15 minutes each in $2 \times$ SSC (*Table 7*), remove the membrane, and dry overnight between two pieces of filter paper. Store flat in a cool dry place as desired until hybridization.

The hybridization reactions require preparation of a radioactively-labelled, single-stranded DNA probe as outlined for example in *Protocol 24*. The nylon membrane with DNA fragments is treated in the pre-hybridization solution (*Table 8*) and the hybridization step is accomplished subsequently. Finally, the location of hybrids is visualized after autoradiography. See *Protocols 31–33*.

Protocol 31. Pre-hybridization procedure

1. Wash the nylon membranes in 1 litre of $(0.1\times)$ SSC, (0.5%) SDS for 10 min at 65 °C.
2. Roll the membranes while still in the washing solution and transfer to a plastic hybridization bag with lower end sealed. Unroll the nylon membranes (2 per bag) such that the sides with DNA on each will face out. Seal the upper end of the bag and make a small diagonal cut in one corner.
3. Using a 25 ml pipette, insert 30 ml of pre-hybridization solution into the bag, chase out bubbles, and seal the bag.
4. Place the bag in a plastic container filled with water at 65 °C, immerse completely in a shaking water bath and shake at 65 °C overnight.

The following day proceed with the hybridization proper (*Protocol 32*).

Protocol 32. Hybridization procedure

1. Cut a corner of the bag and drain off the pre-hybridization solution as completely as possible by rolling over the bag surface with a pipette. Avoid drying the membranes as this may increase background labelling.
2. Mix the radiolabelled probe with hybridization solution (*Table 8*) to give about 2×10^5 c.p.m./ml, and immediately insert the solution into the bag.
3. Seal the bag and incubate as before for 72 h at 65 °C with shaking.

Finally, wash the membranes and process for autoradiography.

Protocol 33. Autoradiography procedure

1. Cut off one corner of the bag, remove and discard the radioactive solution appropriately.
2. Thoroughly wash the membranes with agitation in pre-warmed solutions using the following regimen:
   - 6 times for 7 min each in 500 ml $1 \times$ SSC, 0.1% SDS at 65 °C
   - 2 times for 15 min each in 500 ml $0.1 \times$ SSC, 0.1% SDS at 65 °C
   - once for 10 min in 500 ml $3 \times$ SSC at room temperature

3. Place each membrane between two pieces of Whatman 3M filter paper to remove excess fluid and air-dry for about 2 h.

4. Wrap the membranes with a single fold of plastic wrap and tape to Kodak XAR-5 autoradiograph film with screen.

5. Expose for about 7 days at −60°C and develop according to supplier's instructions.

An autoradiograph with typical result for DNA from 18 human cell lines is provided in *Figure 9*.

4.4 Verifying tissue of origin

The markers used for verification of the source tissues for cell lines are probably as numerous as the types of metazoan cell. The means of major utility include analyses of fine structure, immunological tests for cytoskeletal

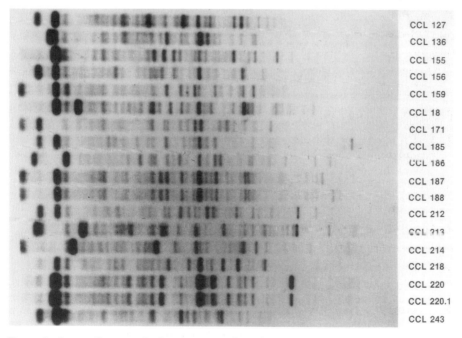

CCL 127
CCL 136
CCL 155
CCL 156
CCL 159
CCL 18
CCL 171
CCL 185
CCL 186
CCL 187
CCL 188
CCL 212
CCL 213
CCL 214
CCL 218
CCL 220
CCL 220.1
CCL 243

Figure 9. Autoradiogram of a Southern transfer using DNA from 18 human cell lines. The DNA was digested with HAE III restriction endonuclease and transferred to a nylon membrane. The membrane was hybridized with a single-stranded ^{32}P-labelled probe to the hypervariable region of the human genome (ref. 41). Note that the profiles for all lines are distinctly different except for those of CCL 187 and 188, and CCL 220 and 220.1. Each of these two pairs are lines from the same individual. CCL 187 and 188 were isolated from a colon tumour cultivated without and with trypsinization respectively. CCL 220 and 220.1 were isolated from a colon tumour of another patient, and 220.1 is a variant with homogeneously staining regions in marker chromosomes. (Photograph courtesy of Y. A. Reid.)

and tissue-specific proteins, and of course any of an extremely broad range of biochemical tests for specific functional traits of tissue cells. Some of the more general methods in these categories will be outlined below.

4.4.1 Fine structure analyses for characteristic markers

The presence of desmosomes and keratin filaments, both of which can be visualized by electron microscopy, is generally accepted as characteristic of epithelia. The so-called Weibel–Palade bodies are specific for endothelial cells of the umbilical vein and other sources. Cells of the islets of Langerhans can be characterized morphologically by demonstrating the presence of their specific secretion granules.

A basic technique to prepare cultured cells for such fine structural analysis is outlined below. Reagents required are listed in *Table 9*. *Protocol 34* describes

---

Table 9. Reagents used in processing cultured cells for electron microscopy. (All reagents can be obtained from Polysciences, Inc.)

Stock solutions
Sodium cacodylate buffer, 0.1 M, pH 7.2–7.3

Na cacodylate	42.8 g
Aqueous phenol red (0.5%)	4 ml

Dissolve in 1 litre double-distilled water and adjust pH with 1 M HCl
Glutaraldehyde: 8% aqueous
Dilute to 4% with double-distilled
Osmium tetroxide 4% aqueous
Uranyl acetate 0.5% aqueous
Reynolds' lead citrate

Lead nitrate	1.33 g
Trisodium citrate, dihydrate	1.76 g
Double-distilled water	30 ml

Dissolve with stirring for about 30 min. Add 8.0 ml of 1 M NaOH (carbonate free) and dilute to 50 ml.
The solution, which is stable for at least 6 months should be passed through a Millipore filter (0.22 μm or less) just before use to avoid small particular precipitates.
Alcohol series to 100% absolute and propylene oxide
Plastic mixtures

A Epon 812	62 ml
Dodecenyl succinic anhydride (DDSA)	100 ml
B Epon 812	100 ml
Methyl nadic anhydride (MNA)	89 ml

These can be made up in advance and stored at 4°C in well-sealed containers for up to 6 months. Accelerator (2,4,6-tridimethylaminomethyl phenol) (DMP 30) is added just before use at 1.5–2.0%.

Working solutions
Buffered glutaraldehyde (2%) 1:1 stock buffer/4% glutaraldehyde stock
Buffered osmium tetroxide (1%) 1:3 4% osmium/stock buffer
Plastic mixture
Thoroughly mix 7 parts of mixture B with 3 parts of mixture A just before use and add the accelerator at 1.5–2.0%.

---

the preparation of a cell monolayer or pellet of cells harvested either from a suspension culture or by dissociation.

Protocol 34. Preparation of cultured cells for fine structural analysis

1. Collect a pellet by centrifugation of not more than $1–3 \times 10^6$ cells in a 15 ml glass or polypropylene test tube. Monolayers can be processed directly on plastic.

2. Remove culture medium and wash the cells once with stock buffer at 4°C. Steps 1–5 are all performed at this temperature.

3. Fix for 1 h in 2% buffered glutaraldehyde. If a pellet is being processed, dislodge it gently from the test tube sides using an orange stick, thus facilitating diffusion of reagents from all sides of the three-dimensional mass. It is important to retain a mass of cells throughout, as opposed to a dispersed population. These larger conglomerates will pass through the more viscous mixtures to be used in later steps.

4. Decant off the fixative and rinse three times with buffer for at least 10 min per rinse.

5. Post-fix with 1% osmium tetroxide for 1 h.

6. Decant with osmium tetroxide fixative and wash three times with double-distilled water (5 min each).

7. Leave for 16–20 h in 0.5% aqueous uranyl acetate. The low pH of this solution removes glycogen but gives excellent membrane preservation and staining.

8. Pass through an alcohol series (10–30–70–95–100–100%) leaving the sample for 10 min in each solution. For pellet fragments, three subsequent 10 min infiltrations with 1:1 100% alcohol/100% propylene oxide, 100% propylene oxide, and 1:1 propylene oxide/complete plastic mixture are recommended. These latter three steps cannot be used with monolayers on plastic. In that case, infiltrate directly from 100% alcohol for 30 min with a 1:1 solution of the complete plastic mixture and 100% alcohol.

9. Decant and begin infiltration with 100% complete plastic mixture (7:3 B/A plus 1.5–2.0% DMP-30, see *Table 9*). For pellet fragments this can best be accomplished by placing a fragment no larger than 1 mm in diameter in the upper part of the mixture in a BEEM embedding capsule (Polysciences). The fragment should settle to the bottom (pointed end) of the capsule within a few hours.

    For a monolayer, the complete plastic mixture is placed in the culture vessel and allowed to infiltrate for 8–18 h.

    Bubbles can be removed by placing the BEEM capsules or open culture vessels in a vacuum.

Protocol 34. *Continued*

10. Polymerize at 65–70°C for 24 h or longer to give a suitable hardness. The B/A mix is rather hard and can be modified depending on the technician's experience and preference (42).

11. For infiltrated monolayers, cut out the area(s) to be examined using a fine jeweller's saw, and cement to a dummy plastic rod. An 'instantly' drying form of contact cement is best. The sample can be oriented for cross-sectioning or to collect sections which are tangential to the monolayer surface (e.g. for desmosomes). For pellet fragments, simply snap off the BEEM capsule, trim as necessary with a single-edged razor blade, and place in the microtome chuck.

12. Section using an ultramicrotome (e.g. Leica). Silver or grey sections are suitable for examination at magnifications above 15 000. Gold sections can be used at lower magnifications.

13. Collect the sections on a suitable grid, stain for 20–60 s in lead citrate, examine and photograph as desired. Consult refs 42 and 43 for further information.

4.4.2 Immunological testing for cytoskeletal proteins

The cytoplasmic intermediate-filament (IF) proteins which form an essential part of the cytoskeleton have emerged as being highly useful in cell identification. Many filament subgroups have been described on the basis of morphology, polypeptide composition, and unique immunogenicity. The distribution of these has been shown to be tissue-type specific. Thus epithelia from most tissues contain cytokeratin(s), and cells of mesenchymal tissues contain vimentin. Desmin is present in myogenic and other cells, neurofilament protein in neurones, and glial fibrillar acidic protein in astrocytes (44).

The filaments may be visualized by electron microscopy or by direct or indirect immunofluorescence staining, or one can demonstrate specific cytoskeletal protein by immunohistochemical means as indicated below for keratin. Reagents (*Table 10*) and staining kits for the latter are available from a number of commercial sources.

To test for the presence of keratin treat coverslip cultures or cells collected on a slide by cytospin as in *Protocol 35*.

Protocol 35. Testing for the presence of keratin

1. Fix for 20 min in methanol: 3% H_2O_2 and wash briefly (30 s) in a bath of running tap water.

2. Place in wash buffer for about 5 min, remove and drain but carefully avoid drying here and throughout the procedure.

3. Cover the monolayer or cytospin 'spot' with buffered normal serum (*Table 10*) and leave at room temperature for 30 min.

4. Drain off serum, add the diluted rabbit anti-keratin antiserum and retain at room temperature for 30 min.

5. Drain, wash gently several times with wash buffer to remove unbound antiserum, place in a coplin jar containing wash buffer, and leave for 15 min.

6. Drain, cover the cells with swine anti-rabbit IgG antiserum, and leave for 30 min.

7. Rinse several times with wash buffer, place in a coplin jar with fresh buffer for 15 min, and drain carefully.

8. Cover cells with the buffered HPAP reagent, leave at room temperature for 30 min, rinse several times in wash buffer, and place in a coplin jar with fresh buffer for 15 min.

9. Drain, cover with the freshly-prepared chromogen solution, and allow to stain for 5 min.

10. Wash gently with double-distilled water, counterstain with haematoxylin, dehydrate, and mount in Permount.

A positive reaction is indicated by the localized deposit of dark brown, precipitated reaction product. Negative controls which are run in parallel with the test samples, consist of slides with cell preparations exposed to buffer in step 4 instead of anti-keratin antiserum. Positive controls consist of carcinoma

---

Table 10. Reagents for the immunohistochemical demonstration of keratin. (A kit adapted from this method (45) is available from DAKO Co.)

Fixative	50 ml 3% hydrogen peroxide
	250 ml anhydrous methanol.
	Prepare fresh just before use
Tris buffer	0.1 M, pH 7.6
Normal serum	10% swine serum in Tris buffer
Rabbit anti-keratin antiserum	1:20 to 1:100 in Tris buffer
Swine anti-rabbit IgG antiserum	1:20 in Tris buffer
Horseradish peroxidase-rabbit anti-horseradish peroxidase-soluble complexes (HPAP)	1:100 in Tris buffer
Wash buffer	1 part Tris buffer (0.5 M, pH 7.6)
	9 parts 0.9% normal saline
Chromogen	6 mg 3,3'-diaminobenzidine
	tetrahydrochloride
Prepare fresh just before use	0.1 ml 3% hydrogen peroxide
	10 ml Tris buffer

cell lines such as SCC4 or SCC15. Alternatively, sections of skin serve well. See refs 44 and 45 for further detail.

4.4.3 Demonstration of tissue-specific antigens

There are many accounts of the isolation and characterization of tissue and tumour-specific antigens. The methodologies vary somewhat among laboratories and with the different cells lines or tissues used. Only a few examples will be summarized here, to provide perspective.

The human lung tumour line 2563 (derived from MAC-21) was used directly by Akeson (46) as the immunogen in rabbits. The resulting antiserum was found by complement fixation, immunofluorescence, and saturation binding assays to contain antibodies specific for antigens of the normal kidney tissue. These could be selectively absorbed out by kidney preparations, showing the existence of a distinct additional antigen shared by the two normal organs.

Ceriani and associates present data demonstrating the existence of mammary-specific antigens on human normal and tumourigenic breast cells. These workers immunized rabbits with a defatted preparation from the washed cream fraction of human milk. Species-specific antibodies were removed from the resultant antiserum by absorption with human erythrocytes. Indirect immunofluorescence studies demonstrated that antibodies in this antiserum bound specifically to cell lines from normal human breast and breast carcinomas. A sensitive radioimmunoassay was developed which permitted quantitative confirmation of this finding. Seven breast carcinoma lines and primary cultures of mammary epithelia bound at least 10–40 times as much antibody as did breast fibroblasts or epithelia from other sources. Trypsinization of MCF-7 or MDA-MB-157 removed 80–85% of the reactivity, indicating that the breast-specific antigens reside on the cell surface (47).

Procedures were developed by Chu *et al.* (48) for the isolation, purification, and quantitation of prostate antigen (PA), from human seminal fluid or prostatic tissue. Crude extracts were treated with ammonium sulphate, the precipitates were solubilized, dialysed, and further purified by column chromatography (DEAE and Sephadex). Fractions in molecular mass regions 26 000–37 000 which contained the PA were concentrated by ultrafiltration, and this final preparation was used to immunize rabbits.

Interestingly, the resultant antiserum showed a single reaction with prostate tissue extracts on immunodiffusion and did not react with similar preparations from other tissues. Furthermore, quantitative studies, accomplished after development of a sensitive enzyme immunoassay for PA, indicated that human cell lines from prostatic carcinomas (LNCaP and PC-3) expressed antigen. In contrast, cell lines from other tissues had no detectable PA.

Hybridomas producing monoclonal antibodies which show a high degree of tissue specificity have been isolated and characterized in a number of laboratories. For example, Koprowski *et al.* (49) reported on the isolation of 19 hybridomas, 15 of which produced antibodies which were specific for cell lines from colorectal carcinomas. Metzgar and associates (50) developed five hybridomas producing antibodies to human pancreatic tumour cell lines. Some cross-reactivity was noted with other tumour tissues or cell lines, but at least in one case (with DU-PAN-1) this was restricted to a transitional cell carcinoma of the bladder. Hybridomas producing antibodies of utility in distinguishing human B cell, T cell, and T cell subsets have also been described (51), and many reagents and kits for this purpose are already available commercially.

Because of the unlimited potential availability of monoclonal reagents, one might anticipate that kits for identification of cells from many tissues will be developed for laboratory use. These may ultimately replace heterogeneous antisera currently in use for such purposes.

4.4.4 Biochemical testing for cell-specific function

A large number of cell lines and strains can be shown to derive from a particular tissue or tumour by the presence of specific synthetic abilities or metabolic pathways. The methodology for such verification of identity is too extensive to be covered in detail. Only a few examples are cited of strains and lines which express specific function.

Yasumura and associates (52) used transplantable mouse and rat tumours as source material for the establishment of four, clonally derived strains of functional cells. Alternating periods of transplantation in host animals and maintenance in culture were required initially to promote growth during isolation. One strain, designated GH_1, obtained from the MtT/W5 pituitary tumour, releases growth hormone into the culture medium and promotes host growth on re-inoculation to either intact or hypophysectomized animals. A second strain, I-10, was derived from the Leydig cell tumour H10119 maintained in BALB/cJ mice. Progesterone and its derivative 20 α-hydroxy-4-pregnen-3-one are the major steroids secreted by these cells in culture. They do not respond to interstitial cell-stimulating hormone, but are stimulated by cyclic AMP. The murine cultured cell strains M-3 and Y-1 were isolated from a Cloudman S91 melanoma and an adrenal tumour, respectively. The M-3 cells produce melanotic tumours upon re-inoculation into suitable hosts (CXDBA animals). The Y-1 adrenal cell strain releases steroid hormones and responds to ACTH.

The clonal isolation technique has also been applied successfully for the derivation of functional cell strains from normal tissues. One excellent example is FRTL, a rat thyroid cell strain that secretes thyroglobulin into the culture medium and concentrates iodide 100-fold (53). Medium F-12M supplemented only with 0.1–0.5% calf serum, insulin, thyrotropin, transferrin, hydrocortisone, somatostatin, and glycyl-L-histidyl-L-lysine acetate was used. The low serum conditions were said to be critical during early stages for selection of the functional cell type. Strains of myogenic cells have also been isolated from normal rat and mouse muscle. In these cases the specific activity of creatine phosphokinase can be used as a biochemical marker (54).

Functional lines are also available from a variety of human tumours. MCF7, for example, was isolated from the pleural effusion of an individual with adenocarcinoma of the breast. The line retains several characteristics of mammary epithelium, including the ability to process oestradiol via cytoplasmic oestrogen receptors and to form domes in confluent monolayers (55). BeWo is a human trophoblastic cell line isolated from a malignant gestational choriocarcinoma of a fetal placenta. Interestingly, the line has been shown to secrete a spectrum of placental hormones including human chorionic gonadotropin, placental lactogen, plus oestrone, oestradiol, oestriol, and progesterone (8). See refs 7, 8, and 56 for additional detail on differentiated lines and biochemical characterizations.

4.5 Identifying and quantitating immunological products

Given the current intense interest and activity in hybridoma technology, no chapter on cell line characterization would be considered complete without reference to this area of endeavour. The specificity of monoclonal antibody produced by a hybridoma is, of course, the ultimate criterion on which identity must be based. If this cell product is absent and its synthesis cannot be restored, the hybrid cell strain is of little use. However, the potential number of different hybridomas which could be generated theoretically equals the number of epitopes which exist times the number of different fusion lines available. Clearly for a hybridoma banking agency or department this presents a dilemma, since no single organization can realistically identify all monoclonal antibodies even of the comparatively narrow range currently available.

Accordingly, a compromise (which has been adopted at the ATCC) may be considered. The banking agency can determine the class, subclass, and quantity of immunoglobulins secreted by a hybridoma and perform the other standard characterizations required (*Figure 1*). This information should prove sufficient to monitor effectively for cross-contamination after the originator verifies monoclonal specificity from progeny of the initial seed stock.

4.5.1 Applying the Ouchterlony test for immunoglobuin isotypes

The Ouchterlony method for determining the isotypes of immunoglobulin produced by hybridomas is relatively straightforward. *Protocol 36* gives the procedure for making the agar plates.

Protocol 36. Preparation of plates for the Ouchterlony test

1. Dissolve agarose to give 0.8% and sodium azide to make 0.025% in PBS by heating in boiling water bath.

2. Add 10 drops of 0.5% aqueous trypan blue per 250 ml of solution and mix gently.

3. Pour 1.5–2 ml aliquots of the hot solution into 60 × 15 mm Petri plates to form a thin bottom layer. While this is congealing retain the rest of the agarose solution in a 56°C water bath.

4. After the primary layer has congealed, add 4 ml of the same agarose solution and allow this to solidify, forming a secondary upper layer.

5. Punch holes in concentric fashion and in the centre of the agar layer using a tubular well cutter, and aspirate the secondary upper layer of the plug to form each well with a pipette attached to any standard vacuum system.

6. Store plates in the refrigerator until use.

Protocol 37. Ouchterlony method

1. Remove a sample from each of the test hybridoma cultures which are close to maximum in cell density and collect the supernatants after centrifugation (200 × g for 10 min).

2. Place 35 µl of the supernatants in the outer wells. If the titre of antibody in the culture fluid is low, it may be necessary to repeat-load the well for that sample up to four times.

3. Place 35 µl of the specific anti-mouse immunoglobulin typing antiserum in the centre well.

4. Maintain the plates at room temperature in a humid environment overnight and read plates the next day. Precipitation lines should form between the well with the appropriate hybridoma supernatant and the specific known anti-immunoglobulin in the centre well. A typical example is shown in *Figure 10*.

4.5.2 Quantitation of secreted immunoglobulin by radial immunodiffusion

The amount of immunoglobulin produced by a hybridoma can be determined once the isotype is known. The appropriate typing antiserum is mixed with agar (*Protocol 38*). A known concentrate of immunoglobulin from the hybridoma supernatant is then added to an antigen well in the agar layer. As the antigen diffuses into the agar a ring of antigen–antibody precipitate forms around the well. The diameter of this ring is measured and compared with that of standards and a standard curve to permit quantitation of the unknown amount placed in the antigen well. After preparation of the agar plates proceed to quantitate immunoglobulin as in *Protocol 39*.

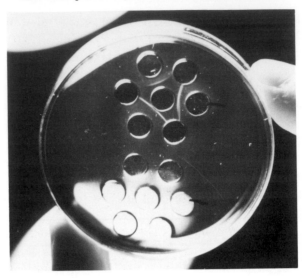

Figure 10. Assay for antibody isotype by the Ouchterlony technique. The precipitin bands between the centre and appropriate outer wells permit identification.

Protocol 38. Preparation of diffusion plates

Buffer
In 100 ml double-distilled water dissolve 0.5 g of K_2HPO_4 and 0.6 g NaCl. Adjust to pH 8.

Agar base
1. Add 1.5 g of Noble agar to 50 ml of buffer and dissolve by heating in a boiling water bath.
2. Store in the refrigerator in 5–10 ml aliquots.

Antiserum-agar mix
1. Dilute the typing rabbit anti-mouse antisera 1:10 in the pH 8 buffer and equilibrate in a 56°C water bath.
2. Liquefy agar base in a boiling water bath and equilibrate at 56°C.
3. Mix equal volumes of the two solutions, dispense immediately on to 2×3 in glass slides and allow to harden.

Protocol 39. Quantitation of immunoglobulin

1. Collect 25 ml of the cell-free supernatant from the test hybridoma culture which is at or close to maximum cell density and cool to 4°C.

2. Precipitate proteins by addition of 25 ml of a saturated solution of ammonium sulfate at 4°C, and centrifuge at 10 000 g for 30 min.

3. Discard the supernatant and dissolve the precipitate in 1–3 ml of distilled water noting the final volume and degree of concentration (typically 8- to 16-fold). Store at −60°C until ready for assay.

4. Using a tubular well cutter, make 12 wells 3 mm in diameter and 12 mm apart in the agar on the diffusion plate(s). Aspirate out the plug with a Pasteur pipette attached to any standard vacuum system.

5. Place 10 μl of antibody concentrate into each test well.

6. For the standard readings, place 10 μl of serial dilutions of the appropriate, purified mouse immunoglobulin into wells to include concentrations of 1, 0.5, 0.25, and 0.125 mg/ml.

7. Place the plates in a sealed, humid chamber at room temperature and leave for 16–18 h.

8. Measure the diameter of the precipitation rings, and calculate the immunoglobulin present by referring to the standards.

Many hybridomas produce 25–75 μg of immunoglobulin per ml of growth medium. *Figure 11* shows results of a typical assay.

5. Computerized cell source and databases

A large number of lines are currently being stored in cell banks and private laboratories throughout the world. In many cases information concerning

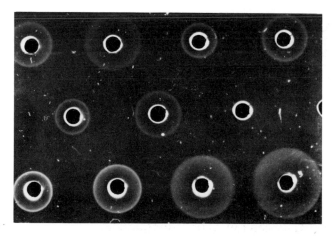

Figure 11. Quantitation by immunodiffusion of immunoglobulins produced by several hybridoma lines. The upper wells contained unknowns and the lower wells a series of standards ranging from 1.25 to 10 μg of immunoglobulin. (Photograph courtesy of W. Siegel.)

availability and special characteristics of these lines has not been widely communicated. This is especially true for hybridomas, both because of the importance of the technology for their development and because so many potentially unique hybridomas can be generated with each fusion.

It would be difficult or impractical for any institution to bank, characterize, catalogue, and distribute all of the cell lines and hybridomas available. Accordingly, groups of interested scientists and organizations have elected instead to develop computerized databases of relevant information on the myriad of lines available for use by the scientific and educational community. The role of such databases is generally that of clearing houses of information relating to the sources and unique characteristics of the specialized cell lines and hybridomas. Often no certification or endorsement of the cells is made. In some cases they are available directly from individual scientists and laboratories participating in the program.

Cell line databases have originated comparatively recently. Their number and evolution with time may differ from the currently conceived plans, depending upon funding, community input, and community requirements. The ATCC initiated a broad-range program of this type in 1977 and currently has source information on some 17 000 metazoan cell lines and hybridomas, many of which are catalogued. The data bases are available internationally for online search through DIALCOM.

Other national cells banks such as the Human Genetic Mutant Cell Repository at The Coriell Institute for Medical Research in Camden, NJ, USA, the European Collection of Animal Cell Cultures supported by the Public Health Laboratory Service, Porton Down, Salisbury, UK, and the Japanese Cell Bank for Cancer Research at the National Institute of Hygiene Services in Tokyo also provide information both on the lines in their collections and on availability of specific cell cultures banked elsewhere.

Acknowledgements

The author wishes to acknowledge the assistance of many members of the Cell Culture Department in developing the various protocols referred to in this chapter. He is especially grateful to M. L. Macy, T. R. Chen, and Y. A. Reid for their suggestions and critical review of the manuscript. The work was supported under a grant (1-R26-CA25635) from the National Large Bowel Cancer Project of the National Cancer Institute, contract No 1-CB-71014 from the National Cancer Institute, and contract (NO 1-RR-9-2105) from the Division of Research Resources of the National Institutes of Health.

References

1. Schaeffer, W. I. (1979). *In Vitro*, **15**, 649.
2. Nelson-Rees, W. A. and Flandermeyer, R. R. (1977). *Science*, **15**, 1343.

3. Nelson-Rees, W., Daniels, W. W., and Flandermeyer, R. R. (1981). *Science*, **212,** 446.
4. Nelson-Rees, W. A., Flandermeyer, R. R., and Hawthorne, P. K. (1974). *Science,* **184,** 1093.
5. Barile, M. F. (1977). In *Cell Culture and Its Application* (ed. R. T. Acton and J. D. Lynn), p. 291. Academic Press, London.
6. McGarrity, G. J. (1982). *Adv. Cell Culture,* **2,** 99.
7. Hay, R. J., Williams, C. D., Macy, M. L., and Lavappa, K. S. (1982). *Am. Rev. Respir. Dis.,* **125,** 222.
8. Hay, R. J., Caputo, J., Chen, T. R., Macy, M. L., McClintock, P., and Reid, Y. A. (1992). *American Type Culture Collection Catalogue of Cell Lines and Hybridomas*, 7th edn. Rockville, MD.
9. Hay, R. J. (1984). In *Markers of Colonic Cell Differentiation* (ed. S. R. Wolman and A. J. Mastromarino), p. 3. Raven Press, New York.
10. Stulberg, C. S. (1973). In *Contamination in Tissue Culture* (ed. J. Fogh), p. 1. Academic Press, London.
11. Macy, M. L. (1978). *Tissue Culture Association Manual,* **4,** 833.
12. Hsu, T. C. and Benirschke, K. (1967–1975). *An Atlas of Mammalian Chromosomes* (9 volumes). Springer-Verlag, New York.
13. Dilworth, S., Hay, R. J., and Daggett, P.-M. (1979). *Tissue Culture Association Manual,* **15,** 1107.
14. Freshney, R. I. (1987). *Culture of Animal Cells*, 2nd edn. Alan R. Liss, New York.
15. Cour, I., Maxwell, G., and Hay, R. J. (1979). *Tissue Culture Association Manual,* **5,** 1157.
16. Macy, M. L. (1980). *Tissue Culture Association Manual,* **5,** 1151.
17. Hay, R. J., Macy, M. L., and Chen, T. R. (1989). *Nature,* **339,** 487.
18. Rovozzo, G. C. and Burke, C. N. (1973). *A Manual of Basic Virological Techniques.* Prentice-Hall, Englewood Cliffs, NJ.
19. LeDuc, J. W., Smith, G. A., Macy, M. L., and Hay, R. J. (1985). *J. Infect. Dis.,* **152,** 1081.
20. Weiss, S. H., Goedert, J. J., Gartner, S., Popovic, M., Waters, D., Markham, P., Veronese, F. M., Gail, M. H., Barkley, W. E., Gibbons, J., Gill, F. A., Leuther, M., Shaw, G. M., Gallo, R. C., and Blattner, W. A. (1988). *Science,* **239,** 68.
21. Hay, R. J. (ed.) (1992). *ATCC Quality Control Methods for Cell Lines*, 2nd edn. Rockville, MD.
22. Barkley, W. E. (1979). *Meth. Enzymol.,* **63,** 36.
23. Richardson, J. H. and Barkley, W. E. (ed.) (1984). *Biosafety in Microbiological and Biomedical Laboratories*. DHHS Publication No. (CDC) 84-8395, U.S. Govt. Printing Office, Washington, DC.
24. Hay, R. J. (1988). *Analyt. Biochem.,* **171,** 225.
25. Povey, S., Hopkinson, D. A., Harris, H., and Franks, L. M. (1976). *Nature,* **264,** 60.
26. O'Brien, S. J., Kleiner, G., Olson, R., and Shannon, J. R. (1977). *Science,* **195,** 1345.
27. O'Brien, S. J., Shannon, J. E., and Gail, M. H. (1980). *In Vitro,* **16,** 119.
28. Wright, W. C., Daniels, W. P., and Fogh, J. (1981). *J. Natl. Cancer Inst.,* **66,** 239.
29. Rutzky, L. P. and Siciliano, M. J. (1982). *J. Natl. Cancer Inst.,* **68,** 81.

30. Halton, D. M., Peterson, W. D. Jr, and Hukku, B. (1983). *In Vitro,* **19,** 16.
31. Nichols, E. A. and Ruddle, F. H. (1973). *J. Histochem. Cytochem.,* **21,** 1066.
32. Stoner, G. D., Katoh, Y., Foidart, J.-M., Trump, B. F., Steinert, P. M., and Harris, C. C. (1981). *In Vitro,* **17,** 577.
33. Pollack, M. S., Heagney, S. D., Livingston, P. O., and Foh, J. (1981). *J. Natl. Cancer Inst.,* **6,** 1003.
34. Seabright, M. (1971). *Lancet,* **2,** 971.
35. Sun, N. C., Chu, E. H. Y., and Chang, C. C. (1973). *Mammalian Chromosome Newsletter,* January, 26.
36. Chen, T. R., Hay, R. J., and Macy, M. L. (1982). *Cancer Genet. Cytogenet.,* **6,** 93.
37. Jeffreys, A. J., Wilson, V., and Thein, S. L. (1985). *Nature,* **314,** 67.
38. Jeffreys, A. J., Wilson, V., and Thein, S. L. (1985). *Nature,* **316,** 76.
39. Jeffreys, A. J., Brookfield, J. F. Y., and Semeonoff, R. (1985). *Nature,* **317,** 818.
40. Gill, P., Jeffreys, A. J., and Werrett, D. J. (1985). *Nature,* **318,** 577.
41. Gilbert, D. A., Reid, Y. A., Gail, M. H., Pee, D., White, C., Hay, R. J., and O'Brien, S. J. (1990). *Am. J. Hum. Genet.,* **47,** 499.
42. Luft, J. H. (1961). *J. Biophys. Biochem. Cytol.,* **9,** 409.
43. Rash, J. E. and Fambrough, D. (1973). *Dev. Biol.,* **30,** 166.
44. Ramaekers, F. C. S., Puts, J. J. G., Kant, A., Moesker, O., Jap, P. H. K., and Vooijs, G. P. (1982). *Cold Spring Harbor Symp. Quant. Biol.,* **46,** 331.
45. Schlegel, R., Banks-Schlegel, S., McLeod, J. A., and Pinkus, G. S. (1980). *Am. J. Pathol.,* **101,** 41.
46. Akeson, R. (1977). *J. Natl. Cancer Inst.,* **58,** 863.
47. Sasaki, M., Peterson, J. A., and Ceriani, R. L. (1981). *In Vitro,* **17,** 150.
48. Papsidero, L. D., Kuriyama, M., Wang, M. C., Horoszewicz, J., Leong, S. S., Valenzuela, L., Murphy, G. P., and Chu, T. M. (1981). *J. Natl. Cancer Inst.,* **66,** 37.
49. Koprowski, H., Steplewski, Z., Mitchell, K., Herlyn, M., Herlyn, D., and Fuhrer, P. (1979). *Somatic Cell Genet.,* **5,** 957.
50. Metzgar, R. S., Gaillard, M. T., Levine, S. J., Tuck, F. L., Bossen, E. H., and Borowitz, M. J. (1982). *Cancer Res.,* **42,** 601.
51. Kung, P. C., Goldstein, G., Reinherz, E. L., and Schlossman, S. F. (1979). *Science,* **206,** 347.
52. Yasumura, Y., Tashjian, A. H., Jr, and Sato, G. H. (1966). *Science,* **154,** 1186.
53. Ambesi-Impiombato, F. S., Parks, L. A. M., and Coon, H. G. (1980). *Proc. Natl. Acad. Sci. USA,* **77,** 345.
54. Kimes, B. W. and Brandt, B. L. (1976). *Exp. Cell Res.,* **98,** 349.
55. Soule, H. D., Vazquez, J., Long, A., Albert, S., and Brennan, M. (1973). *J. Natl. Cancer Inst.,* **51,** 1409.
56. Wigley, C. B. (1975). *Differentiation,* **4,** 25.

5

Separation of viable cells by centrifugal elutriation

D. CONKIE

1. Introduction

In many experimental studies in cell biology, the investigation of cell populations consisting of a single cell-type confers great advantage over the inherently complicated examination of mixed cell populations. As an extension of this concept, studies relating cellular function to the cell division cycle have a requirement for a single cell-type population separated, and so synchronized, with respect to the position of the cells in the cell division cycle.

Between 1965 and 1970, the successful development of isopycnic (buoyant density) and velocity sedimentation methods, alone or in sequential combination, resulted in numerous reports of useful cell separations. However, prior to this period of general interest in cell separation, Lindahl in 1948 (1) described a centrifuge and cell separation chamber for the 'counter-streaming centrifugation' of cells. From 1955 this apparatus was used by Lindahl and colleagues to enrich specific cell types from mixed cell populations and pre-mitotic or mitotic cells from randomly proliferating ascites tumour populations (2). The technique was named *centrifugal elutriation* (3).

2. Cell separation methods

The theory of sedimentation as applied to cells has been reviewed in detail previously (4). Isopycnic or buoyant density sedimentation of cells in continuous gradients requires sufficient force and/or time for cells to relocate to respective density bands within the gradient. Since separation is effected on the basis of respective cell densities, and the densities of many cell types overlap broadly, the application of this method has some limitations (5). A further disadvantage is the high centrifugal force often used and the possible toxicity of the gradient medium employed. Nevertheless, in combination with sedimentation velocity, the method has provided highly enriched populations of viable cells. For example, 80–100% erythropoietic stem cells have been isolated from haemopoietic populations and separated into S phase and non-S phase fractions in Percoll (6).

Velocity sedimentation separates cells on the basis of both cell density and cell diameter. Sedimentation at unit gravity (7) is reliable and technically easy to operate and has the advantage of using inexpensive equipment such as the Staput apparatus (8) or the CelSep chamber (9) (Du Pont).

The apparatus can accommodate a limited number of about 10^8 cells maximum layered on to a gradient of serum, BSA, or Ficoll. Problems arise at the sample/gradient interface at high cell number where cells stream down into the gradient. However, streaming can be minimized by using an intermediate layer between cells and gradient. A major disadvantage of unit gravity sedimentation is the long separation times required.

Sedimentation in an isokinetic gradient (e.g. Ficoll in tissue culture medium) also offers the advantage of reliable cell separation in sterile conditions using standard equipment normally available. As in the case of unit gravity sedimentation, relatively small cell numbers are processed.

The theory of centrifugal elutriation has been discussed by Sanderson *et al.* (10). At a low g force the system separates up to 10^9 cells in a very short time. Throughout the procedure the cells may be suspended in their normal growth medium in sterile conditions and, prior to loading the rotor, there is no requirement for pelleting the cells, thus avoiding minor perturbations and maximizing cell viability. However, the method requires more skill, and the equipment used is more expensive than for some other sedimentation procedures. Hydrodynamic shear stresses may cause problems, although these are alleviated somewhat by elutriating in the presence of serum. Reaggregation of cells is a problem in most forms of velocity sedimentation, more so when separating a cell suspension derived from disaggregated tissue than when fractionating an established cell line. In centrifugal elutriation the formation of cell aggregates interferes with the flow of elutriation buffer. Nevertheless, as a high-resolution separation device, centrifugal elutriation can produce synchronized populations of minimally perturbed viable cells.

Electronic sorting of cells labelled with monoclonal antibodies specifically binding to surface markers is an extremely effective alternative to the use of sedimentation methods (see Chapter 6). For example, Watt *et al.* (11) described the differential binding of two monoclonal antibodies to haemopoietic cells. When used sequentially in conjunction with flow cytometry an enriched population of up to 60% committed erythroid precursors was obtained. However, relative to centrifugal elutriation, only small cell numbers can be separated in a brief process time by this method. In addition the cell sorting apparatus is expensive to purchase and costly to maintain.

3. The centrifugal elutriator

In 1973 Beckman marketed the JE-6 Elutriator Rotor, and around 1980 this model was upgraded to the currently available JE-6B rotor which runs in a standard J2-21 centrifuge.

The complete apparatus for continuous-flow elutriation consists of the JE-6B rotor containing the elutriation chamber, a pump, and additional equipment for loading/elutriating the cell sample or collecting elutriated fractions.

3.1 Elutriator rotor

The elutriation rotor comprises the following main parts:

- a black anodized aluminium body containing a central stainless steel column with entry and exit apertures
- a separation chamber and counterbalanced bypass chamber connecting directly with the central column
- a rotating plastic seal which allows continuous entry and exit of the sample and elutriating liquid to the separation chamber while the rotor is spinning

The assembled rotor is shown in *Figure 1* with the stationary input and output connections sited above the rotating seal. The aperture on the top and underside of the rotor body permits observation of the chamber within and its contents.

Figure 1. An assembled elutriator rotor showing the sample input (A) and output (B), the rotating seal (C), and the aperture (D) for viewing the elutriation chamber and contents by means of a stroboscopic light.

3.2 Elutriation chamber

The standard separation chamber, shown in *Figure 2A*, is constructed from clear epoxy resin. The sample, or elutriating buffer, enters from the side at the base of the chamber such that the flow of liquid directly opposes the centrifugal force. A modified J2-21 centrifuge lid incorporating a viewing port allows the spinning rotor chamber to be seen, illuminated by a synchronized stroboscopic light. The chamber filled with buffer appears as in *Figure 3* which also shows the transmitted image of the strobe lamp filaments.

During loading of a cell suspension, a cell boundary can be observed advancing gradually towards the centripetal exit from the chamber (*Figure 3*) assuming that the sedimentation rate of the cells in the spinning rotor is exactly balanced by the flow rate of the fluid through the chamber. The discrete cell boundary transiently disappears when a fraction of cells is elutriated in response to a small increase in flow rate. Thereafter the boundary is re-established and again visible as equilibrium is restored between the two opposing forces acting on the remaining cells.

The standard chamber's geometry, with continuous taper, produces a gradient of flow rates from one end to the other. Therefore, cells with a wide range of different sedimentation rates can be held within the chamber. As an alternative option, the Sanderson chamber (supplied by Beckman) can be

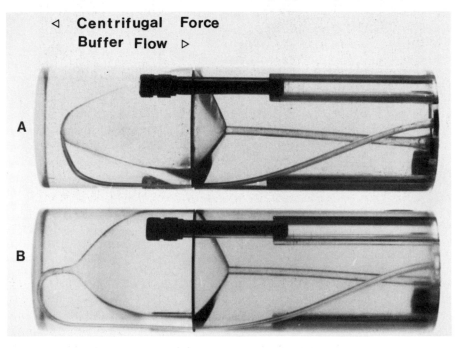

Figure 2. The standard (A) and Sanderson (B) elutriation chamber.

Figure 3. The standard elutriation chamber as viewed *in situ* after loading the cell sample. Arrows indicate advancing cell boundary and inlet at base of chamber.

used to retain cells of differing sedimentation rate in the lower part of the chamber where the walls diverge rapidly (*Figure 2B*). This reservoir of cells advances, as the flow rate is increased incrementally, into the upper chamber where the walls are almost parallel. This results in greater resolution, separating cells differing minimally in physical characteristics. No cell boundary is visible at the centripetal end of this chamber. A further contrast from the standard chamber is in the inlet at the base. In the Sanderson chamber, cells and elutriating fluid enter at the base directly from below rather than from the side. This helps to counteract any tendency for the cells to form clumps.

3.3 Peristaltic pump and additional apparatus

A variable-speed peristaltic pump regulated by a micrometer speed control is required to provide the flow of elutriating buffer. The pump should preferably have automatic compensation for changes in load torque demand and provide a linear relationship between the setting of the pump speed control potentiometer and the observed flow rate over the useful range of 10–60 ml/min. Examples of such pumps are the Watson-Marlow 501 and the Cole Parmer Masterflex 7520.

The pump draws the cell suspension or the elutriation buffer from containers at constant temperature in a water bath or ice bath. Each container is open to the atmosphere via a 0.22 μm filter (Millipore or Gelman), and is connected by silicone tubing (1.5–3 mm internal diameter and 1.5 mm wall thickness, Cole Parmer) and Teflon valves as shown in *Figure 4*. The pump is also connected to the rotor input via a pressure gauge which indicates any line blockage, and a bubble trap which collects stray air bubbles preventing their entry to the chamber and disruption of the suspended cells. The output from the rotor passes via a flow cell (Starna) to the sterile collection area within a laminar flow hood. Observation of the flow cell with a low-power microscope or absorption at 660 nm permits monitoring of the fractions as they are collected and indicates, when free of cells, that a subsequent fraction can be

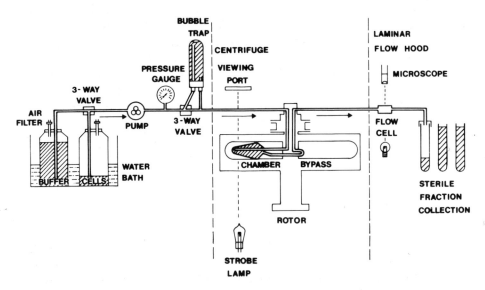

Figure 4. Diagrammatic representation of the complete elutriation apparatus.

elutriated. This also ensures that discrete fractions are collected in a minimum volume.

3.4 Complete elutriation system

The detailed arrangement of the rotor and ancillary equipment is flexible, and can be varied to suit individual experimental requirements. For example, it is possible to assemble the apparatus with the pump on the rotor output line. This tends to reduce the turbulence observed at the base of the separation chamber. In another variation the cell sample can be introduced between the pump and the input to the rotor. This avoids passage of the cells through the pump head. However, in practice, no deleterious effect arising from passage through the pump head is obvious as judged from cell viability studies.

When cells are loaded from an upstream reservoir as shown in *Figure 4*, tangentially acting Coriolis forces affect the initial low cell number in the chamber resulting in random elutriation of cells of all sizes (10). Therefore, as an alternative, a concentrated cell sample may be injected from a syringe into the elutriating medium immediately upstream of the rotor. This bulk sample arriving in the chamber results in an equilibrium between the inertial and hydrodynamic forces.

4. Technical protocols

Elutriation can be performed without sterilization of the apparatus when no subsequent tissue culture of the separated cell fractions is required. Otherwise, the procedure described in *Protocol 1* should be followed. Prior to assembly of the apparatus, autoclave sections of the tubing and connections to be used with the buffer and cell sample or for fraction collection.

The manufacturers recommend sterilization of the complete assembled apparatus by pumping through 6% hydrogen peroxide. As alternatives, either 0.2% diethyl pyrocarbonate or Cidex (Arbrook) are useful. Occasionally, the entire assembled rotor, with the exception of the separation chamber gasket, may be autoclaved at 121 °C for about 1 h. The vinyl gasket can be separately sterilized in 6% hydrogen peroxide. Because of a possible fire hazard from damaged or sparking electrical connections resulting from a system failure, disinfection of the assembled apparatus by pumping through ethanol is inadvisable while the centrifuge circuits are live.

Protocol 1. Sterilization of the apparatus

1. Assemble the apparatus as shown in *Figure 4* but with the buffer and cell sample replaced by vessels each containing approximately 200 ml of Cidex.

2. With the rotor stationary, start the pump and set the speed such that about 40 ml/min of the sterilizing agent is drawn into the system. Vary the three-way valve settings so that all tubing and apparatus are filled.

Protocol 1. *Continued*

3. Invert the pressure gauge to remove air from the neck, then return to normal position.

4. Similarly, invert the bubble trap and purge all air by temporarily withdrawing the small needle to the inner level of the cap. Reposition the bubble trap as in *Figure 4*.

5. To purge air from the rotor, start the centrifuge and accelerate the rotor to 1000 r.p.m. Switch off the pump and then switch off the rotor. Restart the pump when the rotor speed drops below 500 r.p.m. Air bubbles should flow out of the rotor. Repeat the procedure until no further air bubbles emerge from the rotor. The pressure gauge should read less than 70 kPa (approximately 10 p.s.i. or 1.67 bar).

6. Switch off the rotor and pump and leave the apparatus filled with Cidex overnight.

7. After sterilization, clamp both liquid input lines and transfer them with sterile precautions to vessels each containing 1000 ml sterile distilled water. Release the clamps.

8. Pump the water through the system, collecting Cidex for future use.

Protocol 2. Flow rate calibration

1. Start the rotor and accelerate to 2000 r.p.m.

2. When the rotor speed has stabilized, switch on the strobe control and adjust the delay until the elutriation chamber is 'stopped' and clearly visible through the viewing port (*Figure 3*).

3. Obtain from the strobe counter an accurate record of the rotor speed.

4. Set the pump potentiometer to a low arbitrary setting and measure the flow rate by collecting the effluent in a measuring cylinder over 2 min.

5. Repeat for incremental increases in the pump potentiometer and from the data construct a graph or program a microprocessor to indicate the relationship between the pump setting and flow rate.

Protocol 3. Sample loading

1. Using sterile technique, if required, clamp the input lines and remove them from the distilled water containers, taking care to avoid introduction of air.

2. Transfer the input tubes to containers of 1 litre and 100 ml of elutriation buffer. As an example, the buffer for elutriation of cell fractions may be a

balanced salt solution when no further culture is required. Ideally, complete culture medium should be used to elutriate cells which are subsequently to be recultured. The medium may be maintained at 37°C in the water bath as shown (*Figure 4*).

3. Unclamp the tubes and use appropriate settings for the three-way valve to pump most of the 100 ml aliquot into the input tube. This will displace the water in the tube. Rearrange the three-way valve to pump buffer from the 1 litre container through the system until all water is displaced. Retain the bulk of the 1 litre sample as the elutriation buffer.

4. Clamp and remove the input tube from the 100 ml aliquot container. Replace in the cell sample, as in *Figure 4*, and release the clamp. A suitable cell sample may be 2×10^8 cells suspended in 200 ml of complete medium.

5. Set the pump flow rate to 12 ml/min and check that the rotor speed is 2000 r.p.m. Rearrange the three-way valve to pump the cell sample into the separation chamber. This condition will retain in the chamber all particles greater than 8 μm diameter. If peripheral RBCs are loaded at this setting, they will immediately pass through the chamber to the collection vessel.

6. When about 95% of the cell sample has been loaded, adjust the three-way valve to bring the elutriation buffer on line once again. When subsequently all of the cells have been pumped from the bubble trap, adjust the local three-way valve to bypass the bubble trap.

There are two methods of altering the equilibrium of the two opposing forces (sedimentation rate in the centrifugal field and the flow rate) acting on the cells suspended in the separation chamber. Either the centrifugal force is reduced or the flow rate is increased.

Although the modification to the standard speed control potentiometer on the J2-21 centrifuge is available to aid accurate rotor speed adjustment, the rotor deceleration tends to overshoot the set speed. Furthermore, the strobe control must then be readjusted to view the chamber at each new lower speed selected for discrete cell population elutriation. Thus, in practice, it is best to use constant rotor speed and variable flow rate.

Protocol 4. Adjusting the flow rate for elutriation of discrete cell populations

1. By advancing the pump control, increase the flow rate of the elutriation buffer by about 5 ml/min. The exact increment required to elutriate a specific cell fraction is determined empirically in the first instance. However, some guidance is given by the expression

$$F = XD^2(R/1000),$$

Protocol 4. *Continued*

where F is the flow rate at the pump (ml/min), X is 0.0511 (standard chamber) or 0.0378 (Sanderson chamber), D is the cell diameter (μm), and R is the rotor speed (r.p.m.). This equation is an expression of conditions at the elutriation boundary and is used in the Beckman manual to construct a nomogram relating flow rate, cell size, and rotor speed. The same data can be modified, for example, using a microprocessor program, to incorporate a correction factor obtained from the relationship between calculated and actual cell size of cells elutriated at any combination of rotor speed and flow rate.

2. Observe the cell boundary advancing centripetally in the separation chamber and cells passing through to the flow cell. Using sterile conditions if required, collect a cell fraction. About 150 ml of elutriation buffer may be needed to collect all of the cell fraction.

3. When the cell boundary is again visible in the separation chamber and the flow cell is free of cells, a further incremental increase in the pump flow rate will elutriate the next cell fraction for collection as before.

Elutriation of a specific cell fraction at a given flow rate and rotor speed is reproducible provided that the separation is performed using the same elutriation buffer, the same temperature and the same initial cell density in the chamber. Constant rotor speed and constant flow rate may be combined with step increases in elutriating medium density for high-resolution fractionation. Note that the entire chamber contents will be pumped from the chamber if the rotor is switched off. This technique may be useful when concentrating a specific cell type in the chamber such as white blood cells from whole blood. Failure of the pump causes the cells in the chamber to pellet.

5. Collection and interpretation of data

The information which can be obtained from elutriation experiments depends on whether the initial sample is a heterogeneous or single-cell type population.

5.1 Heterogeneous cell populations: assessment of cell fraction purity

Normally the aim of an elutriation experiment in this case would be to separate and collect distinct cell types. For example, host cells may be removed from a tumour cell suspension. Diagnosis of the cell type(s) present in each fraction can be accomplished by classical cytological fixation and staining methods. Morphological studies indicate the degree of cellular integrity, particularly when mechanical or enzymatic disaggregation is used to produce the initial cell suspension from an organ or solid tumour (see Chapters 1 and 6). The enrichment of cell fractions can be further estimated by measurement of cell size with a Coulter particle size analyser, biochemical methods, clonal culture in soft agar or other *in vitro* colony-forming assays, and the ability of cell fractions to colonize sub-lethally irradiated mouse spleen *in vivo*, for haemopoietic precursor cells, or to grow as xenograft in athymic nude mice, for tumour cells.

An example of a simple separation of cell types from a mixture is the elutriation of avian erythrocytes and mouse erythroleukaemia cells. The erythrocytes are elutriated under the conditions used to load the mixed population (2000 r.p.m., flow rate of 14 ml/min), whereas the larger erythroleukaemia cells are retained in the chamber to be elutriated later at 2000 r.p.m. and 25 ml/min. The initial mixed population containing 55% RBCs and 45% erythroleukaemia cells results in an elutriated fraction greater than 99% RBCs with a second fraction containing more than 97% erythroleukaemia cells.

5.2 Established cell lines: recovery and viability

Alternative procedures are required in the classification of cell fractions elutriated from established cell lines of homogeneous cell type. In this case centrifugal elutriation can yield fractions of cells separated on the basis of their position in the cell division cycle. Morphological criteria do not readily distinguish between the different cell cycle stages.

To distinguish S-phase cells from other cell cycle phases, the incorporation of tritiated thymidine into DNA can be used followed by a suitable assay for the isotope such as autoradiography or scintillation counting. However, by far the most informative analysis of elutriated fractions is obtained by flow cytometry (see Chapter 6).

Two examples are provided which demonstrate the resolution of the method and the viability of the separated cells.

5.2.1 HL60 cells

The flow cytometry profile obtained for randomly dividing HL60 cells is shown in *Figure 5*. This profile is fairly typical for established cell lines even though HL60 cells have an inherent heterogeneity of DNA content. 2×10^8 of these cells are loaded at 15 ml/min with a rotor speed of 2000 r.p.m. The elutriation buffer is complete RPMI1640 medium containing 10% FBS at 20°C. At a flow rate of 20 ml/min, flow cytometry reveals a fraction of G1 cells. Both G1 cells and early S-phase cells are collected at 25 ml/min (not shown). A fairly pure mid S-phase fraction is collected at 30 ml/min and at 35 ml/min late S and G2 cells are obtained with a small contaminating peak of G1 cells. The total yield is typically greater than 80%, leaving a proportion of aggregated cells from all cell cycle stages in the elutriation chamber.

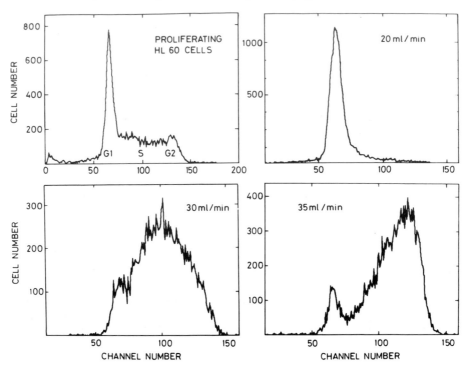

Figure 5. Flow cytometry profiles of HL60 cells randomly proliferating together with separated fractions elutriated at the flow rates shown.

5.2.2 Erythroleukaemia cells

As a measure of viability, elutriated G1 erythroleukaemia cells may be recultured in elutriation buffer (complete medium) at 36.5°C. At various times during culture, samples are prepared for flow cytometry. The profiles (*Figure 6*) reveal a fairly synchronous progression through the cell division cycle obvious between 2 and 7 h and complete by 12 h (one cell cycle time) when the cells re-enter G1 (not shown).

Thus centrifugal elutriation is a procedure producing minimally perturbed synchronous cell populations at low *g* force, avoiding the deleterious effects of pelleting or the osmotic effects of density gradient centrifugation media. Furthermore, brief separation times are possible with up to 10^9 cells processed each time the chamber is loaded.

The applications of centrifugal elutriation are extensive, and a bibliography is available from Beckman Ltd.

6. Future development

Using the standard chamber the total yield of a specific cell type following

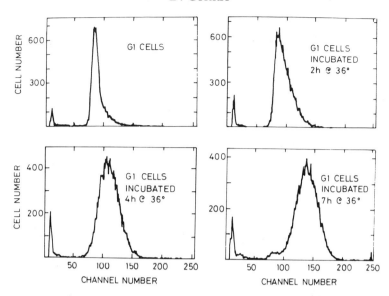

Figure 6. Flow cytometry profiles of G1 erythroleukaemia cells and G1 cells recultured for the times shown. The small peak in the first few channels represents avian erythrocytes added as a marker from which the distance to the G1 peak may be measured. Twice this distance marks the G2 position.

elutriation of 10^9 cells may not always be adequate. Batch processing can increase total yield, but also requires an increased time. A double separation chamber, where the bypass chamber is replaced by a second standard chamber in series, does not provide a significant advantage over two consecutive single chamber elutriations. The use of a double chamber system operating in parallel would require substantial modification to the rotor design. In principle such a modified system need not be limited to two separation chambers.

However, the development of a scaled-up version of the original standard separation chamber has proved encouraging, maintaining the separation characteristics of the original system (12). The new high-capacity JE-10X Elutriator rotor marketed by Beckman has a 10-fold increased chamber volume. Preliminary reports suggest that separation characteristics and fraction purity are maintained, with a 10-fold increase in yield.

A custom-made smaller chamber for applications with small samples has been developed (13), but, as yet, this has not been produced commercially. Further applications are listed in *Table 1*, which provides information on conditions that have been used for a variety of different cell types. In most cases, it is apparent not only that there is a high yield, but also that viability, as measured either by dye exclusion or by more functional criteria such as hormonal stimulation (17) or clonogenicity (19), is maintained.

It is a sign of the success of this method that the technology has remained

Table 1. Applications of centrifugal elutriation

Reference	14	15	16	17
Cell type	Rat biliary epithelium	Guinea-pig keratinocytes	Human squamous carcinoma xenograft	Rat anterior pituitary
Dissociation medium	coll/pron/ DNAse	EDTA/trypsin	coll/pron/ DNase	buff. salt, gluc., BSA
Suspension medium	MEM/10CS/ DNAse	Hanks' BSS	TC medium	—
Cell concentration	—	—	10^7/ml	—
Total load	—	—	2×10^8	1×10^8
Centrifuge	J2-21	Hitachi SCR-20-BA	—	J2-21M
Rotor	JE-6B	—	—	Sanderson JE-6B
Speed (r.p.m.)	2500	2000	4400–2000	1960
(g)	—	—	—	—
Flow rate (ml/min)	11–44	10–30	35–45	18, 28, 40
% Recovery	36–44 BE	85	—	60
% Enrichment	60	—	> 90 host 85–100 tumour	77–90
% Viable	> 97	> 95	—	> 90
Comments	α-naphthalene thiocyanate stimulated	3 size classes: basal, inter- mediate, large	Synchronized population	Percoll 2nd step functionally responsive

Abbreviations: coll, collagenase; pron, pronase; buff, buffered; gluc, glucose; CS, calf serum;

fairly constant over the past five years, with Beckman still dominating the market. The major deterrent of the method is, on the one hand, the capital expenditure involved in setting up, and on the other, the apparent complexity. In practice, given an adequate disaggregation procedure providing a good single-cell suspension, the method is not unreasonably difficult; it requires only a few practice runs for the operator to become familiar with the system before the major parameters may be established. Although somewhat over-shadowed by flow cytometry in the sophistication of the criteria for separation that can be employed and the resultant resolution that can be obtained, elutriation is still the only method that can yield large numbers of cells with reasonable purity and high viability. If this is the requirement, it is well worth overcoming the psychological barrier that deters many people; the complexity may turn out to be more apparent than real.

13	18	19	20	21
Peripheral blood leukocytes	Chick chondrocytes	Tracheal basal epithelium	Rat marrow endothelium	Lymphocyte depletion
—	trypsin/coll/ hyaluronidase	Sigma XIV protease	coll	Needle
Hanks'	F-12/5FB	—	PBS	EDTA/gluc/ RPMI1640
—	10^7/ml	—	$0.2–1.2 \times 10^8$/ml	$2.0–6.7 \times 10^8$/ml
10^9	—	—	—	—
Benchtop	—	—	—	—
CS-1	JE-6	JE-6	—	JE-6B
500–2800	3130–2300	—	—	3000
—	990–530	560	613	900
2.5, 4, 6	20	7.3–11, 14.9–34.5	15–20	15, 25, 29, 33
> 90	60–70	—	$1–2 \times 10^6$ cells	—
80–90	—	90	50–90	> 98 depletion
—	Grew in culture	Clonogenic	> 90	> 95
Custom designed Ficoll 1st step	—	Basal and non-basal both clonigenic	Discontinuous Percoll 1st step	—

TC, tissue culture

Acknowledgement

Original work described was supported by grants from the Cancer Research Campaign.

References

1. Lindahl, P. E. (1948). *Nature,* **161,** 648.
2. Lindahl, P. E. (1960). *Cancer Res.,* **20,** 841.
3. McEwan, C. R., Stallard, R. W., and Juko, E. T. (1968). *Anal. Biochem.,* **23,** 369.
4. Pretlow, T. G., Weir, E. E., and Zettergren, J. G. (1975). *Int. Rev. Exp. Pathol.,* **14,** 91.

5. Pretlow, T. G. and Pretlow, T. P. (1982). In *Cell Separation* (ed. T. G. Pretlow and T. P. Pretlow), Vol. 1, p. 41. Academic Press, London.
6. Nijhof, W. and Wierenga, P. K. (1982). *Exp. Hematol.,* **10,** Suppl. 12, 307.
7. Miller, R. G. and Phillips, R. A. (1969). *J. Cell Physiol.,* **73,** 191.
8. Miller, R. G. (1973). In *New Techniques in Biophysics and Cell Biology* (ed. R. H. Pain and B. J. Smith), Vol. 1, p. 87. Wiley, New York.
9. Wells, J. R. (1982). In *Cell Separation* (ed. T. G. Pretlow and T. P. Pretlow), Vol. 1, p. 169. Academic Press, London.
10. Sanderson, R. J., Bird, K. E., Palmer, N. F., and Brenman, J. (1976). *Anal. Biochem.,* **71,** 615.
11. Watt, S. M., Melcalf, D., Gilmore, D. J., Slenning, G. M., Clark, M. R., and Waldmann, H. (1983). *Mol. Biol. Med.,* **1,** 95.
12. Jemionek, J. F. (1981). *Transfusion,* **21,** 268.
13. Bauer, J. and Hannig, K. (1988). *J. Immunol. Meth.,* **112,** 213.
14. Alpini, G., Lenzi, R., Wei-Rong Zhai, Liu, M. H., Slott, P. A., Paronetto, F., and Tavoloni, N. (1989). *Gastroenterology,* **97,** 1248.
15. Sakamoto, F., Nomura, H., Hachisuka, H., Sasai, Y., and Shiotsuki, K. (1988). *J. Dermatol.,* **15,** 14.
16. Keng, P. C., Allalunis-Turner, J., and Siemann, D. W. (1987). *Int. J. Radiat. Oncol. Biol. Phys.,* **18,** 1061.
17. Scheikl-Lenz, B., Sandow, J., Herling, A. W., Trager, L., and Kuhl, H. (1986). *Acta Endocrinol. (Copenh.),* **113,** 211.
18. O'Keefe, R. J., Crabb, I. D., Puzas, J. E., and Rosier, R. N. (1989). *J. Bone Joint Surg.,* **71A,** 607.
19. Inayama, Y., Hook, G. E., Brody, A. R., Jetten, A. M., Gray, T., Mahler, J., and Nettesheim, P. (1989). *Amer. J. Pathol.,* **134,** 539–49.
20. Irie, S. and Tavassoli, M. (1986). *Exp. Hematol.,* **14,** 912.
21. Wagner, J. E., Donnenberg, A. D., Noga, S. J., Wiley, J. M., Yeager, A. M., and Vogelsang, G. B., Hess, A. D., and Santos, G. W. (1988). *Exp. Hematol.,* **16,** 206.

6

Flow Cytometry

JAMES V. WATSON and EUGENIO ERBA

1. Introduction

Flow cytometry enables scattered and fluorescent light from individual cells in fluid suspension to be quantitated at very rapid rates. Typically up to 5000 cells can be measured per second. The quantitative aspects of the technology take their origins from the work of Caspersson and colleagues in the 1930s, where stained images were projected on to a wall and the amount of light absorbed in different areas of the images was quantitated with primitive photodetectors. Nucleic acid metabolism in *Drosophila melanogaster* salivary gland chromosomes was studied by banding pattern changes using this method (1). In addition to their analytical capability, many instruments have the capacity for electrostatic cell sorting which places individual cells with predetermined characteristics in a test tube for subsequent morphological identification or biological manipulation.

The technology has a number of advantages and disadvantages. The former includes objective quantitation of specific molecules, statistical precision, multiparametric cross-correlated data analysis, distributional information and hence subset identification, dynamic measurements, sensitivity, speed, and the generation of a vast amount of data. The disadvantages include loss of 'geographical' information from solid tissues as a single-cell suspension is mandatory, absence of a direct visual record, and the generation of vast amounts of data. This last point is included under both headings because it is a two-edged sword. Data are meaningless until converted to information, and this may present considerable problems, particularly for multi-parametric data.

It is pertinent at this point to ask why we should wish to make measurements on an individual cell basis and at such rapid rates. If we take a sample of tissue, homogenize it, and perform a given assay we obtain a grand average for that sample. Let us suppose that the answer is 100 units. However, this answer does not tell us if half the cells in the sample have zero units and the other half has 200 units, or if all cells have exactly 100 units each. Individual cell analysis, by whatever means, is the only method of resolving this problem and hence of obtaining reliable data in heterogeneous populations. That in itself is justification, irrespective of the other advantages including statistical precision, that flow cytometry offers. It is possible to use a fluorescence microscope to determine the proportion of fluorescently labelled cells in a population, but the precision of manual counting is highly dependent on the proportion of labelled cells and the number of cells counted. If the labelled fraction constitutes only 5% of the population and 200 cells are counted, there will, on average, be a score of 10 positive cells. Statistically this value could fluctuate between 3 and 17 cells, giving a range of 1.5% to 8.5% at the 95% confidence interval. Hence, the ability to analyse and count large numbers of cells very rapidly has considerable advantages, particularly for analysis of minority subsets. However, the various flow cytometry techniques should not be regarded as replacing existing methods entirely; they should be regarded as an adjunct, although 'classical' techniques just cannot compete in terms of speed. Moreover, there are some things that can be done using flow cytometry that just could not be done in any other way.

The technique relies upon measuring both scattered light and fluorescence from suitably stained constituents in individual cells in the population. The stained cells are streamed in single file in fluid suspension through the focus of a high-intensity light source. As each cell passes through the focus, a flash of scattered and/or fluorescent light is emitted. This is collected by lens systems and filtered before reaching a photodetector which may be either a photo-multiplier or a solid state device. The photodetector quantitatively converts the light flash into an electronic signal, which is digitized by an analogue-to-digital converter into a whole number (integer) that is then stored electronically. The data can subsequently be recalled for display and analysis.

The early commercial instruments were complex and cumbersome, and not very easy to use. However, the recent generation of machines have been simplified considerably, with the built-in computer taking over many of the tasks which the operator previously had to perform manually. The 'user-friendliness' of these modern instruments, together with the relative reduction in initial capital cost, is a considerable advantage, and has made the technology available to many more users. This in turn creates its own problems. Many current bench-top devices require a minimum of operator interaction, so that all that appears to be needed is to stain the cells, shove them in the instrument, and out will come the numbers. This can generate a degree of complacency based on ignorance; the operator may not be sufficiently aware of the basic principles which underlie the considerable complexity of these systems, and where potential problems can arise. This chapter considers some of those basic principles, with the intention of providing an outline of how flow cytometry systems work and where problems can arise.

2. Principles of operation

2.1 Hydrodynamic focusing

The most important essential feature of any flow cytometric instrument is a stable fluid stream which presents the cells one at a time to a sensing volume where the measurements are made. To obtain consistency of measurement, each cell has to be presented to the same volume within the sensor. This is achieved by hydrodynamic focusing, a concept which stems from the work of Bernoulli (2) who showed that pressure and velocity are inversally related in fluid flow, and Euler (3, 4) who showed that the velocity profile is parabolic with the greatest velocity in the centre of the stream. Thus, the pressure is lowest in the centre of the tube so any non-compressible particles will tend to flow coaxially in the centre where the pressure is lowest. A manifestation of hydrodynamic focusing can be seen when you pull the plug out of a basin full of water: the vortex passes down through the centre of the waste pipe.

Crosland-Taylor (5) used these principles to construct a flow chamber which is the fore-runner of all those used in flow cytometry. A representation based on the original is shown in *Figure 1*. It consists of a closed cylinder with an inlet port, forming a sheath through which fluid is pumped, and an exit constriction. Cells are introduced into the flow by a needle whose tip is located just above the exit constriction. The combination of hydrodynamic focusing and the coaxial pressure drop causes cells to pass down the centre of flow through the exit nozzle. By suitably adjusting the nozzle size, the constriction cone, flow rates, and relative pressures it is possible to constrain one

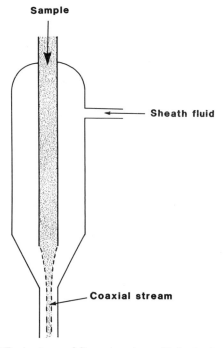

Figure 1. The Crosland-Taylor type of flow chamber with hydrodynamic focusing of the sample in the exit nozzle by the sheath flow.

cell at a time to pass through the nozzle, with the supporting medium containing the cells making up the core of the coaxial stream in the nozzle. Single-sheath systems can attain positional accuracy of the core to within ±2 μm, which is adequate for most applications. However, flow chambers incorporating a double sheath (6, 7) which confer greater stability on the axial stream, have been constructed, and positional accuracy of ±0.5 μm can be achieved. This is necessary for some types of assay (e.g. chromosome slit-scanning) where extreme positional accuracy is required.

The other requirements are a high-intensity light source to elicit fluorescence and light scatter, suitable light collection optics, and electronics to quantitate the response from individual cells. A typical layout of a flow cytometer is shown in *Figure 2*, where the planes of the cell stream, light illumination and light collection are placed orthogonally (at 90° to each other).

2.2 Excitation

Due to the relatively small number of fluorescent molecules per cell and the short time that each cell is exposed to the exciting light (1–5 μs), it is necessary to achieve very high light fluxes at the intersection of the cell stream with the illumination. A high light flux means that a very large number of photons is passing through a small volume of space, and this requires that the source of illumination must be as small as possible and as bright as possible.

The quantity of electromagnetic irradiation emitted from a conventional incandescent source is directly proportional to its temperature, and the wavelength is inversely proportional to temperature. Thus, in order to obtain sufficient higher-energy photons (u.v., violet, and blues) to be useful in flow cytometry, an incandescent source must be very hot, in the region of 6000 K. Most of the energy radiated from such a source is emitted as heat, which has to be disposed of using mirrors which reflect heat but transmit light. The only

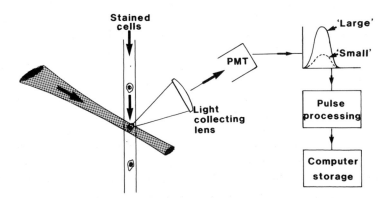

Figure 2. Basic layout of a typical flow cytometer.

conventional source which is practicable in flow cytometry is the mercury arc lamp, where the source size can be as small as 250 μm × 250 μm and can be focused down to a spot size of about 50 μm.

The best illumination source is undoubtedly the laser, for several reasons. Firstly, the light is very bright. Secondly, the beam emitted by a laser is coherent, which means that the light is polarized, the photons are all 'in step', and the beam is essentially parallel. This last attribute means that the effective spot size of the illumination source is *very* small, tending to a point, because the parallel beam appears to be coming from infinity. Finally, lasers give a more stable output than conventional sources, which is essential for quantitative studies. Most lasers can be stabilized to within ±1% over many hours of continuous operation. However, they do have a number of disadvantages, including cost. Furthermore, a laser will generally emit at only one well-defined and specific wavelength at a time, although tuning to a number of individual wavelengths from u.v. to red is possible, depending on the type of laser being used. This specificity is a considerable disadvantage if two fluorochromes with very different absorption spectra need to be investigated simultaneously; the only totally satisfactory solution is to use two lasers. Dye lasers, pumped by part of the output from a medium- to high-power laser, can be tuned within a given wavelength range (8). However, it is not possible to fully stabilize both the pumping and the dye laser simultaneously.

2.3 Focusing

Two systems are used in flow cytometry to focus the illuminating light to the point at which it intersects the cell stream. The first system uses a conventional spherical lens which will focus the illuminating light to a spot size of 30–60 μm. The second system uses a pair of crossed cylindrical lenses to focus the light to a sheet about 120 μm wide and 4–7 μm deep. Cylindrical lenses are constructed with a curved surface in only one plane which focuses the incident light to a line in the focal plane. Two cylindrical lenses, with their focal planes at 90° to each other, can be used in flow cytometry as follows. The lens nearest to the laser has a long focal length, which focuses light at the cell stream in the horizontal plane. The second lens, which is further from the laser, has a short focal length and focuses light in the vertical plane. Hence, the focal length of the lens nearest the laser must have a focal length equal to the distance between the lenses plus the focal length of the second lens.

Generally, there is little to choose beween these two focusing systems for the simpler types of assays. However, the crossed cylindrical lens pair does have the apparent disadvantage of requiring slightly more complicated electronics. The object in this system is larger than the depth of the focal volume, hence the area under the pulse of light emitted from each cell has to be digitized to obtain a measure of the total quantity of light emitted, and frequently the pulse height and width (time of flight through the beam) are also digitized. In contrast, with spherical lens focusing the object is totally enclosed within the focal volume so only pulse height has to be digitized. The crossed cylindrical lens pair has the following advantages.

(a) The sheet-like focusing allows greater latitude in positioning the core within the sheath, a point which will be discussed in the next section.

(b) Depending on the exact design features, there can be a higher light flux at the focus with the crossed pair.

(c) It is possible to focus two very different wavelength beams to the same point, by increasing the light path length of the longer wavelength beam through the system by exploiting spherical and astigmatic aberrations (9).

(d) The increased data obtained with the crossed cylindrical lens pair (pulse height, width, and area) can be used to give some low-resolution object shape information which is very useful in quality control (see later).

2.4 Beam geometry

The beam from a laser can assume various so-called *transverse emission modes* (TEM) which depend on a number of factors including plasma tube construction and mirror alignment. These modes are labelled 00, 01, 01*, 10, 11, etc. depending on the segmentation of the beam. The first three are shown in *Figure 3*. The intensity of the beam is Gaussian distributed in TEM_{00}, which is depicted in *Figure 4*. It is conventional to describe the width of a beam as the diameter where the irradiance falls to 0.1353 of that at the peak. (This is the numerical value of $1/e^2$.) Let us now assume that a typical 1.2 mm diameter beam is focused down to a 50 μm diameter spot at the $1/e^2$ irradiance width and that a core stream diameter of 15 μm containing 10 μm cells is passing through the centre of the focused beam in the most intense section. This is depicted in *Figure 5* with two cells, A and B, in the centre and at the periphery of the core respectively. Cell A, in the ideal position, has a 6% variation in light intensity across it. Cell B, in the worst position at the core periphery, suffers a 15% intensity variation. We can now see that very minor instability in the core position, of as little as ±2 μm, could make a profound

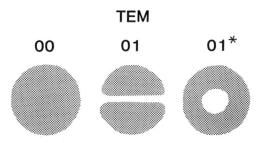

TEM

00   01   01*

Figure 3. Illustration of the segmentation of a laser beam in various transverse emission modes (TEM).

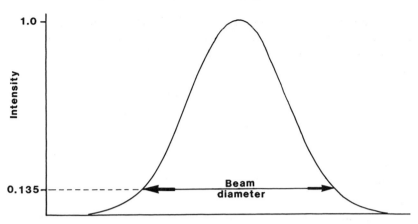

Figure 4. The light intensity profile across a laser beam is Gaussian distributed in TEM$_{00}$. The width of a beam is defined as the diameter across the beam where the irradiance falls to 1/e^2 of that at the peak.

difference to the illumination intensity experienced by individual cells. With this degree of core instability the variation could easily amount to 40%, as depicted in *Figure 5*.

Two options are available to overcome this potential variation in illumination intensity. Firstly, a larger spot size could be used to 'spread out' the highest intensity region with respect to the core, but this also reduces the light flux and is not desirable. The second is to use a crossed cylindrical lens pair as

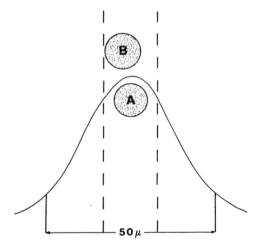

Figure 5. A 15 μm diameter core passing through the centre of the focused Gaussian beam in the most intense section and containing two 10 μm cells. Cell A, in the ideal position, has a 6% variation in light intensity across it. Cell B, in the worst position at the core periphery, suffers a 15% intensity variation. Core position instability of as little as ±2 μm could give rise to a 40% illumination variation.

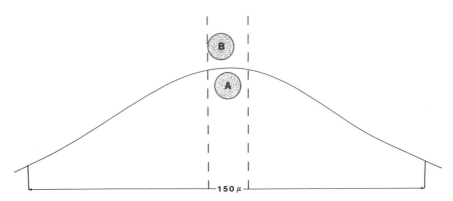

Figure 6. Crossed cylindrical lens pair illumination with compression of the focal spot from top to bottom to form an oval sheet 150 μm wide and 4–7 μm deep. A 15 μm diameter core containing two cells, directly analogous to that in *Figure 5*, is shown and the potential variation in light intensity across a 10 μm cell is now less than about 2% and there is now considerable latitude for any core positional instability.

described above. With this option the focal spot is compressed from top to bottom to form an oval sheet. The intensity profile is still Gaussian-distributed in whichever radial plane the beam is 'sliced' but the standard deviation is 'tightest' in the vertical plane. *Figure 6* shows a representation of the focal spot intensity obtained with a crossed cylindrical lens pair which 'spreads' the beam to 150 μm horizontally together with a 15 μm diameter core directly analogous to that in *Figure 5*. The potential variation in light intensity across a plane of a 10 μm cell is now less than about 2%, and there is considerable latitude for any core positional instability.

2.5 Optical filtration

Some form of optical filtration system is required whenever fluorescence is being observed or measured, and filtration relies on only two properties, absorption and interference. These properties can be used either individually or in combination to give five types of filter: short-pass, long-pass, dichroic mirrors, bandpass, and neutral-density filters. Short- and long-pass filters respectively transmit (pass) light below and above a specific wavelength, the 50% transmission wavelength. Dichroic mirrors are designed to reflect above and below the specified 50% transmission level and bandpass filters transmit light within a given wavelength band, where again the 50% transmission wavelength is specified for both the cut-on and cut-off. Neutral-density filters attenuate light by specified quantities over a given wavelength range.

2.5.1 Absorption filters

Absorption filters are coloured glasses which absorb light of specific wave-lengths. However, it is difficult to construct coloured glass to absorb longer wavelengths without absorbing the shorter wavelengths (short-pass). It is much easier to absorb light below a specified wavelength and to transmit the longer wavelengths (long-pass). Hence, absorption filters tend to be of the long-pass variety.

2.5.2 Neutral-density filters

Neutral-density filters are absorption filters which are designed to absorb all wavelengths of light to an equal degree over a specified wavelength range. The attenuation is specified in OD units, which are logarithmically calibrated. For example, OD filters of 1, 2, and 3 attenuate the incident beam by factors of 10, 100, and 1000. Neutral-density filters can be used to great effect when sequential samples with very different fluorescence intensities are being analysed on the same photodetector and comparisons between samples are required without changing the instrument settings.

2.5.3 Interference filters

Interference filters are designed to reflect some wavelengths and to transmit others using both destructive and constructive interference in multiple layers of material of alternating high and low refractive index, which acts as a resonance cavity to selectively reflect or transmit some light wavelengths and destroy others. The thicknesses of the coatings of high and low refractive index materials are critical, and they vary according to the design wavelength and required angle of incidence. When interference filters are designed for light entering perpendicular to the surface, the respective thicknesses of the high and low refractive index material are exactly one quarter ($\lambda_0/4$) and one half ($\lambda_0/2$) of the design wavelength λ_0.

2.5.4 Bandpass filters

Bandpass filters are constructed as a combination of interference and coloured glass absorption filters to transmit a specific wavelength band as efficiently as possible. However, it is never possible to obtain a bandpass filter with zero probability of transmitting an unwanted photon: there is always a shoulder in both the cut-on and cut-off regions of the spectrum. These filters are relatively expensive, because their structure is complex and they must be made to very tight specifications. *Figure 7* shows a cross-section of a typical, but relatively simple, two-cavity bandpass filter. More complex filters have many more layers and cavities.

2.5.5 Dichroic mirrors

Dichroic mirrors also operate using interference, but they are designed to be used at an inclination of 45° and the thicknesses of the high and low refractive index layers deviate from the $\lambda/4$ and $\lambda/2$ which are used for perpendicular incidence. They are also designed to minimize absorption as both the reflected and transmitted rays are required. The wavelength specified is again

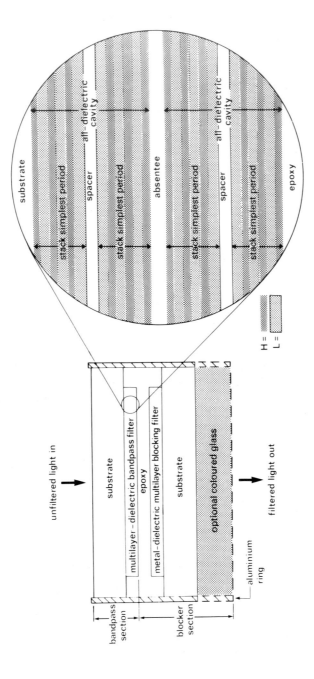

Figure 7. Construction of a relatively simple, two-cavity bandpass filter. More complex filters have many more layers and cavities. Redrawn from the Melles Griot catalogue.

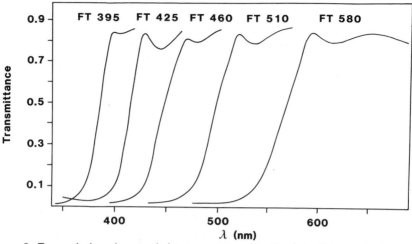

Figure 8. Transmission characteristics versus wavelength of the Zeiss dichroic mirrors. Redrawn from the Zeiss monograph, *Fluorescence Microscopy*.

that of the 50% transmission point and the Zeiss catalogue dichroic filter transmission characteristics are shown in *Figure 8*.

2.5.6 Dichroic combinations

Multifluorescence assays are becoming increasingly necessary, and these require that the spectrum from u.v. to visible light be divided into a number of different wavelength bands by means of a series of dichroic mirrors. Care must be exercised in their arrangement to minimize light loss. One arrangement of four of the Zeiss series of dichroics plus two from Melles Griot, with 50% transmissions centred at 390 nm, 420 nm, 460 nm, 510 nm, 560 nm, and 630 nm respectively, is shown in *Figure 9*. This splits the spectrum into seven primary bands namely: u.v. (<390 nm), violet (390–420 nm), indigo/low-blue (420–460 nm), high-blue/green (460–510 nm), green (510–560 nm), yellow/orange (560–630 nm), and red (>630 nm). With this particular arrangement the red light has to pass through all six filters and will suffer some light loss at

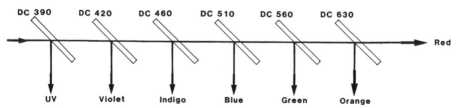

Figure 9. Sequential arrangement of six dichroic mirrors giving seven spectral bands: u.v. (<390 nm), violet (390–420 nm), indigo/low-blue (420–460 nm), high-blue/green (460–510 nm), green (510–560 nm), orange (560–630 nm), and red (>630 nm).

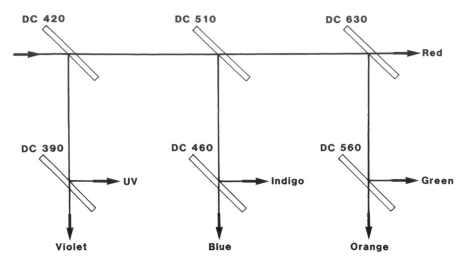

Figure 10. Alternative dichroic arrangement to give the same seven spectral bands as in *Figure 9*, but with fewer light losses, see *Table 1*.

each transmission. At the other end of the spectrum the u.v. light is reflected off the first surface of the first dichroic, and experiences no transmission. Generally, relatively more light is lost in transmission than in reflection, and another arrangement of the dichroics is shown in *Figure 10*. Here, the red light is transmitted through only three mirrors and *Table 1* shows a summary of the numbers of reflections and transmissions for each colour band with both configurations. Clearly, the arrangement in *Figure 10* is superior overall.

2.6 Light scatter

Theoretically, Maxwell's equations of electromagnetism (10) could be solved to describe the propagation of light scattered by objects of any size and refractive index. In practice, exact solutions have only been worked out for particles much smaller than the wavelength of the illuminating light (11) and uniform spheres which approximate to the sizes of cells (12). No satisfactory

Table 1. Number of dichroic reflections (R) and transmissions (T) for each wavelength band for the mirror configurations shown in *Figures 9* and *10*.

Figure	T:R	U.v. (<390)	Violet (390–420)	Indigo (420–460)	Blue (460–510)	Green (510–560)	Orange (560–630)	Red (>630)
9	T	0	1	2	3	4	5	6
	R	1	1	1	1	1	1	0
10	T	0	1	1	2	2	3	3
	R	2	1	2	1	2	1	0

solutions have been found for the reality of real cells, which present almost insuperable problems for the theorist. Scatter depends on a number of optical processes including diffraction, reflection, refraction, and Rayleigh scattering. In a biological cell these optical processes in turn depend on total size, refractive index of the medium, the number of dielectric interfaces within the cell (granularity) that the wave front passes through, plus their sizes, light wavelength, and angle of observation. Furthermore, any intracellular refractive index changes may also give rise to both constructive and destructive interference. Light scatter from intact cells is extremely complex and ill-understood; however, under specific conditions, it can give information about cell size and some morphological data which is low resolution at present.

2.6.1 Diffraction

An opaque particle in a beam of light, which is large compared with the light wavelength, will absorb the light incident upon it and thus create a 'hole' in the wavefront. Each point on the circumference of this hole will act as a focus for secondary irradiation (Huygens' principle, 13) and diffraction can take place where these secondary waves interact. The diffraction pattern of 632 nm light (helium–neon) produced by a circular disc corresponding to a 10 μm sphere suspended in medium with refractive index of 1.33 is shown in *Figure 11* which is redrawn from Salzman *et al.* (14).

2.6.2 Refraction and reflection

Both of these phenomena contribute to the total quantity of light scattered from an object into a detector at any given angle. The exact magnitudes of their respective contributions from unstained cells are not known. However, cells stained with immuno-gold scatter very large quantities of light to 90° although their forward scatter characteristics are similar to unstained cells. Very granular cells, e.g. polymorphonuclear leucocytes and macrophages, also exhibit large scatter signals to 90°, and it is now generally accepted that reflection and refraction play a major role in scattering light to larger angles.

2.6.3 Anomalous diffraction

Classical diffraction occurs with opaque objects in the path of the light beam where there are large differences in refractive index between two media. With biological material the refractive index difference between supporting medium (water or saline) and the object of interest is frequently small. Anomalous diffraction is a term which describes interference due to a phase shift induced by an effective path-length difference when an object is large compared with the incident light wavelength and where the refractive index difference is small. The phase-shifted wave emerging from the object then interferes with the wave passing round the object to give rise to a scatter pattern.

177

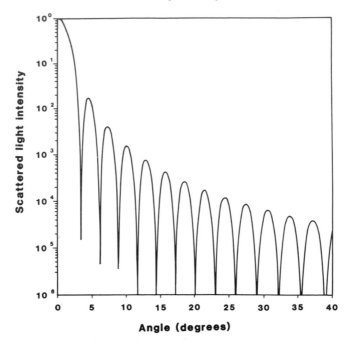

Figure 11. Diffraction pattern of 632 nm light (helium–neon) produced by a circular disc corresponding to a 10 μm sphere suspended in medium with refractive index of 1.33. Redrawn from ref. 14.

2.6.4 Rayleigh scattering

This behaviour is exhibited by particles, such as macromolecules or small organelles, which are smaller than the wavelength of the incident light. The latter causes displacement of atomic charges in the particle, inducing a dipole which oscillates in phase with the illuminating light. This oscillation now acts as a point source which re-irradiates light in all directions.

All of these processes contribute to the amount of light scattered into a detector at any given angle. Brunsting and Mullaney (15) have compared the total light scattered by CHO cells at various angles with homogeneous and coated microspheres, and their data are reproduced in *Figure 12*. Up to an angle of about 7.5° all three types of objects gave very similar results, compatible with diffraction making the dominant contribution. Compatible with Mie theory (12), at angles greater than 7.5° the CHO cells gave results intermediate between those from the artificial objects.

2.6.5 Forward light scatter

This is also referred to as FALS, or forward angle light scatter. It is very rarely specified adequately and almost invariably is equated directly with cell size. Forward scatter is proportional to particle size at narrow forward angles,

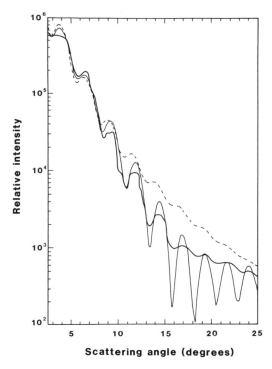

Figure 12. Comparison of the total light scattered by Chinese hamster ovary cells (heavy solid line) at various angles with homogeneous (dashed line) and coated microspheres (light solid line). Redrawn from ref. 15. Up to an angle of about 7.5° all three types of objects gave very similar results which were compatible with diffraction making the dominant contribution.

but the relationship is strictly only valid over collection angles between about 0.5° and 1.5°. At wider angles the correlations tend to break down. This is illustrated in *Figure 13* by work of Mullaney *et al.* (16) where total light scatter intensity at 632.8 nm (helium–neon) is plotted against diameter for particles with refractive indices of 1.533 and 1.373, corresponding approximately to those of fixed and unfixed cells respectively, and the collecting angle was between 3° and 5°. It will be noted that the scatter intensity between these particular angles is very insensitive for diameters of 7 μm to 13 μm, which is the range of most interest in biology.

Another problem encountered with all scatter measurements, irrespective of the collection angle, is the dependence on cellular orientation. A spherical cell with a centrally placed spherical nucleus constitutes no problem, as the light-scattering pattern is orientation independent. However, cells which are not radially uniform in all orientations, e.g. nucleated discs such as chicken RBCs or elongated fibroblasts, can give rise to artefactual light-scatter

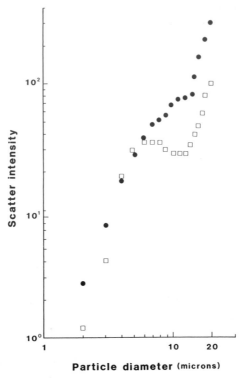

Figure 13. Light scatter intensity (ordinate) versus diameter for particles with refractive indices of 1.373 (●) and 1.533 (□) at a collection angle between 3° and 5°.

distributions. Loken *et al.* (17) found a bimodal distribution with light-scatter measurements from fixed chicken RBCs. The two peaks were sorted, and each was re-run through the instrument. The same bimodal pattern was obtained, indicating that the bimodality of the distribution was instrument-generated and due to changes in cellular orientation.

The major problem with biological material is non-uniformity. Scattered light in the forward direction seems to be dominated by diffraction (regular and anomalous) at the medium/cytoplasmic and cytoplasmic/nuclear interfaces. Light scattered to 90° tends to be related more to the number of dielectric interfaces through which the photon passes, and to the magnitude of those changes in refractive index. Thus, 90° scatter is more related to reflection and refraction from intracellular structures, though diffraction and Rayleigh scattering will also contribute.

It has become part of the dogma of flow cytometry that forward light scatter is proportional to particle size. In most cases this is true, but the exact position of the detector in relation to the particle, the light collection angle, the type of flow chamber and focusing system used, the laser wavelength and illuminating power, as well as the composition of the particle, can all influence the total quantity of light scattered into the detector. Thus, forward light scatter should not be related indiscriminately to the size of a particular cell or cell type unless this has been proven to be true by an independent method.

3. Data output and analysis

3.1 Data display

Any system which produces large quantities of data, particularly if these are multidimensional, must have efficient data presentation methods. This is an important step in converting data, which is just a series of numbers, into information.

3.1.1 Monodimensional histograms

The simplest data set in flow cytometry is the monodimensional histogram, where frequency is scored on the *y*-axis versus the magnitude of the measurement (fluorescence or scatter intensity) on the *x*-axis. *Figure 14* shows an example of a DNA histogram which has two peaks separated by a trough. The first peak corresponds to cells with G1 DNA content and the second, recorded at double the *x*-axis scale reading, corresponds to cells with G2 + M DNA content. Cells between the peaks are at various stages of DNA synthesis. This type of histogram can be analysed by a number of computer programs to give the proportions of cells in G1, S, and G2 + M (18).

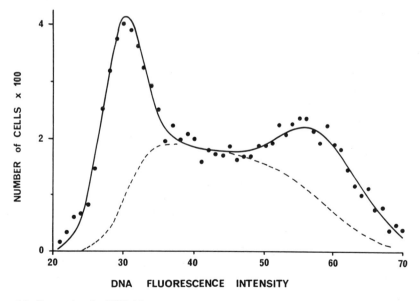

Figure 14. Example of a DNA histogram.

3.1.2 Bivariate data

A bivariate data set is one in which two measurements were made on each cell in the sample and a number of methods have been developed to present the data within these two-dimensional data spaces.

i. Dot-plots

The simplest method of presenting bivariate data is to display the two measurements from each cell as a dot with x and y coordinates proportional to the two measurements. This is shown in *Figure 15*, where forward scatter is plotted on the y-axis versus 90° scatter on the x-axis with cells from mouse bone marrow. The associated monodimensional histograms are presented adjacent to the respective axes. The problem with this type of display is that it gives little depth or height perception. Once a pixel on the oscilloscope screen is lit up with the first cell to be scored in that location it cannot be lit up again if another cell has exactly the same coordinates. This can be overcome to some extent in systems with grey-scale capability, but many earlier instruments do not have this feature.

ii. Contour maps

A second method is to present the data as a contour plot where lines are drawn to connect points of the same frequency. This is directly analogous to a terrestrial contour relief map showing distances above sea level. The data in the previous figure are redrawn in this form in *Figure 16*. It can sometimes be

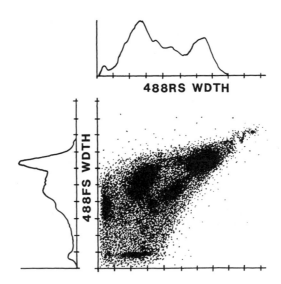

Figure 15. Dot-plots of 90° scatter on the y-axis versus forward scatter on the x-axis with cells from mouse bone marrow. The associated monodimensional histograms are presented adjacent to the respective axes.

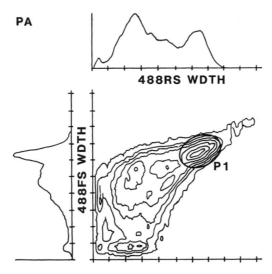

PA

488RS WDTH

488FS WDTH

P1

Figure 16. Contour map displays of the data shown in *Figure 15*.

a little difficult to appreciate the data fully in this form, as hills and troughs are not always immediately apparent.

iii. Stereo-perspective graphics

This type of display presents the data as a solid model in stereo and perspective (colour can also be included) with frequency on the *y*-axis versus the two measurements on the horizontal *x*- and *z*-axes; an example is shown in *Figure 17*. This is the same data set presented in the previous two figures, and it is clear that this is a far superior display system where the depth and height perception is self-evident. However, we must also be able to rotate these displays in order to see hidden surfaces and lines which may also mask some data.

3.1.3 Trivariate display

Three-parameter data is a little more difficult to display; ideally, stereoscopic views should be used. An example from bone marrow for the hydrolysis of two esterase substrates (horizontal axes) versus time (vertical axis) is shown in *Figure 18*. Four populations can be identified, and line vectors have been drawn through the medians of each of the 10 contoured time-slices. This figure should be inspected with a stereo viewer.

3.1.4 Multiparameter data

The world in which we live is three-dimensional, and conceptually it is difficult to visualize the geometry and meaning of a data space containing four or more dimensions. It is not possible to display four-dimensional data within

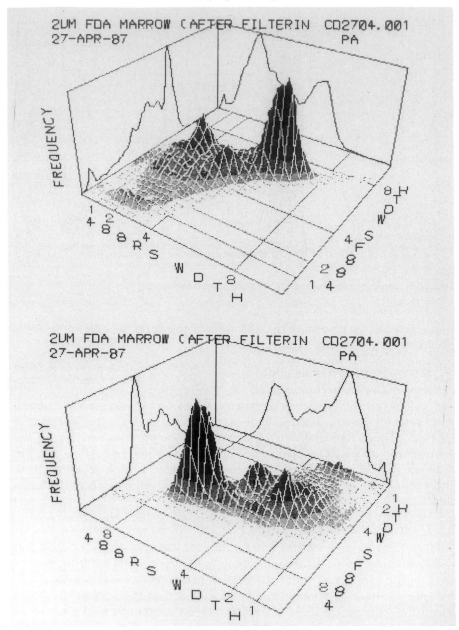

Figure 17. Two perspective 'hidden surface elimination' views with of the same data as in the previous two figures. The 'dip' in the centre of the data is now clearly apparent.

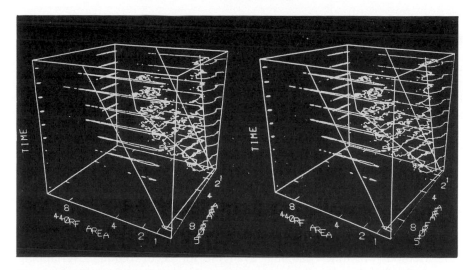

Figure 18. Stereo-perspective three-dimensional display for the hydrolysis of two ester-ase substrates (methyl umbelliferyl acetate, 440RF AREA, versus fluorescein diacetate, 520RF AREA) versus time (vertical axis). The four vectors have been drawn through the medians of the populations at each of the 10 time-slices.

a geometrical space as shown for three dimensions in *Figure 18*, and multi-dimensional data are best displayed in a series of related two-dimensional data spaces.

3.2 Data analysis

Data analysis is too large a subject to be covered in the space available here. However, there is room to make some comments. Firstly, a distinction must be made between data handling and data analysis. Gating and counting the number of cells within that gate is commonly regarded as data analysis but, in reality, this is data handling. Data analysis, the means by which information is extracted, starts after gating procedures have been carried out.

One of the frequently voiced advantages of flow cytometry is that it produces 'good statistics' because large numbers of cells have been analysed. However, confidence limits are rarely placed on results, and hence the reader has little or no feel for the inherent variability in the information produced. This variability is important and has three major components. The first is due to the measuring system, and applies not just to flow cytometry but to every measurement system. The second component is due to variability in the processes involved in making the measurement possible, and in flow studies this includes variability in fluorescence staining procedures including the various reagents as well as the technical competence with which the procedures are carried out. The last, and most important, source of variability is within the biological system being studied, and it is from this that we might gain some extra information.

Each of these components has many sources of variation which give rise to a distributed response. This has one very important implication which does not appear to be generally appreciated. When the distributions of two over-lapping immunofluorescence populations, for example, are present in the same sample it is not merely possible, but in fact *inevitable*, that there will be specifically labelled cells of low intensity which elicit a lower 'response' from the instrument than do some non-labelled cells. **This is a direct consequence of the simple fact that the data are distributed**. This cannot be emphasized too strongly. The same phenomenon is seen in the deconvolution of DNA histograms, where early S-phase cells overlap some of those in the G1 distribution. In the early days of DNA histogram analysis this was an article of faith derived from logical deduction; however, absolute confirmation has been provided with the advent of the bromodeoxyuridine technique (see *Protocol 11*). Methods are now becoming available for deconvoluting overlapping immunofluorescence histograms using curve fitting and analytical procedures of the type used for many years in DNA histogram deconvolution.

4. Quality control

Quality control is of profound importance, particularly with increasing use of flow cytometry technology in clinical practice. The potential for litigation (in the USA in particular) has added increasing urgency to the quest for standardization of quality control. The commonest reasons for poor-quality data are poor preparation and partial nozzle blockages, and the latter is often due to the former. A number of techniques are available to monitor the preparation and hence quality of the data.

4.1 Inspection

Visual inspection frequently seems to be forgotten. The availability of a sophisticated and expensive instrument does not mean that the fluorescence microscope should be abandoned. Newcomers to flow cytometry should always be encouraged to carry out a quick check before presenting the sample for flow analysis. If the preparation looks bad under the fluorescence microscope, with shredded cells and clumps, it will look much worse in the cytometer. These instruments are very sensitive, but they have no inherent intelligence or recognition capability. A small chunk of garbage scatters light just as effectively as a cell, and it will always be seen and recorded. The human eye and mind recognize rubbish and clumps and exclude them automatically; this editorial capacity will always lead the observer to believe that the preparation is better than it is. If it looks bad under the microscope it really is too bad to be worth running through the instrument. No worthwhile results will be obtained, and there is a risk of plugging up the tubes and nozzle.

4.2 Coincidence correction

Even if the preparation looks perfect under the microscope there is always a chance that two cells will pass through the focus either very close together or simultaneously. The probability of this occurrence increases with increasing cell concentration. Two methods are available to correct for coincidence, using time in slightly different ways.

The data acquisition computer can be programmed to calculate the average time interval between arrival of cells at the analysis point from the rate at which the digitized signals are being presented to it for storage. It can then calculate the Poisson probability of coincidence, and set the data acquisition logic so that only events spaced at greater than specific intervals are recorded. Events arriving 'too close' to each other are ignored.

The second method uses a double electronic threshold triggering system. *Figure 19* shows pulses from two closely spaced cells passing through the beam. Threshold 1 (T1) is set lower than threshold 2 (T2), and nothing at all will happen unless the voltage from the photodetector exceeds T1. This enables the electronic acquisition system, which begins to integrate the area under the pulse, to start timing the width of the pulse, and record the peak height when this is reached. In the top panel the voltage drops below T2 as

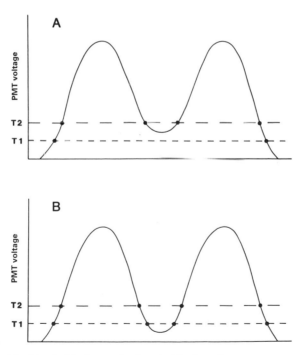

Figure 19. Double-threshold electronic threshold triggering system to identify closely spaced pulses.

the first cell passes out of the beam, but does not drop below T1 before rising again to cross T2 as the second cell enters the beam. This aborts the data acquisition logic, which is then reset; the recording is not completed. The difference between the upper and lower panels of *Figure 19* is that in the lower panel the voltage drops below T1 before it rises again through T2. In this case a recording is made; the width of the pulse is timed when the voltage drops below T1, and pulse height, width, and area are digitized and stored. With the double threshold system the sequence T1, T2, T2, T2 means *stop*, events are too closely spaced. In contrast, the sequence T1, T2, T2, T1 means *go*, and the data are acquired. There is a third possibility, shown in *Figure 20*, where the threshold sequence is T1, T2, T2, T1. The data would be recorded although, clearly, this is a double event. However, this can be excluded after acquisition using pulse shape analysis as described in the next section.

4.3 Pulse shape analysis

Some types of preparations, e.g. nuclei extracted from paraffin-wax-embedded archival material, inevitably contain a variable quantity of debris, clumps, and nuclear fragments as well as intact single nuclei. After DNA staining with propidium iodide (red fluorescence) any small debris and large clumps can be gated out on a combination of forward and 90° scatter signals. Large fragments of nuclei and small clumps not identifiable in the forward versus 90° light scatter data space can be identified and gated out using the shape of the pulse from the red (DNA) photomultiplier which is usually the master triggering detector in these types of assay. This involves displaying pulse height versus pulse area, and a schematic of two pulses, the first a single pulse and the second from a doublet, is shown in *Figure 21*. Each of these pulses has the same peak height, but the area under the doublet is double that of the singlet. Thus, when pulse height is plotted against pulse

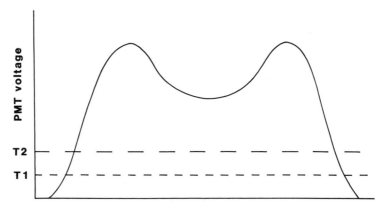

Figure 20. Illustration of two closely spaced pulses which would not be recognized by the double-threshold system.

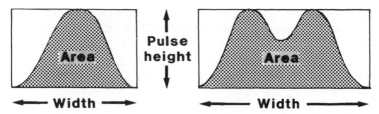

Figure 21. Relationship between pulse height and area for a single event and a doublet.

area for the whole population we get the display shown in *Figure 22* and the doublets can be excluded. This type of quality control procedure can be carried out with conventional spherical lens focusing if pulse width and area are digitized together with pulse height, but it is considerably less efficient than with crossed cylindrical lens pair focusing.

4.4 Time

Continuous-time recording was introduced in flow cytometry parameter by Martin and Schwartzendruber (19) to follow the kinetics of fluorescein diacetate (FDA) hydrolysis in populations of intact cells. However, time can also be used as a quality control parameter by continuously monitoring versus time an independent parameter which does not vary with time (20). This is shown in *Figure 23* where a sample of isolated nuclei stained for DNA with ethidium bromide was introduced into our instrument and the red fluorescence associated with each event was digitized and the time at which each event occurred in relation to the start of the run was recorded from the computer time clock with 50 ms resolution. The sheath feed tubes were partially occluded on three

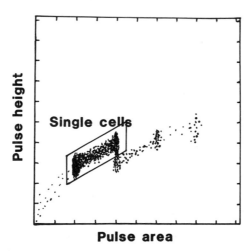

Figure 22. Pulse height plotted against pulse area for a population stained for DNA which contains a proportion of doublets. The single events are within the gate.

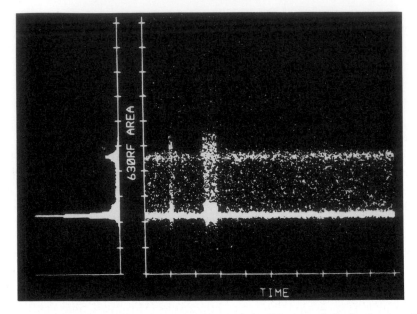

Figure 23. Perturbation induced by partial occlusion of the sheath feed tubes on three occasions at 5 s for about 1.0 s and twice for about 0.5 s at approximately 13 and 14.5 s into the run. This was responsible for the loss of definition particularly noticeable at the base of the G1 peak of the DNA histogram.

occasions, once at 5 s for about 1.0 s and twice for about 0.5 s at approximately 13 and 14.5 s into the run. The perturbations are readily appreciated where on each occasion there was the expected 'smearing' of the fluorescence data and consequent loss of definition, particularly noticeable at the base of the G1 peak of the DNA histogram. These data were redisplayed from the list-mode data file commencing at 15 s (beyond the disturbance) to give a histogram indistinguishable from that in a non-perturbed run.

5. Cell sorting

Sorting or separation of cells can be carried out using a number of physical and biological agencies. Perhaps the most extensively and frequently used, but least appreciated, is gravity. RBCs settle out to form the haematocrit layer in heparinized blood, leaving the white cells in the buffy coat. Variations on this theme include density gradient sedimentation (Ficoll and Ficoll-paque) and cell electrophoresis. The latter uses charge in combination with gravity in the separation process. Magnetism can also be used to sort macrophages which have ingested iron particles. Biological procedures include complement lysis of specific antibody-tagged unwanted cells, leaving the cells of interest intact. Also, magnetic beads coated with antibody can be used to select positively for the cells of interest. Some of these procedures do not yield particularly pure sub-fractions and they will not be discussed further because they are not relevant to flow cytometry. However, they should not be forgotten altogether because they can, and frequently should, be used in conjunction with flow technology.

5.1 Electrostatic sorting

5.1.1 Ink-jet writing

This was developed in the early and mid 1960s (21) in an attempt to cut down on the noise generated by mechanical printing and to make it possible to write on fragile objects, for example, date-stamping individual eggs. This is notoriously difficult with traditional mechanical stamping devices which inevitably result in a significant proportion of breakages. In ink-jet writing the ink is forced at high speed through a fine nozzle and the jet breaks up into droplets. Normally, the droplet break-off point occurs at random distances from the nozzle and the droplet sizes are not constant. The trick is to make this break-off point a constant distance from the nozzle. This was achieved by introducing a high-frequency perturbation into the jet by oscillating the whole nozzle assembly with piezoelectric diodes. With jets of 50–70 μm in diameter an oscillating frequency between 40 000 and 50 000 Hz will stabilize the droplet break-off point and form constant and regular-sized droplets at the oscillating frequency. The stream of droplets can be guided in the X- and Y-axis planes by placing a variable charge on the droplets as they break off from the jet. Downstream from the droplet break-off point there are pairs of charged deflection plates at right angles to each other which deflect the charged droplets by defined distances proportional to the charge placed upon them. Thus, it is possible to guide the stream of ink droplets to trace out almost any geometrical shape, including lettering.

5.1.2 Electrostatic cell sorting

This was developed from ink-jet writing, and a typical layout shown in *Figure 24* was developed initially by Fulwyler at Los Alamos (22) then by the Herzenbergs at Stanford University in the late 1960s and early 1970s in conjunction with Sweet (23, 24, 25). The jet issues from the nozzle of the Crosland-Taylor type flow chamber and the laser intersects the jet just below the nozzle (the analysis point). The whole of the flow chamber assembly is oscillated at about 40 000 Hz with the formation of regular droplets which break off the jet at a constant distance from the analysis point at this frequency. Fluorescence and/or 90° light scatter is collected by the optics at 90° to the intersection of the jet with the laser, and solid state detectors collect the forward scattered light. The system operates as follows. Cells are 'interrogated' at the analysis point, and their fluorescence and light-scatter characteristics are determined. The jet velocity is constant, as is the distance between

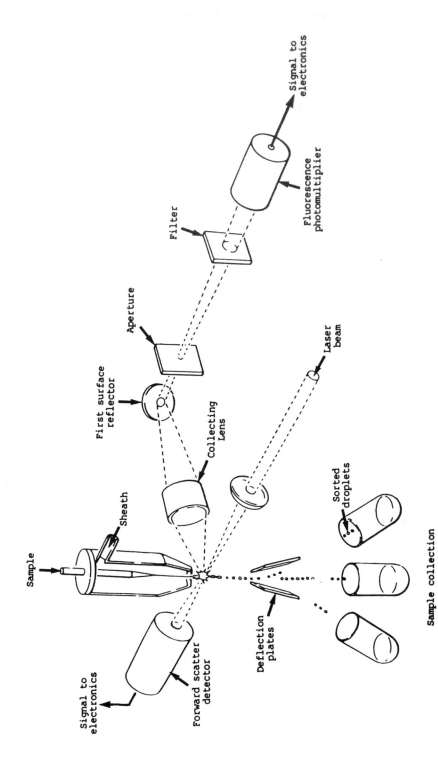

Figure 24. Typical layout of an electrostatic cell sorter.

analysis and droplet break-off point. Thus, the time for a cell to travel from the analysis point to the break-off point, Δt, is also constant. If a given cell is analysed and is found to have characteristics in which the investigator is interested, then the jet containing the cell is charged at a time Δt after it is analysed. Thus, the jet will be charged as the cell of interest is contained within the tip of the jet just before the droplet, which is about to surround the cell, breaks free from the jet. A split second later the charged droplet containing the cell breaks off from the jet; it is now isolated and in free fall. Further downstream the stream of droplets passes between charged plates which deflect charged droplets left or right, depending on the sign of the charge, into suitable collecting vessels.

Figure 25, reproduced from Herzenberg *et al.* (25), is a time-lapse photograph

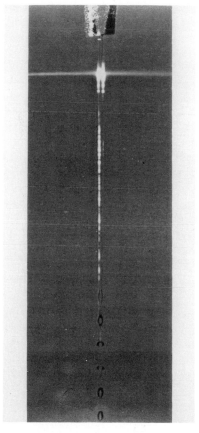

Figure 25. Time-lapse photograph of the jet issuing from the nozzle located at the top with the laser cutting across the jet just below the nozzle. The illuminating light for the exposures was stroboscopically flashed at exactly the same frequency as the droplet break-off frequency which 'freezes' the jet and droplets.

of the jet issuing from the nozzle which is located at the top with the laser cutting across the jet just below the nozzle. This and the following picture are quite remarkable as the illuminating light for the exposures was stroboscopically flashed· at exactly the same frequency as the droplet break-off frequency. These photographs, therefore, represent multiple superimpositions (40 000 per second) of the jet and the droplets. No trace of 'fuzz' can be seen because each superimposition was virtually exact and the droplet breakoff point is a constant distance from the laser intersection analysis point. The second picture (*Figure 26*) shows the droplet stream between the deflection plates where three droplets are being deflected to left or right. Again, there is exact superimposition of droplets on this multiple exposure photograph. Incidentally, data such as those in *Figure 25* can be used to calculate the surface tension of water from the shapes of the droplets as they break off from

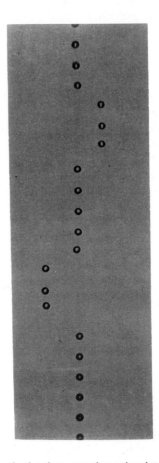

Figure 26. Downstream from the jet there are three droplets being sorted left and right.

the jet, a technique originally suggested and used by Niels Bohr at Cambridge in 1909 (26).

The time it takes for a cell to travel from the analysis point to the droplet break-off point, however, is never *exactly* constant because the velocity is not exactly constant. Very minor variation is possible, as too is variation in the electronic timing, although this is also extremely small. Furthermore, the electrostatic charging of the jet cannot be absolutely instantaneous as it takes finite time (usually only nanoseconds) for the charge to build up to maximum. The timing of the charging must be to within ± 10 μs at a droplet break-off frequency of $40\,000$ s^{-1} and it is possible, therefore, that the cell of interest may not be exactly at the droplet break-off point when the charge is applied to the jet. These various factors can be described by Poisson statistics. Because of these potential sources of variation it is frequent practice to charge three droplets in succession, with the timing arranged so that the cell is expected to reside in the middle droplet of the three. This minimizes the chance of missing a cell of interest.

5.2 Sorting efficiency

Sorting efficiency means that the majority (should be $>95\%$) of the cells are collected in the intended pot. This can be checked by running a small sample from the sorted population through the instrument to check that all the cells/objects appear within the sorting selection window. An example of sorted chromosomes is shown in *Figure 27*, where Davies *et al.* (27) were sorting the X chromosome from the flow karyotype shown in the upper panel. The lower panel shows a sample of the sorted fraction which confirms that almost all of the sorted fraction was chromosome X. In a long sorting run it is well worth doing this quality control check at intervals. The yield can be checked either by haemocytometer or by Coulter counting.

If the purity or yield is not what it should be, the problem is most likely to be due to instability in the jet and droplet formation. This can be checked by looking at the droplet break-off point. Most cell sorters are equipped with either long-focal-length stream-viewing microscopes or video cameras for this purpose. The illuminating light for the microscope is stroboscopically flashed at the droplet formation frequency, and patterns similar to those in *Figures 25* and *26* should be visible.

The most likely cause of jet and droplet formation instability is a partial blockage of the chamber outflow nozzle, which is usually between 50 μm and 70 μm in diameter. A very little material in the nozzle can completely wreck the jet stability. The cause of the problem is usually cell debris (loose, spewed-out DNA is bad) or clumps of cells in a badly prepared specimen. However, non-cellular materials that may block the nozzle can be introduced at all stages in sample preparation. In my experience these include fibres from pipette plugs, short segments of hair or fur introduced during splenic removal

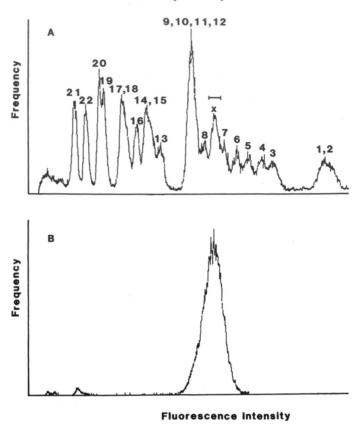

Figure 27. Chromosme sorting. The top panel shows the selection gate for the X chromosome. The bottom panel shows an analysis of the sorted fraction which confirms that the X chromosome was obtained. Data taken from ref. 27.

from rodents, or little slivers of plastic. These can arise from two sources. First, Petri dishes or wells where the cells have been harvested with a glass Pasteur pipette. Any scraping of the pipette tip over the plastic surface planes off little curls and slivers of plastic. These are very difficult to see not only because they are small, but also because the refractive index of plastic is fairly close to that of water and they are therefore almost invisible. The second source of plastic slivers is polystyrene universal containers. In badly made batches (and they do vary) very fine wispy threads of plastic are still attached to the mould injection point. In some containers, where the injection mould sections have been forced slightly apart during manufacture there is a sharp ridge of plastic which can cut into the cap and shower flakes of plastic from the cap into the sample.

6. Preparation of samples

Flow cytometry measures one or more signals from each individual particle passing through the light path: therefore cells must be singly suspended and not aggregated, and, as discussed in the previous section, no non-cellular debris must be present. The stain must be specific and not leach out into the carrying fluid.

6.1 Cell suspension from different tissues

Protocol 1. Blood samples

Reagents and materials

- heparinized tubes
- HBSS or Dulbecco's PBS
- Ficoll-Hypaque or equivalent, 1.077 g/ml (Flow Laboratories, Pharmacia, Nygaard)
- universal containers (Sterilin) or 25 ml centrifuge tubes

Method

1. Blood cells do not need to be disaggregated but fibrin may form clots interfering with the flow system or aggregating single cells. Freshly drawn blood must therefore be collected into heparinized tubes.

2. Dilute sample with four volumes of PBS or HBSS and layer 10 ml of diluted blood over 5 ml of Ficoll-Hypaque in a universal container or 25 ml centrifuge tube to eliminate RBCs.

3. Centrifuge at 400 g. for 20 min.

4. At the end of the centrifugation the mononucleate cells form a visible interface between the PBS and Ficoll. Aspirate the interface and re-suspend in 10 ml PBS.

5. Centrifuge twice at 400 g. for 10 min.

6. Finally, re-suspend the pellet in HBSS at 10^6 cells/ml and analyse by flow cytometry.

6.1.1 Disaggregation of solid tissues

i. Enzymic disaggregation

The enzymes used for disaggregation of tissue may be divided into three classes: enzymes active on fibrous structures such as collagen or elastin (28), enzymes that hydrolyse mucopolysaccharides, e.g. hyaluronidase, and non-specific proteolytic enzymes, such as trypsin, pronase, and dispase. Many factors must be taken into account in using enzymatic disaggregation, such as the medium in which the enzyme is dissolved; duration of the exposure of the tissue to enzyme; temperature, pH, and concentration of enzyme used; and the means by which the enzymatic activity is stopped (29).

(a) *Trypsin*. A concentration of 0.25% crude trypsin in PBS is effective with a wide variety of tissues (30–35) with a 90–95% viability even after prolonged exposure at 37°C (35). Trypsin is often used in combination with chelating agents (see below).

(b) *Pronase*. Obtained from *Streptomyces griseus*, pronase is another enzyme widely used on a wide range of tissues (36–40). There is no concrete evidence for the best concentration to be used for optimal tissue disaggregation (41), although a concentration of 0.1% of pronase is routinely employed in the technique described (38). Many data have shown that the effects of high concentrations of pronase for a short time are similar to those of low concentrations for a relatively long period. Like trypsin, pronase has been used in combination with enzymatic agents such as DNase (40).

(c) *Collagenase*. Used on many kinds of tissues such as mouse myeloma, mouse mast cell tumours, human Hodgkin's tumour, human spleen, and human ovarian cancer (34, 42, 43).

Other proteases may be present in crude preparations of collagenase, and even purified preparations have been found to contain hyaluronidase (28). Purified collagenase may not yield a satisfactory single-cell suspension, particularly with epithelial cells, and may require supplementing with trypsin or pronase. Collagenase may be used at different concentrations, from 100 to 1000 units/ml in culture medium for 15–45 min at 37°C (34, 43, 44). Bacterial collagenase, like pronase, is not inhibited by serum, so prolonged digestion in complete medium with serum is possible, even up to several days' duration.

ii. Chelating agents

Another approach used to obtain cell suspensions from solid tissue is to treat the tissue with chelating agents, usually in combination with enzymes. When used alone, chelating agents have been found to be less efficient than enzymatic treatment in producing intact cells. The most commonly used are EDTA, ethyleneglycol (2-aminoethylether)-*n,nl*-tetraacetic acid (EGTA), and citrate.

EDTA acts by complexing both Ca^{2+} and Mg^{2+} ions; it is used on different tissues at concentrations of 0.01–0.10 M. EGTA is used at a concentration of 0.1 mM. Both are dissolved in PBS or BSS free of Ca^{2+} and Mg^{2+}. Citrate is used at a concentration of 40 μM (45, 46).

iii. Mechanical disaggregation

Soft tissue containing small amounts of fibrous tissue can sometimes be disaggregated mechanically, although the number of viable cells obtained is always smaller than the number obtained using other methods. Another disadvantage of mechanical disaggregation is that many cells released from the tissue remain in clusters rather than being fully dispersed as single cells. Mechanical disaggregation can be achieved by forcing fragments of tissue through sieves of progressively smaller size (e.g. 250 μm, 100 μm, 20 μm) or through hypodermic needles of progressively smaller size (18G, 21G, 23G, etc.) (29, 47, 48).

It is difficult to recommend one single method for disaggregation of cells from solid tissues (especially for solid tumours) as one method may be used with good results, in terms of amount of debris and type of damage sustained by cells, for one particular tissue, but may not be suitable for another. *Protocol 2* is offered as a starting point. Further development may be necessary, based on experience and reports in the literature.

Protocol 2. Disaggregation of cells

Reagents and materials (sterile)
- scissors or pair of scalpels
- magnetic stirrer
- nylon or stainless steel gauze, 100 μm and 20 μm
- HBSS
- enzyme solutions: 0.25% trypsin in PBS or PBS with 1 mM EDTA, 0.1% pronase (Sigma) in PBS or complete culture medium; DNase 100 μg/ml (1–5 μg/ml final concentration) (Sigma, Worthington). Collagenase crude CLS grade (Worthington) or grade 1 (Sigma), 2000 units/ml stock, 100–1000 units/ml final

Method
1. Place tissues in HBSS at 4°C. Mince the tissue into 1 mm^3 pieces with scissors or crossed scalpels and wash twice with HBSS.
2. Remove HBSS and replace with enzyme solution for disaggregation at a ratio of 10–15 ml/1 g tissue.
3. At the end of digestion, stop the enzyme activity by placing on ice or by addition of serum. Centrifuge the cells, decanted from the tissue fragments, at 1000 g. for 10 min. Filtration through nylon or steel gauze, 100 μm followed by 20 μm, will help to remove aggregated cells and larger debris at this stage.

Note: Never add DNase to these preparations if you plan to evaluate DNA content in the cells. Otherwise 1–5 μg/ml may be used to minimize reaggregation due to DNA released from lysed cells.

6.1.2 Fixation of cell suspension

A cell suspension can be used immediately after preparation as living intact cells, or it can be fixed if staining and flow cytometry are to be carried out later, or if fixation is required for effective staining. Ethanol fixation is used most frequently.

Protocol 3. Fixation procedure

Reagents and materials
- HBSS or PBS, free of Ca^{2+} and Mg^{2+}, with 1 mM EDTA
- 95% ethanol

Method
1. Bring cells to a concentration of about 1×10^6 cells/ml in suspension in cold HBSS, Ca^{2+} and Mg^{2+} free, containing 0.5 mM EDTA.
2. Keeping the suspension at 4°C add 95% ethanol dropwise with constant mixing in the ratio of three volumes for each volume of cell suspension, to give a final ethanol concentration of 70%.
3. Cells can be kept in this suspension for a few months.
4. Before being used for cytophotometry, cells must be freed of ethanol by centrifugation and re-suspended in PBS or whatever reagent is to be used for the procedure.

---

7. Examples of types of assay

7.1 Light scatter of unstained cells

The presence of a particle, for example a cell, in a light beam, perturbs the beam and the incident light is re-distributed. The energy is re-emitted by the cells in a complex pattern where diffraction, refraction, and reflection interact together in a function of the volume, surface configuration, and internal structures of the cell. Commercial instruments normally measure light scatter in two fixed positions, a small-angle forward scatter, about 10°, and a 90° scatter.

The 90° scatter is considered to be affected mostly by reflection and re-fraction due to the internal structures of the cell, while the low-angle forward scatter is considered as more representative of cell volume (16, 49, 50) (*Figure 28*).

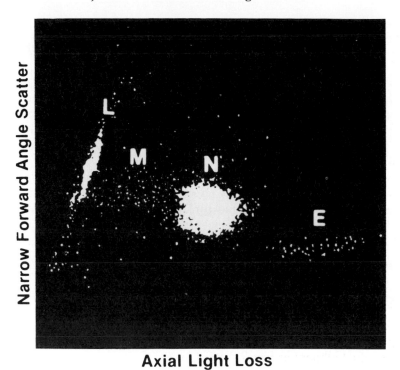

Narrow Forward Angle Scatter

Axial Light Loss

Figure 28. Light scatter of unstained cells. Differential counting of neutrophils, eosinophils, lymphocytes, and monocytes in human blood by non-fluorescent method. (Photograph from Ortho Instruments.)

Protocol 4. Scatter measurements

Reagents and materials

● HBSS

Method

1. Suspend cells in HBSS to obtain a concentration between 1×10^5 and 1×10^6 cells/ml.

2. Fill the reservoir of the instrument containing the sheath fluid with the same balanced salt solution as used for the cells.

3. Set the flow of cells through the instrument to 500 cells/s to obtain the best performance.

7.2 Staining procedures

7.2.1 Suspension of nuclei and DNA staining with propidium iodide

Propidium iodide is an analogue of ethidium bromide and, like ethidium bromide, intercalates in the nucleic acid helix with a resultant increase in fluorescence of about 20-fold. Propidium iodide is about 1.8 times more fluorescent than ethidium bromide. At 488 nm excitation the emission maximum for the DNA–propidium iodide complex occurs at approximately 615 nm (47, 51, 52).

A typical histogram is shown in *Figure 14*, where the major peak corresponds to the G1 phase of the cell cycle, the minor peak to G2/M, and the intervening saddle to the S phase. Absolute values for DNA content can be obtained by using a standard such as chicken erythrocytes, run simultaneously with the sample.

Protocol 5. DNA content of Lewis lung carcinoma of mouse (3LL)

Reagents
- HBSS
- staining solution: 100 ml of sodium citrate 0.1%, 5 mg of propidium iodide, 900 μl of Nonidet P40 1% in water

Method

1. Remove tumours from C57BL/6 mice and wash in a Petri dish with PBS.
2. Remove obvious connective tissue and fat and mince with scissors.
3. Force the small pieces through a 1.20 × 38 mm (18G × 1.5 in) needle and re-suspend in HBSS maintained at 4°C.
4. Stain 3LL cell suspension by adding 3 ml of propidium iodide solution to 200–300 μl of cell suspension (500 000 cells/ml).
5. Store at 4°C for 20–30 min before analysis.
6. Measure fluorescence emission between 580 nm and 750 nm to exclude the overlapping region of excitation and emission spectra of unbound propidium iodide.

Protocol 6. Application to cells in culture

Reagents: as for *Protocol 5*

Method

1. Remove medium from culture dish and rinse twice with HBSS.

2. Add 5 ml of propidium iodide solution and refrigerate the dishes for 10 min at 4°C.

3. Dislodge the nuclei by repeating pipetting and analyse by flow cytometry.

Protocol 7. DNA staining with propidium iodide in whole cells

Reagents
- 70% ethanol
- propidium iodide, 50 μg/ml, containing 40 g/ml RNase

Method
1. Fix cells in 70% ethanol (see *Protocol 3*).
2. After centrifugation remove fixative, stain an aliquot of cells (10^5–10^6 cells/ml) with propidium iodide for 30 min at room temperature.

In the combination staining technique using the fluorescent antibiotic mithramycin and propidium iodide, the energy transfer occurs from the mithramycin donor molecule to the propidium iodide acceptor molecule. This technique is useful for staining and analysing solid tumour samples, sperm cells, and gynaecological material, as well as tissue culture cells (43).

Protocol 8. DNA staining by mithramycin and propidium iodide

Reagents
- 10 μg/ml mithramycin
- 10 μg/ml propidium iodide
- 15 mM magnesium chloride
- 0.1% Nonidet P40
- Tris–HCl pH 7.5

Method
1. The cell suspension can be used immediately after disaggregation or up to 1 week after fixation in 70% ethanol.
2. Stain with propidium iodide/mithramycin solution for 1 h at 4°C.
3. After staining run the sample through the instrument excited by a mercury 100 W lamp, with barrier filter BG12 3 mm plus one step filter K590.

7.2.2 Differential staining of RNA and DNA

The metachromatic property of acridine orange is used for the differential identification of DNA and RNA. Acridine orange has been used as fluorescent stain for nucleic acids in fixed and unfixed cells, or as a marker of lysosomes, or in observing the metachromatic changes of fluorescence in dead cells, or to differentiate between cycling and quiescent cell populations. Despite the technical difficulties in obtaining a reproducible measurement of DNA and RNA, this method has been extensively used by some groups (54–56).

Protocol 9. DNA and RNA staining by acridine orange

Reagents

Solution A: Stable for about 2 weeks, must be kept cold

- 0.1 ml Triton X-100 (0.1%)
- 8 ml 1 M HCl
- 15 ml I M NaCl
- 76 ml distilled water
- 100 ml total volume, pH 1.5

Solution B: Stable for several months, but best stored as a sterile solution (autoclave or filter).

- 10 ml 0.01 M EDTA
- 15.5 ml 1 M NaCl
- 31.5 ml 0.4 M Na_2HPO_4
- 18.5 ml 0.2 M citric acid
- 24 ml distilled water
- 99 ml total volume, pH 6.0

Acridine orange stock solution

- 1 mg/ml in distilled water.

Acridine orange is carcinogenic: handle with care.

Acridine orange working solution

- 0.1 ml of stock solution plus 9.9 ml of solution B.

Note: the sheath flow system of the instrument must be kept cold (4°C).

- Argon laser excitation: 488 nm
- Red fluorescence is DNA ($F > 600$ nm)
- Green fluorescence is RNA ($F > 530$ nm)

Method

1. Add 0.2 ml of cells (8×10^6 cells/ml) suspended in PBS with 15% serum to 0.4 ml of solution A and keep at 4°C for 45–60 s.

2. Add 1.2 ml of solution B for 2 min at room temperature.

3. Analyse by flow cytometry within 10 min from the addition of solution B.

Note: every step of the procedure is critical for obtaining reproducible results. It is particularly important to have a constant number of cells to be stained; depending on the ratio between nucleic acids and acridine orange, RNA metachromasia will be affected. Time of exposure to staining solution and temperature will also strongly affect the reaction.

7.3 Application of staining procedures

Protocol 10. Differential staining of denatured and double-stranded DNA

Reagents

- HBSS containing 1000 u/ml RNase A, 0.2 M KCl, pH 1.35.
- acridine orange, 5 μg/ml (16.7 μm) in 0.1 M citric acid, 0.2 M Na_2PO_4 buffer, pH 2.6

Method

1. Suspend 2×10^5 cells in 1 ml HBSS/RNase.

2. Incubate at 37°C for 1 h.

3. Mix 0.2 ml of cell suspension containing 4×10^5 cells (still in HBSS/ RNase) with 0.5 ml of 0.2 M KCl pH 1.35 at 20°C for 30 s.

4. Stain by adding 2 ml acridine orange solution for 2 min.

5. The green (530 nm) and red (600 nm) fluorescence represent, respectively, the amount of denatured and double-stranded DNA of the cells.

Note: see note to *Protocol 9*.

Hoechst 33258 stain fluoresces between 410 nm and 580 nm and needs to be excited in the u.v. at 330–360 nm. It binds preferentially to adenine– thymidine base pairs and is therefore specific not just for DNA but for the presence of thymidine. In the analysis of the cell cycle, the incubation of cells in the presence of BrdU substitutes BrdU for thymidine in DNA. Cells in the synthetic phase will therefore remain apparently diploid and after division will give rise to a new peak in the histogram of half the G1 peak.

This technique permits direct measurement of the G2 plus mitosis time to be made (57–59).

Protocol 11. Cell cycle analysis by bromodeoxyuridine (BrdU) and Hoechst 33258

Reagents
- medium containing BrdU at 33 µg/ml and deoxycytidine at 26.4 g/ml (equimolar)
- staining solution: Hoechst 33258 in PBS. Cells are run through the instrument about 2 h after staining; stability of the stain is reported to range from 30 min to 24 h.

Method
1. Add BrdU medium to the growth medium 1:10.
2. Incubate cells for different lengths of time appropriate to the predicted cell cycle time.
3. After incubation dissociate cells as for subcultivation.
4. Re-suspend the cells directly in the staining solution.

Hoechst 33342 can be used as vital stain for DNA. It is known to produce relatively good stoichiometric DNA staining while preserving cell viability. Excitation is in the u.v. range of 350–363 nm, while emission is recovered over 450 nm (60–62).

Protocol 12. DNA staining of viable cells by Hoechst 33342

Reagent
- Hoechst 33342, 0.25 M in distilled water

Method
1. Prepare cell suspension at 10^6/ml and stain with Hoechst 33342 at the A concentration of 5–10 µM for 20 min at room temperature.
2. Do not wash the cells before analysis.

Protocol 13. Protein staining by fluorescein (FITC)

Reagents
- 70% ethanol
- fluorescein isothiocyanate: 0.1–1.0 µg/ml in PBS containing 40 g/ml RNase

Method

1. Fix cells in 70% ethanol for at least 18 h before staining.
2. Centrifuge fixed cells to remove fixative.
3. Stain cells for protein at room temperature for 30 min in fluorescein isothiocyanate solution.
4. Analyse by flow cytometry with argon laser excitation at 488 nm and fluorescence emission 515–555 nm.

7.3.1 Double staining for DNA and proteins

For two-colour staining, fixed cells are stained with a solution containing 18 μg/ml propidium iodide in 0.1% sodium citrate, 0.05 μg/ml FITC and 40 μg/ml RNase in PBS for at least 20 min at room temperature (*Figure 29*).

7.4 Fluorescence antibody technique using fluorescein

This technique is used to detect cells having specific membrane antigens by treating a cell population with monoclonal antibodies conjugated with fluorescein or rhodamine.

Protocol 14. Direct immunofluorescence staining of cell surface

Reagents

- FITC-conjugated antibody diluted in PBMEM, containing 0.1% sodium azide and 2% calf serum. To test for appropriate concentration add 50 μl of different dilutions to cells in a microtitration plate and check for optimal staining by fluorescence microscopy. In analysis with anti-human LEU-1, dilute to 5 μg/ml or 0.25 g/50 μl. Mouse or human cells, at a concentration of 2×10^7, must be 90% viable.

Method

1. Add 50 μl of the cells to a microtitration plate containing 50 μl diluted antibody per well and mix.
2. Incubate for 45 min on ice.
3. Centrifuge the plate at 500 g.
4. Discard the supernatant and wash the cell pellet twice with 100 μl of PBMEM. Centrifuge cells after each washing at 500 g for 3 min and aspirate supernatant from the pellets.
5. For flow cytometry analysis re-suspend the cells in 1 ml of cold medium PBMEM to have a concentration of 1×10^6 cells/ml; keep cold before analysis (63, 64).

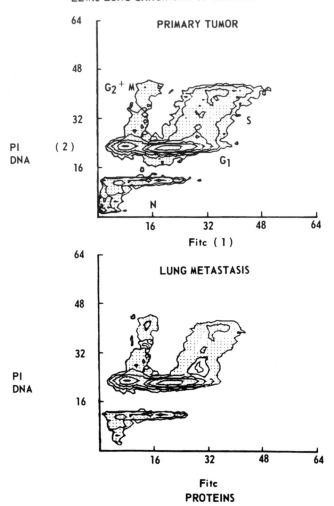

Figure 29. Double staining for DNA and proteins (from E. Erba, Mario Negri Institute, Milan).

8. Discussion

Flow cytometry provides a large set of instruments, similar but not identical, optimized for specific situations. Limiting factors are the sources of excitation light, the sensitivity of photomultipliers, and the computerized functions available.

High resolution is always easily obtained with coherent light from low-intensity lasers in a narrow beam of light focused on the flow of cells. High sensitivity can only be achieved with very costly high-power lasers requiring the availability of a very high power supply (a multiline 18 W output argon laser absorbs almost 50 kW/h and the commonly used 4 W argon laser requires more than 12 kW/h). High excitation power is required, for example, when immunofluorescence of a small number of membrane receptors must be detected but also when fluorochromes require excitation in the u.v., where both argon and krypton lasers have low efficiency. Again, if a wide spectrum of wavelengths is required with a dye laser included in the system, initial costs and power required increase. To these apparently negative comments it must, however, be added that fluorochromes like propidium iodide permit analysis of nucleic acid content in vertebrate cells with the power given by a very small 100 mW laser, and that the mithramycin propidium iodide method works perfectly with a 100 W mercury lamp.

The performance of cell sorting must be considered separately from the analytical use of the instrumentation. This is characterized by the need of more sophisticated equipment from the electronic point of view but dedicated to a preparative function and not expected to perform at the limits of sensitivity and resolution.

Cell sorting provides cell suspensions of a good grade of purity in relation to the discriminating parameter chosen. All the methods described above can be applied in sorting. It must be noted, however, that only a few of them can be applied to living cells. This area requires further development. Further limitations in cell sorting that must be discussed are of technical origin: first of all cells are sorted in drops of fluid interspaced with about a 15-fold excess of empty droplets of the flow sheath. This means that sorted cells are strongly diluted and normally must be concentrated by further manipulation, producing an unavoidable cell loss.

Speed of analysis greatly reduces the purity of preparations, and the 'official' speed given by commercial firms of 5000 cells/s must be considered a maximum giving fair quality preparations. A more efficient speed of analysis could be considered to be less than 2000 cells/s. This speed of analysis applies to all the cells passing through the nozzle, and the real recovery of sorted cells depends on the ratio between cells to be sorted and cells which are discarded because the computer was still elaborating information about previous cells in real time. In practice a low frequency of droplets containing cells, 1:15 empty droplets, is required to avoid this problem and reduce the statistical chance of more than one cell occupying a droplet.

The sorting time required to select a sufficient number of cells of the desired type may therefore become very long, increasing the risk of poor performance linked to the probability that the cell suspension will contain debris, aggregates, or any type of foreign material able to clog the nozzle or eventually modify the flux of fluid through the nozzle. This last possibility is the most serious in terms of cell loss and purity of the sorted cells prepared.

All these drawbacks of the methodology are presented in the light of our own experience. The advantages of flow cytometry and cell sorting are well known to most scientists, and enthusiasm for the new possibilities offered by this technique tends to underestimate the difficulties and the limitations of the technology or of the type of equipment available.

References

1. Caspersson, T. O. and Shultz, J. (1938). *Nature,* **142,** 294.
2. Bernoulli, D. (1738). *Hydrodynamica.* Argentorati.
3. Euler, L. (1755). Principes généraux du mouvement des fluides. *Hist. Acad. Berlin.*
4. Euler, L. (1759). *Novi Comm. Acad. Petrop.,* **15,** 1.
5. Crosland-Taylor, P. J. (1953). *Nature,* **171,** 37.
6. Eisert, W. G. and Dennenloehr, M. (1981). *Cytometry,* **1,** 249.
7. Eisert, W. G. (1981). *Cytometry,* **1,** 254.
8. Arndt-Jovin, D. J., Grimwade, B. G., and Jovin, T. M. (1980). *Cytometry,* **1,** 127.
9. Watson, J. V. (1981). *Cytometry,* **2,** 14.
10. Maxwell, J. C. (1891). *Treatise on Electricity and Magnetism,* 3rd edn. Oxford University Press.
11. Rayleigh, J. W. S. (1871). *Phil. Mag.,* **41,** 447.
12. Mie, G. (1908). *Ann. Physik.,* **25,** 377.
13. Huygens, C. (1690). *Traité de la lumière.* Reprinted as *Treatise on Light* (1962), Dover, New York.
14. Salzman, G. C., Mullaney, P. F., and Price, B. J. (1979). In *Flow Cytometry and Sorting* (ed. M. R. Melamed, P. F. Mullaney, and M. L. Mendelsohn), Chapter 5, p. 105. Wiley, New York.
15. Brunsting, A. and Mullaney, P. F. (1972). *Rev. Sci. Instr.,* **45,** 1514.
16. Mullaney, P. F., Crowell, J. M., Salzman, G. C., Martin, J. C., Hiebert, R. D., and Goad, C. A. (1976). *J. Histochem. Cytochem.,* **24,** 298.
17. Loken, M. R., Parks, D. R., and Herzenberg, L. L. (1977). *J. Histochem. Cytochem.,* **25,** 790.
18. Watson, J. V., Chambers, S. H., and Smith, P. J. (1987). *Cytometry,* **8,** 1.
19. Martin, J. C. and Swartzendruber, D. E. (1980). *Science,* **207,** 199.
20. Watson, J. V. (1987). *Cytometry,* **8,** 646.
21. Sweet, R. G. (1965). *Rev. Sci. Instr.,* **36,** 131.
22. Fulwyler, M. J. (1965). *Science,* **150,** 910.
23. Hulett, H. R., Bonner, W. A., Barrett, J., and Herzenberg, L. A. (1969). *Science,* **166,** 747.
24. Bonner, W. A., Hulett, H. R., Sweet, R. G., and Herzenberg, L. A. (1972). *Rev. Sci. Instr.,* **43,** 404.
25. Herzenberg, L. A., Sweet, R. G., and Herzenberg, L. A. (1976). *Sci. American,* **243,** 108.
26. Bohr, N. (1909). *Phil. Trans. R. Soc. (London),* **A209,** 281.
27. Davies, K. E., Young, B. D., Elles, R. G., Hill, M. E., and Williamson, R. (1981). *Nature,* **293,** 374.
28. Waymouth, C. (1974). *In Vitro,* **19,** 97.
29. Pretlow, T. G., Weir, E. E., and Zettergren, J. C. (1975). In *International Review*

of Experimental Pathology (ed. M. A. Epstein and G. W. Richter), Vol. 14, p. 91, Academic Press, New York.

30. Brattain, M. G., Kimball, P. M., Pretlow, T. G. II, *et al.* (1977). *Br. J. Cancer,* **35,** 850.
31. Kreisberg, J. I., Pitts, A. M., and Pretlow, T. G. II (1977). *Amer. J. Pathol.,* **86,** 591.
32. Willson, J. K. V., Luberoff, D. E., Pitts, A., *et al.* (1975). *Immunology,* **28,** 161.
33. Dow, S. H., and Pretlow, T. G. II (1975). *J. Natl Cancer Inst.,* **54,** 147.
34. Morasca, L., Erba, E., Vaghi, M., Ghelardoni, C., Mangioni, C., Sessa, C., Landoni, F., and Garattini, S. (1983). *Br. J. Cancer,* **48,** 61.
35. Morasca, L., Erba, E., Vaghi, M., Amato, G., Pepe, S., Mangioni, C., Colombo, N., Landoni, F., and D'Incalci, M. (1984). *J. Exp. Clin. Cancer Res.,* **3,** 305.
36. Helmes, S. R., Brazeal, F. I., Bueschen, A. J., *et al.* (1975). *Am. J. Pathol.,* **80,** 79.
37. Pretlow, T. G. II, Dow, S. R., Murad, T. M., *et al.* (1974). *Am. J. Pathol.,* **76,** 95.
38. Pretlow, T. G. II, Jones, J., and Dow, S. (1974). *Am. J. Pathol.,* **74,** 83.
39. Pretlow, T. G. II, Scalise, M. M., and Weir, E. E. (1974). *Am. J. Pathol.,* **74,** 83.
40. Bowman, P. and McLaren, A. (1970). *J. Embryol. Exp. Morphol.,* **23,** 163.
41. Bowman, P. and McLaren, A. (1970). *J. Embryol. Exp. Morphol.,* **24,** 331.
42. Lasfargues, E. Y. and Moore, D. H. (1971). *In Vitro,* **7,** 21.
43. Kraehenbuhl, J. P. (1977). *J. Cell Biol.,* **72,** 390.
44. Bashor, M. M. (1979). *Meth. Enzymol.,* **58,** 119.
45. Ham, R. G. and McKeehan, W. L. (1979). *Meth. Enzymol.,* **58,** 44.
46. Jacob, S. T. and Bhargava, P. M. (1972). *Exp. Cell Res.,* **27,** 453.
47. D'Incalci, M., Torti, L., Damia, G., Erba, E., Morasca, L., and Garattini, S. (1983). *Cancer Res.,* **43,** 5674.
48. Erba, E., Ubezio, P., Colombo, T., Borggini, M., Torti, L., Vaghi, M., D'Incalci, M., and Morasca, L. (1983). *Cancer Chemother. Pharmacol.,* **10,** 208.
49. Meyer, R. A. and Brunsting, A. (1975). *Biophys. J.,* **15,** 191.
50. Gledhill, B. L., Lake, S., Steinmetz, L. L., *et al.* (1976). *J. Cell. Physiol.,* **87,** 367.
51. Erba, E., Vaghi, M., Pepe, S., Amato, G., Bistolfi, M., Ubezio, P., Mangioni, C., Landoni, F., and Morasca, L. (1985). *Br. J. Cancer,* **52,** 565.
52. Crissman, H. A., Mullaney, P. F., and Steinkamp, J. A. (1975). In *Methods in Cell Biology* (ed. D. M. Prescott), Vol. 9, p. 179. Academic Press, New York.
53. Starace, G., Badaracco, G., Greco, C., Sacchi, A., and Zuppi, G. (1982). *Eur. J. Cancer Clin. Oncol.,* **18,** 973.
54. Darzynkiewicz, Z., Traganos, F., and Sharpless, F. (1976). *J. Cell. Biol.,* **68,** 1.
55. Kapuscinski, J., Darzynkiewicz, Z., and Melamed, M. R. (1982). *Cytometry,* **2,** 201.
56. Darzynkiewicz, Z., Traganos, F., Xue, S., Coico, L., and Melamed, M. R. (1981). *Cytometry,* **1,** 279.
57. Beck, H. P. (1982). *Cytometry,* **2,** 170.
58. Steel, G. C., Adams, K., and Barret, J. C. (1977). *Growth Kinetics of Tumours.* Clarendon Press, Oxford.
59. Bohmer, R. M. (1979). *Cell Tissue Kinet.,* **12,** 101.
60. Loken, M. R. (1980). *Cytometry,* **1,** 136.

61. Arndt-Jovin, D. J. and Jovin, T. M. (1977). *J. Histochem. Cytochem.,* **25,** 585.
62. Fried, J., Doblin, J., Takamoto, S., Perez, A., Hansen, H., and Clarkson, B. (1983). *Cytometry,* **3,** 42.
63. Zeile, G. (1980). *Cytometry,* **1,** 37.
64. Colotta, F., Peri, G., Villa, A., and Mantovani, A. (1984). *J. Immunol.,* **132,** 936.

7

Organ culture

ILSE LASNITZKI

1. Introduction

Organ culture means the explantation and growth *in vitro* of organs or part of organs in which the various tissue components, such as parenchyma and stroma, and their anatomical relationship and function, are preserved in culture, so that the explanted tissue closely resembles its parent tissue *in vivo*. The outgrowth of isolated cells from the periphery of the explants is discouraged and minimized by suitable culture conditions, and the 'new' growth is composed of differentiated structures. Thus, in glands, new glandular structures are formed, or, in lung tissue, new small bronchi develop at the periphery of the explant. They consist of alveoli lined with secretory, cuboidal, or columnar glandular or bronchial epithelium.

In tissues lined with squamous epithelium, such as skin or oesophagus, or in bladder lined with transitional epithelium, the epithelium follows a similar pattern of differentiation as in the organs *in vivo*. Hormone-dependent tissues remain hormone sensitive and responsive, and endocrine organs continue to secrete specific hormones. Finally, in fetal tissues, morphogenesis *in vitro* closely resembles that seen *in vivo*.

2. Survey of organ culture techniques

Loeb (1) was the first to culture fragments of adult rabbit liver, kidney, thyroid, and ovary on small plasma clots inside a test tube and found that they retained their normal histological structure for 3 days. His work antedates the introduction of tissue culture by Harrison (2) by 10 years.

Loeb and Fleischer (3) showed that the tube must be filled with oxygen to prevent central necrosis of the explants. Parker (4) also emphasized the necessity for oxygen and grew fragments of several organs in a shallow layer of medium in a flat-bottomed flask filled with 80% oxygen. This method proved unsatisfactory as most tissues sank to the bottom of the flask except skin which floated on the surface of the medium. Skin is not wettable, and this property was used by Medawar (5) to grow slices of rabbit ear skin on a serum–saline mixture in a flask filled with 70% oxygen. However, a disadvantage of growing skin in fluid medium is that the fragments tend to curl up and the epithelium migrates and covers the dermal surface.

It had been recognized earlier that, with the exception of skin, most organ rudiments or organs could be better maintained growing on a solid support than in fluid medium.

2.1 Clotted plasma substrate

Fell and Robison (6) introduced the 'watchglass technique' by which organ rudiments or organs were grown on the surface of a clot consisting of chick plasma and chick embryo extract, contained in a watchglass. This became the classical standard technique for morphogenetic studies of embryonic organ rudiments. The method has later been modified to investigate the action of hormones, vitamins, and carcinogens in adult mammalian tissues (7) and will be described in detail below.

Another type of culture vessel consisted of an embryological watchglass containing a plasma clot and closed with a glass lid sealed on with paraffin wax. This was first introduced by Rudnick (8) and later adopted by Gaillard (9). He used a clot consisting of two parts of human plasma, one part of human placental serum, and one part of human baby brain extract mixed with six parts of a saline solution.

The plasma clot, although it supported the growth and development of fetal and adult organs, had several disadvantages. It usually became liquefied in the neighbourhood of the explants so that they came to lie in a pool of medium. Moreover, because of the complexity of the medium no biochemical investigation was possible.

2.2 Agar substrate

The problems encountered using plasma clots could be eliminated by the use of agar gels. The agar gel technique was first introduced by Spratt (10, 11). Wolff and Haffen (12) modified Gaillard's technique and used an agar gel contained in an embryological watchglass. The agar method has been successfully used for developmental and morphogenetic studies and will, like the watchglass technique, be described in detail below.

Although the agar does not liquefy, it cannot be added or analysed without transplanting the cultures. This disadvantage was overcome by the use of fluid media combined with a support which prevented the cultures being immersed.

2.3 Raft methods

Chen (13) found that lens paper used for cleaning microscope lenses (Gurr) is non-wettable and will float on fluid medium. He explanted 4–5 cultures on a 25 × 25 mm raft of lens paper which floated on serum in a watchglass. Richter (14) improved on this by treating the lens paper with silicone which enhanced its flotation properties. Lash *et al.* (15) combined the lens paper with Millipore filters. They punched a small hole in the centre of the lens paper raft and covered this with a strip of Millipore filter. Different types of tissues were cultured on either side of the filter and their interaction with each other studied. Shaffer (16) replaced the lens paper with rayon acetate. The rayon acetate strips were made to float on the fluid medium by treating the four corners with silicone. The rayon acetate has the advantage over lens paper that it is acetone soluble and can be dissolved during the histological procedures by immersing it in acetone.

2.4 Grid method

The use of rafts floating on a fluid medium did not provide ideal conditions. The rafts often sank and the tissues became frequently, and to different depths, immersed into the medium. This difficulty was overcome by the grid technique devised by Trowell (17). He introduced metal grids, made at first of tantalum wire gauze. This was replaced later by the more rigid expanded metal, obtainable as a continuous sheet of stainless steel (18), or by titanium (19). The grids were square with a surface of 25×25 mm, with the edges bent over to form four legs, and about 4 mm high. Skeletal tissues could be cultured directly on the grid, but softer tissues such as glands or skin were first explanted on strips of lens paper and these deposited on the grids. The grids with their explants were placed in the culture chamber filled with medium up to the level of the grid. The original Trowell technique was aimed at maintaining adult mammalian tissues which have a higher requirement for oxygen than fetal organs. To achieve this, the culture chambers were enclosed in containers which were perfused with a mixture of carbon dioxide and oxygen. The method succeeded in preserving the viability and histological structure of the adult tissues, such as prostate glands, kidney, thyroid, and pituitary. The technique, particularly the application of the gas phase, has since been simplified and in this modified form has been and still is widely used.

2.5 Intermittent exposure to medium and gas phase

More recently, a method which provides intermittent exposure to medium and gas phase has been successfully used for the long-term culture of human adult tissues, including bronchial and mammary epithelium, oesophagus, and uterine endocervix (71–74). In this technique, the explants are attached to the bottom of a plastic culture dish and covered with medium. The dishes are enclosed in an atmosphere-controlled chamber which is filled with an appropriate gas mixture. The chamber is placed on a rocker platform and rocked at several cycles per minute during cultivation.

3. Watchglass technique

The technique was introduced by Fell and Robison (6), originally to study the development of avian limb bone rudiments, but was extended to investigate

Organic culture

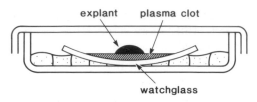

Figure 1. Watchglass technique (6).

growth and differentiation of other avian and mammalian tissues. By this method, avian tissues are placed on a clot, consisting of chick plasma and chick embryo extract in equal proportions, which is contained in a watchglass. One or two such watchglasses are enclosed in a Petri dish carpeted with moist cotton wool or filter paper (*Figure 1*) to prevent evaporation of the clot and explants. The Petri dishes are transferred to an incubator and incubated, usually at 37.5°C

Protocol 1. Preparation of embryo extract

Materials

All reagents and instruments must be sterile.

- fertilized hen eggs incubated for 7–8 days
- Petri dishes
- straight forceps
- curved forceps
- curved scissors
- homogenizer, loose fitting, Potter-Elvejheim, glass with Teflon pestle
- centrifuge tubes, 50 ml (Corning)
- Pasteur pipettes
- wide-mouthed pipettes made by cutting off the tops of ordinary pipettes
- storage vials
- HBSS
- 70% alcohol

Method

1. Before removing the embryos, wipe the blunt end of the eggs with cotton wool soaked in 70% alcohol to sterilize the surface of the shell. Remove the upper end of the shell carefully, by tapping the shell to crack it and peeling outwards, and tear open the inner shell membrane covering the embryo with a pair of straight forceps. Lift embryos out by their heads

216

using a pair of curved forceps and drop them gently into a Petri dish. Wash them thoroughly with HBSS, remove the eyes, and mince the embryos by hand with a pair of curved scissors.

2. Transfer the minced tissue with a wide-mouthed pipette to a homogenizer and grind by hand for a few minutes.

3. Mix the resultant homogenate with an equal amount of HBSS and centrifuge the mixture at 20 g for 10 min at 4°C. Remove the supernatant and if necessary re-centrifuge to remove suspended cell debris. The final supernatant should be free of cells and appear opalescent.

4. The extract can be used immediately or transferred to vials in 1 ml aliquots for storage at −20°C. Use within 3 months.

Protocol 2. Preparation of chicken plasma

Commercial preparations of plasma are not recommended as they usually do not produce a firm clot and may not support the same degree of growth and differentiation as freshly prepared plasma.

Materials
- domestic chicken 6–12 months old
- syringe, 30 ml; hypodermic needles
- centrifuge tubes
- heparin

Method
1. Prepare chicken plasma by the standard technique described by Paul (74). To lower the lipid content of the plasma, starve the animals for at least 24 h before bleeding.

2. To prevent premature clotting of the plasma, siliconize all glassware and keep cold, rinse the collecting syringe with a solution of heparin (0.1–1%). The amount of heparin should be kept as low as possible to permit clotting of the plasma when mixed with embryo extract.

3. Test plasma for sterility and store in siliconized tubes at −20°C. Use within 6 months.

Protocol 3. Explantation

Materials
- watchglass 3.5 mm in diameter
- Petri dishes 91 mm in diameter

Protocol 3. *Continued*

- cotton wool or filter paper to fit Petri dishes
- sterile distilled water or saline solution
- graduated pipettes 10 ml
- Pasteur pipettes
- wide-mouthed pipettes prepared by cutting off tips of Pasteur pipettes
- extra-fine pipettes prepared by drawing out Pasteur pipettes to a fine point in flame
- pipette rack (*Figure 2*).
- cavity slides 6–7 mm deep
- glass slides approximately 8 cm^2
- chicken plasma
- chick embryo extract (50%)
- heat-inactivated (56°C for 30 min) sera; rat, horse, calf, for use with mammalian tissues
- glass rods for mixing plasma and extract
- HBSS
- cataract knives, Swann-Morton blades, No. 11
- dissecting needles
- watchmaker's forceps, straight forceps 10 cm
- medium scissors
- fine scissors for dissection

Method

1. Cover the bottom of the Petri dish with a layer of cotton wool, or preferably, three layers of filter paper which have one or two 3.5 cm holes punched into them, and sterilize by dry heat.
2. On the morning of explantation spread 10 ml of sterile distilled water or HBSS with a graduated pipette on cotton wool or filter paper and place the watchglass over the punched holes. The holes will allow transmission of light for a dissecting microscope and facilitate macroscopic observations of the explanted tissues during cultivation.
3. For the cultivation of avian tissues, prepare clots by mixing 10 drops of chicken plasma and 10 drops of chick embryo extract (6). Stir the mixture quickly and thoroughly with a glass rod.
4. For mammalian tissues, heat-inactivated rat, horse, or calf serum is added to chicken plasma and chick embryo extract, usually in a proportion of chicken plasma:serum:extract of 2:1:1, but this may be modified accord-

ing to the particular tissue used and the problem to be investigated. In experiments in which the effect of chemical agents on growth and differentiation is studied, the compounds to be investigated are incorporated into the mixture before clotting (25).

5. Using fine scissors and watchmaker's forceps, remove organs or part of organs, rinse them with HBSS, and transfer them to a glass plate. Using cataract knives, dissection needles, or Swann-Morton blades, trim them and cut them to an appropriate size, not exceeding $2 \times 2 \times 2$ mm. Place the glass plate on a piece of dark paper, to provide a good background for the tissues and facilitate dissection. It is essential that during dissection the tissue is kept moist with HBSS.

6. Transfer one to three explants with cataract knives or a wide-mouthed pipette to the clot. In the latter case, the explants are sucked up with HBSS and deposited on the clot and the excess fluid is sucked off with the extra-fine pipette.

7. Transfer the Petri dishes containing one or two watchglasses to an incubator kept at 37.5°C. Do not place Petri dishes directly on the metal shelves but interpose a piece of wood. This will prevent condensation of water on the inner side of the Petri dish lid which may drop on the explants and damage them.

The medium has to be renewed every 2–3 days for avian tissues and every 3–4 days for mammalian tissues. For this purpose, fresh clots are prepared as described above.

Protocol 4. Renewal of medium

1. Gently detach the explants from the old plasma clot and lift off with the aid of two cataract knives.

2. Transfer them to, and wash them in, HBSS containing 2% serum contained in the cavity of a cavity slide.

3. After washing, suck the explants up with a wide-mouthed pipette and deposit them on the fresh plasma clot.

4. Separate and spread them out on the clot using a dissecting needle and cataract knife.

5. Suck off the excess fluid with the extra-fine pipette.

3.1 Combination of clot and raft techniques

In this modification, the explants are first placed on strips of lens paper or rayon acetate moistened with HBSS and these strips deposited on the clot.

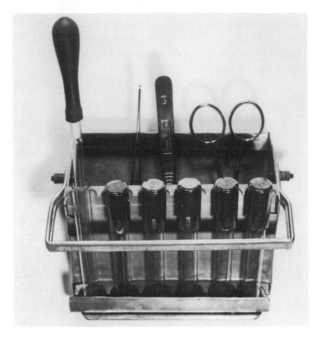

Figure 2. Stainless steel rack to hold Pasteur pipettes and dissecting instruments.

This method does not impair the viability or growth of the explanted tissues. It prevents the tissue from sinking into the pool of liquefied plasma and simplifies the medium changes. Instead of lifting off the explants individually, the rafts carrying the explants can be lifted *in toto*, washed with HBSS, drained on pieces of sterile filter paper, and then placed on the fresh clot.

The watchglass technique has proved eminently suitable for developmental studies on avian bone rudiments (6, 22, 23), and for studies on differentiation of avian epidermis and oesophageal epithelium and its modulation by vitamin A (24, 25). Further, it has proved very useful for investigating the action of carcinogenic agents, steroid hormones, and vitamin A and their interaction in embryonic and adult, human, murine, and rat tissues. Using this technique it could also be demonstrated that steroid hormones were necessary for the maintenance of their target organs and that carcinogenic agents induced precancerous changes which could be inhibited by vitamin A (26–28).

The technique has certain limitations. The explants often liquefy the clot and sink into a pool of liquefied medium which may impair their oxygen supply. Consequently, the medium has to be changed frequently. The complexity of the medium rules out biochemical analysis, and the duration of culture is somewhat limited. Although in many experiments the answer to a particular problem may be obtained within a short time, others may require a

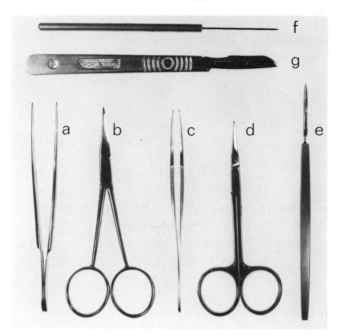

Figure 3. Dissecting instruments. (a) medium size straight forceps, (b) medium-size straight scissors, (c) watchmakers' forceps, (d) smaller fine scissors, (e) cataract knife, (f) dissecting needle, (g) Swann-Morton knife.

longer period of cultivation. Using the plasma clot technique the period of cultivation has not usually exceeded 4 weeks (29).

4. The Maximow single slide technique

This is a modification of Fell and Robison's watchglass technique (6). It is derived from the double cover slip technique by which cells were grown on a plasma-extract clot (30) and it has been successfully adapted by Hardy (31) for developmental studies in fetal organs or organ rudiments (*Figure 4*). Hardy recommends the methods for workers on a limited budget and with limited facilities and when the main effects to be examined are changes in morphogenesis or ciliary and secretory activity. Any changes can be observed by conventional light microscopy or by polarized or u.v. light microscopy.

plasma clot

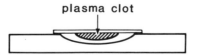

Figure 4. Maximow single slide technique (35).

Protocol 5. The single slide technique

Materials

In addition to those described in the previous protocols, the following are required:

- microconcavity slides 75 × 45 mm, 6–7 mm deep
- square cover slips, No. 3, 40 × 48 mm
- cover slip forceps
- safety razor blades, and a scalpel
- Vaseline and Paraplast: four parts of Paraplast mixed with one part of Vaseline, for sealing cover slips to the chambers
- fine natural bristle paint brush, for applying the Paraplast–Vaseline mixture

Setting-up of cultures

1. Using fine knives and dissecting needle, prepare organs or organ rudiments to pieces of appropriate size (not exceeding 2 mm × 1.5 mm × 1 mm) in HBSS.
2. With cover slip forceps, place 40 × 48 mm cover slips into Petri dishes.
3. With a Pasteur pipette place three drops of chicken plasma into the centre of each cover slip. Stir the plasma with a glass rod to cover a circle of 25 mm.
4. With a second pipette add one drop of chick embryo extract (50%) to the plasma, mix the extract and plasma thoroughly and quickly with the stirring rod.
5. When the mixture has clotted, place 1–3 explants on the clot, usually with the stromal surface attached to the clot.
6. Transfer the explants with a scalpel blade or with the use of a wide-mouthed pipette. In the latter case, suck off the explant and HBSS and deposit on the clot. Remove the excess HBSS with the extra-fine pipette.
7. Invert the cover slips carrying the clot and explants with a cover slip forceps over a Maximow slide. Seal the chambers with three coats of the Paraplast–Vaseline mixture, heated to 60°C.
8. Incubate the slide chambers at the desired temperature.

Renewal of medium

1. Prepare fresh clots according to the method described above. Remove the paraffin seal from the slide chamber with a warmed scalpel or razor blade.

2. Invert the cover slip on the stage of a dissecting microscope. Using fine forceps and a dissecting needle remove explants from the clots and transfer them to the slide well for rinsing with HBSS (two changes, 5 min).

3. Transfer each explant to fresh clot. Invert the cover slip over the Maximow slides and seal chambers as before.

Hardy (31) found the method suitable for studies on hair growth and differentiation of fetal mouse skin (32), for the study of hormonal effects on newly born mouse vagina (33), and the differentiation of fetal mouse gonads (34).

The method was less suitable for highly motile tissues such as fetal intestine and highly secretory organs. Although the method allows examination by light microscopy of the living explants, the optics are not perfect and the tissues are usually too thick for examination by phase contrast.

5. Agar gel techniques

These techniques eliminate some of the disadvantages of the plasma clot, avoiding clot liquefaction and allowing the use of defined media. The original technique was introduced by Spratt (10) who used a 3% Ringer solution albumen gel and later a 4% gel consisting of bicarbonate buffered saline solution, glucose, 11 amino acids, 10 vitamins, and agar (11). Wolff and Haffen (12) modified Gaillard's technique (9) to study development and differentiation of amphibian, avian, and mammalian organs or organ rudiments. They used an agar gel contained in an embryological watchglass (*Figure 5*). Initially, their medium consisted of 1% agar, dissolved in Gey's solution mixed with 50% chick embryo extract and Tyrode solution. In later versions, chicken or horse serum was added to the agar, or the embryo extract was replaced by certain amino acids and vitamins (35).

Wolff (36) found that fetal organ rudiments often became encapsulated by cells migrating from the cut surface of the explants. To prevent this he wrapped them in a piece of vitelline membrane obtained from hen eggs.

The method has also been applied to studies of tumour growth. Wolff *et al.* (37) succeeded in growing cells from human tumour cell lines or fragments of fresh human tumours on agar clots associated with chick mesonephos or on clots reinforced with chicken liver dialysate.

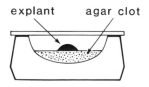

explant   agar clot

Figure 5. Embryological watchglass with agar clot (12).

The method has been and still is widely used, although the composition of the agar gel has undergone some changes and some workers have replaced the saline solutions used to dissolve the agar with defined media.

In a later section a modification of the technique using such defined medium is described.

Protocol 6. Embryological watchglass technique

Materials

- embryological watchglasses, outer diameter 3.9 cm, inner diameter 3.5 cm
- glass lid 3.9 cm in diameter to cover watchglass
- paraffin wax heated to 60°C to seal lid on watchglass
- small natural bristle paint brush to seal lid with paraffin agar 2% (Gibco)
- double-strength Morgan, Morton, and Parker's medium 199 based on Hanks' salts
- single-strength medium 199, based on Hanks' salts
- newborn or fetal bovine serum, inactivated at 56°C for 30 min
- HBSS
- Pasteur pipettes
- wide-mouthed pipettes
- extra-fine Pasteur pipettes made in the laboratory by drawing out Pasteur pipettes to a fine point in flame
- pipette rack
- cataract knives
- watchmaker's forceps, fine curved scissors, dissecting needles
- glass plates, approximately 8 cm square, for dissecting organs

Preparation of agar gel

1. Heat 2% agar in a water bath kept at 90°C until liquefied, then mix quickly in a glass tube with equal amounts of double-strength medium 199, kept at 37°C.

2. When thoroughly mixed, place 0.5 ml in the bottom of the watchglasses, add 0.1 ml fetal bovine or newborn calf serum and 0.4 ml single-strength medium 199, again mix thoroughly and quickly to avoid premature gelling. The mixture begins to gel at 36°C and is firmly set at 30°C.

If a smaller clot is desired the proportions of agar, serum, and medium 199 can be modified accordingly. In experiments which involved a study of hormonal action or that of other agents, the compounds to be investigated are incorporated into the gel by adding them to the single-strength medium 199.

Setting-up of cultures

1. Remove the organs or organ rudiments from the embryos with the aid of a dissecting microscope. During dissection, keep the embryos moist with HBSS containing 2% serum.

2. Transfer the tissues to be cultured and wash them in HBSS contained in a cavity slide.

3. Transfer them to the agar gel, either with the aid of two cataract knives or by gently sucking them up in a wide-mouthed pipette with HBSS, and deposit them on the agar.

4. Orientate the explants on the agar with two needles and suck off the excess fluid with a fine pipette. Usually, one explant is accommodated on each gel. The optimum size of the explant is $1.5 \times 1.5 \times 10$ mm, the maximum size $2 \times 2 \times 1.5$ mm. However, if the tissue is very thin, say less than 0.5 mm thick, the other dimensions can be larger.

5. Using the small paint brush, seal the glass lids on to the chambers with at least two coats of warm paraffin wax ($60\,^\circ$C). Transfer the chambers to the incubator and incubate at $37\,^\circ$C or any other desired temperature. It is advisable not to place them directly on the metal shelves but to insulate from the metal with a polystyrene foam tile.

Many experiments involving studies of morphogenetic changes and cyto-differentiation do not require a lengthy culture period and the answer to the question posed may be available within 3–7 days. In this case, a change of medium is not necessary. If the experiment requires a longer period of cultivation, the explants have to be moved to a fresh agar gel every 5–7 days. A fresh gel is made in a fresh chamber, prepared with the same proportion of agar, serum, and defined medium as before.

Remove the lids from the old embryological watchglasses with a warmed razor or scalpel. Lift the explant off the old gel with the aid of two cataract knives or one cataract knife and a dissecting needle, wash in HBSS contained in a cavity slide, transfer to the fresh gel, and seal the watchglasses with a fresh sterile glass lid, using a small paint brush and paraffin wax.

The viability of the explants and macroscopic changes can be monitored during cultivation by viewing them by daylight or with the aid of a light source from the dissecting binocular. Healthy and growing tissues usually appear translucent with a shiny surface; opacity suggests loss of viability or the beginning of necrosis of the explanted tissue.

This modification, in which the saline solution was replaced by defined medium, has been found suitable for studying the induction by testosterone of the prostate gland from its *anlage*, the urogenital sinus, from mouse or rat embryos, and the role of mesenchyme in this process (38, 39). The agar supported the development of the prostate gland from the whole sinus, as well as that of its components.

Agar is also used for the preparation of combination cultures to study the interaction of various tissue components, in particular that of mesenchyme and epithelium during the development of fetal organs. Using this method it has been found that the mesenchyme plays an essential role in the differentiation of the epithelium, and that in hormone-dependent organs it mediates the effect of the hormone on the epithelium (39). The result was obtained by using cultures in which the mesenchyme and epithelium of fetal organs were separated and re-associated in various recombinations. The method can be used in developmental studies of many organs, but here its application by Mizuno (39) to the study of prostatic development is described.

Protocol 7. Combination cultures

1. Use 15–18 day embryos of mouse (15) or rat (16–18).

2. Kill pregnant animal and remove uterus.

3. Wash it well with cold HBSS to which 100 U/ml penicillin G have been added, and transfer to Petri dish.

4. Remove embryos from uterus with fine scissors, wash them well with cold HBSS, and transfer them to second Petri dish.

Preparation of dissection dish

1. First prepare a Petri dish with black paraffin (*Figure 6*). Melt paraffin in a beaker at 60°C, add activated charcoal powder, and mix well with a glass rod.

2. Pour mixture into the Petri dish and leave at room temperature overnight.

3. Wrap it in paper and dry sterilize it for 2 h at 120°C.

4. Let it cool until paraffin is set.

5. Transfer fetus with drilled spatula to black paraffin and secure it there with fine pins.

Dissection of urogenital sinus

1. Remove the urogenital sinus with its appendages of urinary bladder, urethra, and genital tubercle (*Figure 7*) from the fetus with a pair of fine forceps and scissors (Pascheff-Wolff, Moria MC 19) under a dissecting microscope (magnification × 10) and transfer the whole to cold HBSS in a Petri dish.

2. Remove urinary bladder, urethra, genital tubercle, and blood vessels with two fine forceps under the binocular (magnification × 10).

Separation of epithelium and mesenchyme by treatment with collagenase

1. Immerse sinuses in 0.03% collagenase (CLS grade, Worthington Biochemical Co.) in BSS in an embryological watchglass at 37°C for 20, 30, or 40 min depending on the size of the sinuses.

2. Transfer them to BSS in a second embryological watchglass and separate the epithelium (E) from the mesenchyme (M) with two very fine forceps under the binocular (magnification × 16) (*Figure 8*).

Inactivation of collagenase

Wash epithelium and mesenchyme in several changes of medium 199 and FBS in equal proportions for 2 h, and again in medium 199.

Recombination

1. Place the isolated mesenchyme on to the surface of an agar gel with the aid of a transplantation spoon (Moria MC 10), and forceps under the binocular (magnification × 10).

2. Place the isolated epithelium on to the mesenchyme in such a way that the basal lamina of the epithelium becomes attached to the upper surface of the mesenchyme.

3. Remove excess saline with small pieces of filter paper under binocular (magnification × 16, *Figure 9*).

4. Cover embryological watchglass and seal it with paraffin wax.

5. Incubate it for at least 2 h at 37°C to ensure the coherence of the two tissue fragments.

6. Transfer recombinates to strips of Millipore filter, 0.5 μm pore size (Millipore) and place these either on freshly prepared agar gels in embryological watchglasses and seal these with glass lids and two coats of paraffin wax, or by a modified Trowell method on grids of expanded metal (Expanded Metal Co.) (18 mm square, 4 mm high cut out from a larger piece with a pair of scissors) resting in a small culture dish filled with medium. Place two such culture dishes in one Petri dish.

Medium

• agar gel as described previously or medium 199 (based on Earle's salts)

• 5–7.5% heat inactivated newborn calf serum, 100 U/ml of penicillin G

1. Culture for a period of 4 days, incubating at 37.5°C.

2. Prepare serial sections for light microscopy and stain these by standard techniques. If positive, projections of epithelial buds into the mesenchyme can be recognized.

5.1 Development of embryonic mouse molars *in vitro* using flotation techniques

Various authors have studied the development of tooth germs using a type of Trowell organ culture (40–42). However, although suitable media and gas

Figure 6. Black paraffin dish for dissection of early mouse or rat embryos.

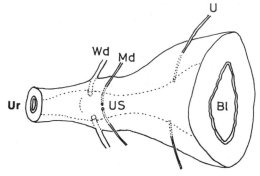

Figure 7. Rat urogenital sinus Bl = urinary bladder; Ur = urethra; Wd = Wolffian duct; Md = Mullerian duct; U = genital tubercle.

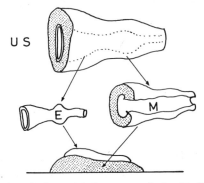

Figure 8. Rat urogenital sinus before and after removal of epithelium (E) and mesenchyme (M) and association of both tissues.

228

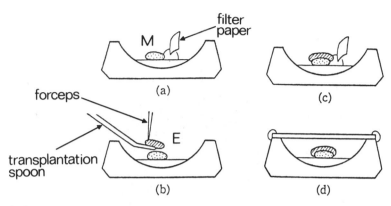

Figure 9. Procedure to combine epithelium with mesenchyme on agar clot.

tension were used, the histological structure of the developing tooth germs was frequently impaired.

Sakakura (43) devised an ingenious flotation method which ensured three-dimensional growth while maintaining the histological structure of the tooth germs. He used a small glass dish and an agar chamber to float the molars to the gas medium surface, where they were maintained by the surface tension of the medium. The molars developed in three dimensions, and it was easy to determine their anterior–posterior and buccolingual orientation. Normal cytodifferentiation of ameloblasts and odontoblasts occurred and a thick layer of enamel was formed.

Protocol 8. Flotation method for growth of tooth germs

Tooth germs. Mandibular first molar tooth germs of 17-day mouse embryos were removed and stored in HBSS solution within 1 h prior to explantation. At the start of explantation the embryonic molars represented the cap stage of tooth morphogenesis, and no dental matrix was detected at the epithelio-mesenchymal interface.

Medium
- DMEM (see Appendix A2) supplemented with 10% heat-inactivated (56°C for 30 min) calf serum (Flow Laboratories), 1m M L-proline, 150 μg/ml L-ascorbic acid, 100 μg/ml streptomycin sulphate, and 100 U/ml penicillin G
- gas mixture: 50% oxygen, 5% carbon dioxide, 45% nitrogen

Culture method
1. Prepare an agar plate by pouring 1.5% agar (Agar Noble, Difco) into a 90 mm Petri dish. Cut an agar block about 12 mm square and 8 mm thick and place it in a 22 mm glass dish. *Figure 10* illustrates the various steps of the original flotation method.

Protocol 8. *Continued*

2. Make a deep depression in its centre.

3. Place the embryonic molar at the bottom of this depression.

4. Fill the space between the agar and the culture dish with medium.

5. Transfer dish to incubator perfused with the gas mixture (see above).

6. With the flow of medium into the depression through the agar the explants float up until equilibrium between medium levels inside and outside the agar chamber is reached. The tooth is then maintained at the level of the gas medium interface by the surface tension of the medium.

7. Cultivate tooth germs for 10 days and change medium every second day.

Recently, Sakakura (44) introduced an improved and simplified flotation technique with the use of disposable 24-multiwell dishes (Nunc) and plastic culture chambers (Millicell-HA PINAO 200, Millipore) containing a membrane filter. The inner surface of the filter is coated with 1% agar, one or two drops per culture chamber, to prevent the molars from sticking to the filter. The chambers are then inserted into the multiwell dishes, and the space between the culture chamber and the multiwell dish is filled with culture medium, about 1 ml per well (*Figure 11*). The explants gradually float up with the slow inflow of medium into the culture chamber through the filter and

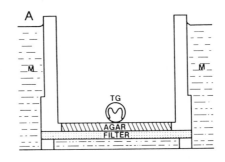

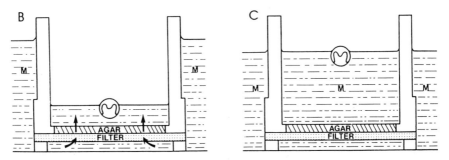

Figure 10. Flotation technique for cultivation of tooth germs (43).

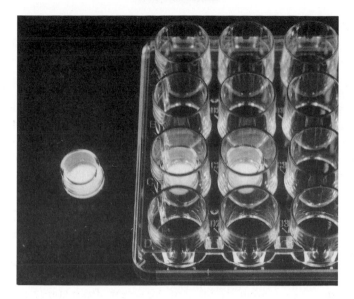

Figure 11. Simplified flotation technique using multiwell dishes and plastic chambers (44).

agar layer. The explant is maintained at an equilibrium level of the medium inside and outside the culture chamber by the surface tension of the medium as with the original flotation technique. The medium is renewed every 2 days, and the molars are cultivated for a period of 10 days. The same gas mixture is used.

6. Embryo culture for teratological studies

New (45, 46) and Cockcroft (47) have developed a method by which whole mouse and rat embryos can be grown in culture. As very early pre-implantation embryos have a great capacity for recovery (44) they are less suitable for teratological studies than older embryos, which are more suscept-ible to the influence of toxic agents.

New and Cockcroft chose, therefore, embryos of two later stages of gesta-tion: 9.5 days and 11.5 days. The 9.5 day embryos consist of a pre-somite cylinder, 1.5 mm long and 0.6 mm in diameter. The 11.5 day embryo has 27–31 somites and is 3–3.8 mm in length

---

Protocol 9. Explantation procedure

1. Remove the pregnant uterus, open in saline solution along the antimeso-metrial side, and cut away the conceptus with a cataract knife.
2. Transfer to fresh saline, remove the maternal decidua, and open Reichert's membrane. The ectoplacental cone, or the allantois in older

Protocol 9. *Continued*

embryos, is left attached to the embryos which are now ready to be placed into their culture chambers. Two types of chamber have been used:

- *cylindrical 60 ml bottles*, filled partially with medium and partially with a gas mixture, laid horizontally on rollers and rotated at 30–60 r.p.m. during incubation to ensure maximum and even exposure to the agent under investigation; *or*
- a *rotator* consisting of small culture bottles attached to a hollow rotating drum. This provides a constant level of carbon dioxide and oxygen with a minimum variation of pH of the medium.

Gas mixture

- for embryos explanted at 9.5 days' gestation: 5% carbon dioxide, 5% oxygen, and 90% nitrogen for the first 24 h; subsequently 5% carbon dioxide, 20% oxygen, and 75% nitrogen
- for embryos explanted at 11.5 days' gestation: 5% carbon dioxide and 95% oxygen throughout the culture period

Medium

- rat serum from ether-anaesthetized rats

i. Centrifuge the blood immediately, before it has formed a clot.

ii. After centrifugation, gently squeeze the clot formed in the supernatant to release the serum underneath.

iii. Re-centrifuge serum, decant, and store at $-10°C$.

iv. Before use, thaw, heat-inactivate at $56°C$ for 30 min, gas to remove any residual ether, and centrifuge to remove any further fibrin clots.

Protocol

1. Place six 9.5 day embryos in each 60 ml serum.

2. Explant three 11.5 day embryos per bottle in 9 ml serum.

At both ages, they are incubated at 37–38°C. The embryos are explanted either floating freely in the serum or placed on collagen-coated fabric attached to a cover slip.

A high proportion of the explanted embryos develop well in culture. The younger embryos form 25–28 somites and their protein content rises significantly, close to the pattern of development seen *in vivo*. Embryos explanted at 11.5 days' gestation develop at a slightly slower rate than *in vivo*, but they also show increased somite number and protein content. Embryos anchored in position can be inspected during cultivation and their development followed. At the same time, the influence of toxic agents on normal development can be monitored. In this way, the effect of carbon dioxide, excess glucose, and hyperthermia has been successfully studied (46). The method in *Protocol 9* is equally suitable for investigating the influence of other toxic or suspect agents on normal development.

7. Grid technique

This method combines the use of fluid defined or semi-defined media with a firm support of the explanted tissues which prevents them being submerged into the medium and thus becoming deprived of oxygen. The importance of oxygen for the growth of adult tissues was recognized by Trowell, who designed an apparatus in which cultures explanted on a grid were enclosed in a chamber under an atmosphere of carbon dioxide and oxygen (17–19). As currently used, the metal grid for supporting cultures in fluid medium is basically unchanged but the original gassing chamber introduced by Trowell has been considerably modified and simplified.

This modified Trowell technique is now widely used for embryonic and adult tissues and is described here (*Figures 12* and *13*).

Several media have been examined as to their efficacy in maintaining structure and function of the explanted organs. These include Ham's F12 medium, Waymouth's medium MB 752/1, medium CMRL 1066, medium 199, and Trowell's T8 medium.

Trowell's medium seems suitable only for relatively short-term maintenance of adult tissue. It contains a high concentration of insulin (50 µg/ml) which may interfere with the action of hormones and mask the effects of other growth factors to be studied, so it is unsuitable.

Medium 199 with 15% NBCS 1.25 µg/ml Insulin

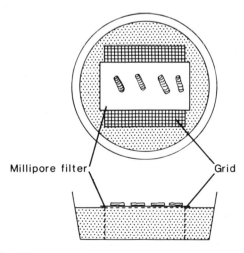

Millipore filter   Grid

Figure 12. Modified Trowell technique.

Figure 13. Modified Trowell technique: Petri dish with culture chambers, grids and Milli-pore filter strips.

As no basic difference has been observed between other media, medium 199 is used routinely, either alone or supplemented with serum and small amounts of insulin as follows.

Medium 199 based on Earle's salts:

- heat-inactivated newborn calf serum, 5% for hormonal studies, 10–15% for studies on carcinogenesis
- insulin 0.5–1.0 μg/ml
- antibiotics and antifungal agents: penicillin 250 U/ml
- streptomycin 100 μg/ml, fungizone 1.4 μg/ml

Protocol 10. Explantation procedure: preliminaries

Materials
- sterile distilled water or HBSS
- Petri dishes, 91 mm in diameter
- round sterile filter papers to fit into the bottom of the Petri dishes with two 3.5 cm holes punched in them.

- plastic culture chambers 32 mm in diameter (Sterilin)
- lens paper (Gurr), degreased in ether and alcohol before use (see Section 2.3 for method of preparing lens paper)
- Millipore filter 0.5 μm pore size (Millipore) cut into strips measuring 1 × 2.2 cm before sterilization

Note: it may be worthwhile trying polycarbonate membranes (Nuclepore, Sterilin) which are transparent and detergent free, instead of Millipore filters. Explants grown on such membranes could be inspected during cultivation. They may be particularly suitable for the cultivation of small whole organs such as fetal mammary glands which could be fixed, stained, and examined on the membrane, thus eliminating sectioning.

- grids of expanded metal (Expanded Metal Co.), 18 mm square, 4 mm high, cut out from a larger sheet with a pair of scissors. The ends are bent down to form a bridge or table. After use, immerse grids in concentrated nitric acid and leave for 24–48 h. Remove grids from acid and transfer them to running tap water for at least 24 h. Rinse in three changes of distilled water and dry in drying oven. The nitric acid can be used several times.
- medium 199 based on Earle's salts
- medium 199 based on Hanks' salts (holding medium)
- insulin (Sigma)
- newborn calf serum, heat-inactivated
- universal containers, glass or plastic (Sterilin), 20 ml capacity
- bijou bottles, glass, 5 ml capacity
- graduated pipettes
- Pasteur pipettes
- pipette rack
- penicillin (Glaxo)
- streptomycin (Glaxo)
- fungizone (Flow Laboratories)
- 2 straight forceps, approximately 10 cm long
- 2 straight scissors, approximately 10 cm long
- 2 straight fine scissors
- 2 watchmaker's forceps, No. 3 or 4 (Weiss)
- 2 cataract knives
- 2 Swann-Morton blades No. 10, degreased with absolute ethanol before use
- glass cavity slides
- glass plate, approximately 8 cm square, for dissection
- 70% alcohol

Protocol 10. *Continued*

- stainless steel rack
- anaerobic Macintosh and Fildes Jar (modified)
- gas cylinder containing 5% carbon dioxide and 95% oxygen

Preparation of lens paper for culture

Lens paper is normally covered with a fine layer of grease which has to be removed before it can be used for culture.

1. Cut lens paper into pieces approximately 5×8 cm and put them in a Petri dish.
2. Immerse in ether for 1 h, change ether, and leave for a further hour.
3. Take ether off and immerse into absolute ethanol for 1 h, change absolute ethanol, and leave for a further hour. Take absolute ethanol off and immerse into glass-distilled water overnight. Dry in drying oven (37 °C).
4. Before use, cut into strips of required size and sterilize with dry heat or by autoclave.

Protocol 11. Explantation procedures: general (1)

1. Assemble the apparatus necessary for cultivation and prepare the media before dissecting the organs.
2. Moisten three layers of filter paper inserted into each Petri dish before sterilization with 10 ml of sterile distilled water or HBSS.
3. Place two plastic culture chambers into each dish over the holes punched into the paper, and place the expanded metal grids into the chambers.
4. Fill another Petri dish with holding medium 199 (based on Hanks' salts) containing 2% newborn calf serum for washing and keeping the tissues moist before and during dissection.

The exact procedures vary with different tissues and those used for the explantation of prostate gland, trachea, and skin are described in some detail in *Protocols 12, 13*, and *14*. The description of general procedures is resumed in *Protocol 15*.

Protocol 12. Prostate gland

This is derived from mice or rats, 6 weeks to 6 months old.

1. Secure the animals with map pins to a cork mat and thoroughly wash their fur with 70% alcohol.

2. Open the abdominal skin with a pair of straight scissors with one and two cuts. Secure the resulting skin flap with a map pin, leaving the area around the bladder exposed.

3. Lift up the bladder with a pair of watchmaker's forceps and cut its adhesion to the underlying muscle. Turn it over to reveal the prostate gland underneath. The gland is distinguishable from the surrounding tissues by its shiny surface and mother-of-pearl colour.

4. Lift the gland up with a pair of watchmaker's forceps and remove it from the bladder with a pair of fine scissors.

5. Rinse the gland with ice-cold holding medium, transfer to a cavity slide filled with the same medium, and keep on ice until required for dissection. At least six glands are required for one group of experiments.

6. Transfer the glands to a glass slide for dissection.

7. First free them from the surrounding fat and connective tissue with a pair of Swann-Morton knives.

8. Separate the two lobes of the gland by cutting through the connective tissue between them.

9. Then subdivide them, not by cutting them as this would result in regenerative hyperplasia at the cut edges, but by teasing them gently apart into smaller lobules. The small lobules measuring $2 \times 2 \times 2$ mm are now ready for explantation.

    Care should be taken to keep the tissue moist during dissection, otherwise it will dry out and stick to the glass.

Protocol 13. Trachea

This is obtained from neonatal mice or rats, 2–5 days old.

1. Secure the animals on a small cork mat with pins, and wash thoroughly with 70% alcohol.

2. Open the skin and rib cage with a pair of fine scissors and expose the trachea by pulling the ribs apart. The trachea is easily recognized by its cartilage rings.

3. Loosen it with the aid of two watchmaker's forceps. One holds the organ while the other slides underneath it and separates it from the underlying oesophagus by moving the forceps to and fro.

4. When this is accomplished, lift it up and remove from the rat with a pair of fine scissors and watchmaker's forceps.

5. Rinse the organ with ice-cold holding medium, deposit in a cavity slide filled with the same medium, and keep on ice until further dissection.

Protocol 13. *Continued*

6. Before explantation, transfer the tracheas to a glass slide, trim at the proximal and distal ends, and halve with the aid of two Swann-Morton knives. The resulting halves measure approximately 1.5–2 mm in length and are explanted as tubes.

Protocol 14. Skin

This is usually obtained from 17–18 day mouse or rat embryos.

1. Secure the embryos on a small cork mat with pins and moisten their skin with cold holding medium.
2. With the aid of a pair of watchmaker's forceps and a pair of fine scissors remove the dorsal skin entirely, rinse with cold holding medium, and transfer to a glass slide. Once removed from the embryo the skin contracts, tends to curl up, and has to be straightened repeatedly during dissection.
3. Scrape the underside free of fat, if present, so that the skin now consists of two main layers: dermis and epidermis.
4. Subdivide it with the aid of two Swann-Morton knives into fragments measuring $3 \times 3 \times 4$ mm.

The explanation of skin is a little difficult as it tends to curl up. This can be overcome by pressing a moistened strip of Millipore filter on to the dermal side of the skin explants. This makes them flatten out and adhere tightly to the substrate. The strips are inverted before placing them on the grids so that the skin is growing with the epidermis uppermost.

Protocol 15. Explantation procedures: general (2)

1. Place the strips of lens paper or Millipore filter on a glass slide or the bottom of a Petri dish, and moisten with 1–2 drops of holding medium.
2. With the aid of two cataract knives, transfer 6–8 fragments of prostate gland or four halved tracheas to the strips; lens paper for adult prostate or Millipore filter for neonatal trachea. If lens paper is not available, Millipore filter can be used for the prostate glands as well.
3. Transfer the strips holding the explants to the grids in the culture chambers with a pair of watchmaker's forceps. Fill the chambers with approximately 3 ml of medium, which should reach the upper level of the grids and keep the surface of the explants moist without submerging them.

Compounds to be investigated can be added to the medium, at the desired concentration, at this stage. Alternatively, they can be incorporated into the medium first and the mixture kept at $-20\,^{\circ}C$ until used. This latter method ensures a tighter control of experimental conditions and is more labour saving.

Protocol 16. Incubation and perfusion

1. Stack the Petri dishes holding two culture chambers each in a rack of stainless steel and place into a Macintosh and Fildes anaerobic jar (Fisons) (*Figure 14*). The attachments meant for anaerobic control are discarded, and the lid fitted with two valves.

2. Place the jar in the incubator and connect to the gas cylinder with silicone rubber tubing running through a small opening at the top of the incubator.

3. Perfuse the jar with a mixture of 5% carbon dioxide and 95% oxygen via sterile filter. For cultures of adult prostate perfuse for 25 min at a flow-rate of 150 ml/min; for neonatal trachea and skin, for 20 min at a flow-rate of 125 ml/min. Check the flow-rate and quantity of the gas mixture with the use of a flow meter.

4. Seal the jar tightly so that the concentration of the gas mixture inside the jar remains constant until it is opened for medium renewal.

5. Incubate the cultures at $37.5\,^{\circ}C$.

A mixture of 5% carbon dioxide and 95% oxygen has been found suitable for the cutlivation of prostate glands, trachea, and skin, and prevents the central necrosis of the tissue which may occur if it is cultured in air. However, in other tissues, a lower proportion of oxygen may be as effective as long as its concentration is above that in air.

Protocol 17. Renewal of medium

The medium is renewed every 2–3 days.

1. Remove the Petri dishes holding the cultures from the Macintosh jar.

2. Suck off the old medium with a Pasteur pipette and replace it with the same quantity of fresh medium. The culture chambers can remain in the same Petri dish if the period of culture is relatively short. For prolonged cultivation, it is recommended to transfer the chambers to a new set of sterile Petri dishes, moistened with sterile water or HBSS.

3. Following the change of medium, return the Petri dishes holding the culture chambers to the Macintosh jar.

4. It should again be tightly sealed and re-perfused with the same gas mixture as before.

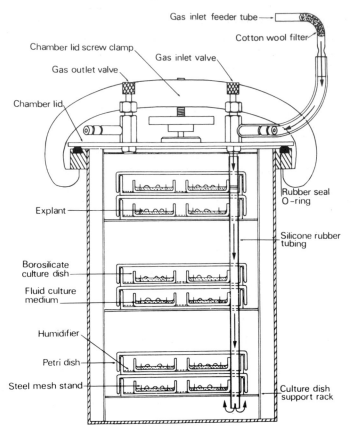

Figure 14. Modified anaerobic Macintosh and Fildes jar for gassing cultures.

The grid technique has proved successful in studying the growth and differentiation of embryonic and adult tissues, in investigating the effect of steroid hormones on their target organs, and in the interaction of carcinogens and smoke condensate with vitamin A.

Studies using rat prostate glands have shown that the organs remain androgen dependent and responsive *in vitro* and that they metabolize testosterone to the active androgen, dihydrotestosterone, as measured by steroid chromatography of tissue and medium (49). Human benign prostatic hyperplasia could be well maintained in organ culture (50), and testosterone was found to increase RNA synthesis of its epithelium and stroma (51). Receptors for labelled androgen could be demonstrated by steroid autoradiography in the same system (52). Using the grid technique, rat and human bladder have been maintained for prolonged periods and the urothelium was seen to differentiate in a manner similar to that *in vivo* (53, 54).

In mouse or rat prostate glands, human and murine embryonic lung, and embryonic and neonatal rat trachea, carcinogenic hydrocarbons and cigarette smoke condensate induced changes of a pre-cancerous nature within a short time of exposure. These could be inhibited or reversed by vitamin A and its structural analogues (55–58).

8. Organ culture of tissues: examples

8.1 Neural cells

Cell or explant cultures of neural cells have been valuable tools in neuro-biological research to investigate their morphological and biochemical properties. In these systems, however, the histiotypic structure of the tissue is artificially disturbed. In contrast, organ culture offers a three-dimensional system where the interaction of the various cellular components and their development can be explored.

Steinsvag and Laerum (59) successfully developed such a system by which fetal rat brain tissue could be successfully grown in stationary organ culture in multiwell dishes coated with semi-solid agar and its growth and development monitored for up to 50 days.

Protocol 18. Organ culture of brain tissue

Medium: DMEM (Flow Laboratories), supplemented with 10% heat inactivated newborn calf serum, four times the prescribed concentration of non-essential amino acids, L-glutamine, penicillin 100 U/ml, and streptomycin, 100 µg/ml

1. Remove fetuses from 18 day pregnant rats by Caesarean section.
2. Dissect whole brains from the fetuses and place them in sterile Petri dishes containing PBS.
3. Dissect brain lobes free and remove meninges.
4. Cut lobe tissue into approximate 800 µm cubes.
5. Explant fragments in multiwell dishes (Nunc), base coated with 0.5 ml. 0.75% semisolid agar (Agar Noble, Difco).
6. Transfer one fragment to each well and add 1 ml of medium.
7. Maintain cultures at 37°C in 5% carbon dioxide in air with 100% relative humidity.
8. Change medium every second day and observe cultures daily under a phase contrast microscope.

Test for viability

1. At days 0, 1, 2, and 5 and every fifth day thereafter until the end of the 50-day culture period, transfer some fragments to wells without agar. Cellular

Protocol 18. *Continued*

outgrowth at bottom of wells within 24 h can be regarded as expression of viability.

2. Monitor progress by light, scanning, and transmission electron microscopy.

The brain tissue fragments grown in culture mimicked some of the events in the developing rat brain *in vivo*, but some of the features observed such as shrinkage and cell loss did not correspond to the *in vivo* situation.

8.2 Postnatal rat liver

As the Trowell method has not proved successful for the cultivation of postnatal liver, Hart *et al.* (60) have developed a method by which postnatal rat liver could be kept viable for periods up to 72 h in Conway microdiffusion units. The criterion of viability used was the incorporation of tritiated orotate and leucine.

Protocol 19. Organ culture: postnatal liver

Medium: medium 199 containing 0.35 g/l sodium bicarbonate with 50 μg/l streptomycin sulphate and 50 u/ml penicillin mixed in equal proportion with FBS

1. Remove liver from a 13-day rat and place in beaker filled with synthetic medium.

2. Transfer to laminar flow hood.

3. Blot lobes on Whatman No. 1 filter paper and trim to give rectangular cutting surface.

4. Using a dermatome, cut slices of approximately 3 mm in thickness.

5. Transfer them to synthetic medium in a Petri dish and swirl them gently around to randomize them.

6. Transfer to No. 1 Conway microdiffusion units (*Figure 15*). Fill central well with culture medium and cover it with No. 1 Whatman filter paper disks.

7. Place 2–3 liver slices on the filter paper.

8. Place 9–10 units into an airtight box. Perfuse it with 95% oxygen and 5% carbon dioxide, and place it in an incubator. Incubate at 37°C.

9. Change medium and renew gas mixture every 24 h.

10. Test viability after 72 h incubation by incorporation of tritiated orotate and leucine into trichloroacetic acid precipitable material measured by autoradiography or the method of Mans and Novelli (61).

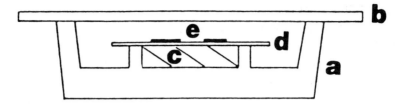

Figure 15. Microdiffusion unit for cultivation of liver slices (60). (a) Base of unit, (b) ground glass lid, (c) medium in centre well, (d) sterile filter paper, (e) liver slices.

8.3 Rabbit aorta

Pederson and Bowyer (62) devised an organ culture system to study the healing of superficial intimal injury using segments of rabbit aorta. The segments could be maintained *in vitro* for several weeks, and the repair of endothelial injury followed after placing a series of precisely located injuries, 100 μm wide, on each aortic segment. It was seen that the pattern of healing was similar to that observed *in vivo*. The cells within the wound underwent a burst of cell division and then became quiescent. The wound area was covered by migrating cells and underwent a pattern of remodelling which restored the normal morphology of the endothelium.

Protocol 20. Organ culture: rabbit aorta

Materials
- New Zealand white rabbits (Redfern strain), killed by sodium pentobarbitone (May & Baker Ltd) (0.5 ml/kg).
- culture support: 12 × 20 × 3.5 mm pieces of silicone rubber topped with a 12 × 20 mm sheet of coarse polyester cloth (P 500, Henry Simon Ltd)
- *culture vessels*: made from 60 × 20 mm Petri plates (Schott Glass) into which a 4 mm layer of Sylgard 184 encapsulating resin (Dow Corning) is cast to form a central well (*Figure 16*), which is of the correct size to accept the silicone rubber support
- *medium*: RPMI 1640 supplemented with garamycin sulphate (10 μg/ml) and up to 30% newborn calf serum. The medium is changed every 48 h and at each change glutamine and amphotericin B are added to produce final concentrations of 0.68 mM and 3 μg/ml respectively (Flow Laboratories, Gibco, Northumbria Biologicals)

1. Expose the aorta and tie a ligature around the aortic arch.
2. Transfer the aortic arch to a silicone rubber tray, immerse in perfusion fluid, and open it along its dorsal midline with iridectomy scissors.

Protocol 20. *Continued*

3. Stretch the opened aorta, with its luminal surface uppermost, over six culture supports, and re-establish the *in vivo* length.
4. Pin the aorta to the culture support using staples fashioned from 0.1×10 mm stainless steel minute pins (Watkins & Doncaster).
5. To detect any areas which might have been injured, stain the vessel with trypan blue (0.5 ml/l) for 30–45 s after a brief rinse with DPBS.
6. Divide into six segments, each 1 cm in length, by cutting the aorta with a scalpel blade along the edges of each culture support.
7. Place cultures in a modular incubation chamber (Flow Laboratories) which is gassed with humidified 5% carbon dioxide in air, and place the incubator chamber on a slowly rocking platform (6 cycles/min) in a hot room kept at 37°C.

Under the conditions described in the protocol the endothelium is well maintained for up to 6 weeks, and the method can therefore be considered a useful tool to study the effects of injury and the recovery taking place. The various stages of recovery are monitored by scanning electron microscopy and by measurements of thymidine uptake using autoradiography. Because the descending thoracic aorta from a single rabbit provides enough material for at least six segments, it is easy and economical to perform comparative experiments using material from the same animal.

8.4 Human trabecular meshwork

Trabecular cells may play a major role in the pathogenesis of glaucoma and loss or decrease of their function may be a key factor in producing glaucoma.

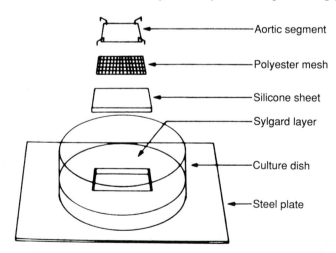

Aortic segment

Polyester mesh

Silicone sheet

Sylgard layer

Culture dish

Steel plate

Figure 16. Method for cultivation of aortic segments (62).

Human trabecular cells have been studied in cell and tissue culture, but the technique has limitations in that cells grown in a single layer on a plastic surface may react differently from cells growing in their normal position on the collagen of trabecular beams. Johnson and Tschumper (63) developed a successful method for organ culture of human trabecular meshwork. In this technique the anterior portion of an entire eye is used, including cornea, meshwork, scleral spur with a 5 mm wide scleral rim, using a modified Petri dish. Intraocular pressure, i.e. the normally directed flow of fluid and corneal clarity, could be maintained, for at least 4 weeks.

Protocol 21. Organ culture: human trabecular meshwork

Medium: DMEM with a mixture of penicillin, streptomycin, and amphotericin B (1 ml/100 of media) (A 9909, Sigma)

1. Obtain human eye bank eyes within 12 h of death from donors 73 ± 12 years of age and culture them within 24 h.
2. Trim globe of excess conjunctiva and bisect at equator.
3. Remove vitreous body and lens.
4. Rinse cornea and sclera slightly with mixture of polymiyxin, bacitracin and neomycin (Burroughs Wellcome).
5. Place anterior section on sterile Petri plate and rinse with PBS.
6. Use a culture dish made of acrylic plastic with a tight fitting O-ring to seal the scleral rim to an inner island, thus creating a closed eye.
7. Place the ring over the eye and tighten using four nylon screws.
8. Two 21G cannulas built into the base allow access to the eye. Perfuse one cannula with DME.
9. Sterilize entire dish with ethylene oxide gas but aerate for 48 h before use to remove traces of the gas.
10. Infuse at a constant rate of 2.5 µl/min using a Harvard microinfusion pump (Harvard Apparatus) fitted with sterile syringes and polyethylene tubing connected to the cannula of the culture dish.
11. Culture explants at 37°C in an atmosphere of 5% carbon dioxide, and remove spent media daily from the dish.

The system maintained the eye in a closed pressurized state for at least 4 weeks. There was no leakage and 54 eyes were cultured for 4 weeks, which showed a 78% success rate as judged by light microscopy and electron microscopic criteria.

8.5 Human gastrointestinal mucosa

As Howdle (64) points out, the method provides a useful tool to study diseases of the gastrointestinal mucosa, in particular coeliac disease and of ulcerative colitis. Coeliac disease is caused by a mucosal abnormality provoked by the gluten fraction of wheat (65). Falchuck *et al.* (66) found that in untreated coeliac mucosa *in vitro* the rise in alkaline phosphatase seen in the normal mucosa is inhibited by the presence of gluten. Organ culture studies of mucosa from patients with ulcerative colitis have demonstrated a change in the control of DNA synthesis followed by an increase in the rate of cell proliferation (67). Rachimelewitz *et al.* (68) and Hawkey and Truelove (69) have shown that in the same system inflamed ulcerative colitis mucosa produces increased amounts of prostaglandin E_2 *in vitro* which could be prevented by addition of prednisolone, sulphasalazine, or flufemio acid to the culture medium.

Protocol 22. Organ culture: human gastrointestinal mucosa

Medium: commercially available media of known constitution supplemented with FCS

1. Obtain mucosal biopsies using an intestinal mucosa biopsy instrument.
2. Cut them into pieces of about 3 mm in diameter.
3. Place them in the culture system within 5 min of excision.
4. Orientate the pieces of mucosa, villous or mucosal side uppermost, on a stainless steel grid.
5. Place grid with pieces of mucosa into the central well of a sterile culture dish.
6. Fill central well with culture medium so that it just touches the under surface of the mucosa. In this way, a thin layer of medium is drawn over the mucosal surface by capillary action.
7. Place felt pad saturated with saline into the outer well. Keep dish at 37°C in an atmosphere of 95% oxygen and 5% carbon dioxide.
   Culture period: 24–48 h.

---

9. Long-term cultivation of human adult tissues

With many organ culture methods, growth and differentiation of the explanted organs can usually be maintained for only limited periods *in vitro*. In many experiments the answer to a particular problem can, in fact, be obtained within days or weeks of cultivation, but in others a longer-term cultivation may be necessary. This is particularly true for studies of carcinogenesis in human tissues which require a prolonged exposure to carcinogenic agents.

With this aim in mind, a method has been introduced by which human tissues are intermittently exposed to fluid medium and an appropriate gas phase. Using this technique, the characteristic growth pattern and differentiation can be maintained in culture for prolonged periods.

The tissues include bronchus (70), breast (71), oesophagus (72), and uterine endocervix (73).

The medium found to be most suitable for the long-term maintenance of human tissues in organ culture is medium CMRL 1066, supplemented routinely with fetal bovine serum, hormones, and antibiotics. The composition of the medium is as follows:

- medium CMRL 1066 (Gibco)
- 5% heat-inactivated FBS
- 2–4 mM L-glutamine (Flow Laboratories)
- 1 µg/ml bovine insulin (see *Table 1*) (Sigma)
- 0.1 µg/ml hydrocortisone hemisuccinate (see *Table 1*) (Sigma)
- penicillin 100–300 U/ml (Glaxo)
- streptomycin 100–300 µg/ml (Glaxo)
- fungizone (Squibb) 5 µg/ml or Gentamicin (Schering) 100 µg/ml

For the cultivation of explants of hormone-dependent organs specific hormones should be added to the medium (e.g. for breast tissue, progesterone,

Table 1. Long-term culture of human explants

	Length of cultivation (months)	Results
Bronchus (70)	4	After 6 weeks, reduction in height of columnar epithelium, some loss of goblet cells, some metaplasia.
Breast (71)	1	Lobular architecture and typical ductal epithelium preserved. Secretory activity present with changes related to menstrual cycle.
Oesophagus (72)	4	Characteristics of squamous epithelium retained as demonstrated by light and electron microscopy.
Endocervix (73)	5	Epithelium viable as seen by mitotic activity and ultrastructure. After 4 weeks, partial loss of mucus secretion with focal metaplasia.

Insulin is a growth factor and in experiments where changes in cell proliferation are a criterion of effect the presence of insulin may mask the effect. It is recommended to omit it for such experiments or reduce the dose (e.g. 10 ng/ml).
Hydrocortisone at the concentration recommended (0.1 µg/ml) inhibits the action of carcinogenic agents (57) and is not suitable for studies of carcinogenesis. It should be omitted or used at a 20-fold lower concentration.

oestradiol, aldosterone, and prolactin have been applied, either continuously or as a 28-day regimen to mimic the menstrual cycle: see ref. 69).

Protocol 23. Long-term cultivation: explantation and cultivation

1. Obtain specimens from immediate autopsies or surgical resections. Dissect them from surrounding tissues and transport to the laboratory, usually immersed in ice-cold Leibowitz medium or other well-buffered medium, to which antibiotics have been added to prevent bacterial infection of the tissues. After arrival in the laboratory, transfer the tissue to a similar medium for further washes. The concentration of antibiotics should be at least twice that used for cultivation (e.g. 200 U/ml penicillin, 200 μg/ml streptomycin).

2. Trim the tissues further, free from fat, stroma, and underlying muscle, and divide into fragments of a size suitable for explanation. The measurements used vary for different tissues. For instance, bronchial explants measure 2 cm^2, breast and endocervix 0.5–1.0 cm^2, and oesophagus 1 cm^2.

3. Place the explants on the bottom of 60 mm tissue-culture-grade plastic dishes to which they adhere and then submerge them in 3–5 ml medium. Place organs composed of epithelium and underlying stroma into the dish with the stromal side attached to the bottom of the dish and the epithelium at the free surface.

The number of explants per dish varies with different tissues and authors. Thus 6–9 explants of endocervix, 2–6 explants of breast tissue, and 6–18 explants of oesophagus can be grown in each dish. However, it may not be necessary to follow this regimen exactly and other workers may find a different explant size and number per dish more suitable for their particular requirements.

Protocol 24. Incubation

The explants are incubated at a temperature of 36.5°C.

1. Place the dishes containing the explants in 3.5 ml medium on individual trays which can hold up to 60 dishes, and transfer these to an atmosphere-controlled chamber (Bellco).

2. Perfuse the chamber with a gas mixture of 5% carbon dioxide, 45% oxygen, and 50% nitrogen.

3. Place on a rocker platform and rock at 3–10 cycles per minute, depending on the tissues used. This causes the medium to flow intermittently over the surface of the explants. The assembled unit will fit into most standard incubators.

4. Change the medium two or three times weekly. After each change, re-place the gas mixture and return the atmosphere controlled chamber to the rocker platform and the incubator.

Using this technique the tissues could be maintained in culture for pro-longed periods with only minor alterations of their normal growth pattern (*Table 1*).

10. Preparation of organ cultures for light microscopy

10.1 Preparation of histological sections

The basic procedures used to prepare histological sections for light micro-scopy are the same as for other tissues, but the timings are shorter.

Protocol 25. Fixation, dehydration, and embedding

Materials
- fixative: Bouin's or 3% acetic Zenker
- ethanol series 50–100%
- 1% eosin in 70% ethanol
- cedarwood oil
- soft wax, melting point 45–56 °C (Raymond Lamb)
- hard wax, melting point 63 °C (Raymond Lamb)
- 2 needles
- 2 cataract knives
- glass watchglasses 5 cm in diameter
- wide-mouthed pipettes
- hotplate

Fixation and dehydration
1. At the end of the culture period, lift the strips carrying explants from medium. Rinse them with cold HBSS.
2. Drain on filter paper.
3. Immerse the strips carrying cultures in fixative for 35–40 min.
4. Following fixation in Bouin, transfer the strips to several changes of 70% ethanol.
5. Following fixation in Zenker, transfer to tap water for 1 h, 50% ethanol

249

Protocol 25. *Continued*

for 1 h, 70% ethanol for 1 h, 70% ethanol overnight, 90% ethanol for 1–1.5 h.

6. Remove explants from the strips with cataract knives.

7. Add a few drops of 1% eosin to 95% ethanol. Transfer the explants to 95% ethanol with eosin for 30 min.

8. Two changes of absolute ethanol, 1 h each. Cedarwood oil at least 1 h; cedarwood oil overnight; soft wax 1–1.5 h; soft wax 1.5–2 h. Two changes of hard wax, 30 min–1 h each.

Embedding

1. Warm the watchglass on the hotplate.

2. Partially fill with hard wax.

3. Transfer the explants and wax to the watchglass with a wide-mouthed pipette.

4. Orientate the explants in wax with two needles before it solidifies. When solidified, plunge the watchglass with embedded explants into cold tap water. This helps the wax block to detach from the watchglass.

5. Block up and cut serial sections for at least 1 day before staining.

10.2 Staining methods

Before staining, the sections have to be de-waxed in xylene and taken down to distilled water via immersion in alcohols of descending concentrations. **As xylene vapour is toxic and may be carcinogenic, it is best handled in a fume cupboard.** Wash in xylene until the wax is completely removed, then transfer to: absolute ethanol; 90% ethanol 5 min each; 70% ethanol; 50% ethanol; distilled water.

Protocol 26. Haematoxylin and Eosin (H & E) staining

This is a routine stain for histology and histopathology, and much of the present knowledge of normal and morbid histology has been gained by the study of H & E stained sections.

1. Ehrlich's haematoxylin 5–15 min.

2. Distilled water, quick dip.

3. Remove excess stain by differentiating in acid alcohol (1% hydrochloric acid in 70% ethanol) for a few seconds. The blue stain of haematoxylin is changed to red by the action of the acid.

4. Distilled water, quick dip.

5. Regain blue colour and stop decolorization by washing in alkaline running tap water for 5–15 min at room temperature, stain in 1% aqueous eosin for 1–3 min.

6. Wash off surplus stain in distilled water, dehydrate in alcohols, and clear in xylene, bearing in mind that eosin is removed from the tissues by water and low-grade alcohols and less readily by absolute alcohol. The degree of staining is thus easily controllable, and any overstaining can be remedied during the passage through the alcohols. The recommended times are: 2–3 s each for 70% ethanol and 90% ethanol; 30 s for absolute alcohol; 1–2 min for xylene; 1–2 min or longer for xylene.

7. Mount in a synthetic resin such as DPX.

Protocol 27. Periodic acid–Schiff (PAS) staining

This method is used to demonstrate polysaccharides (glycogen), neutral mucopolysaccharides, mucoproteins, glycoproteins, and glycolipids.

1. After de-waxing bring sections to distilled water via alcohols in descending concentrations.

2. Oxidize for 15 min in 1% periodic acid.

3. Wash for 15 min in running tap water and rinse in distilled water.

4. Place in Schiff's reagent in the dark for 10–20 min.

5. Wash for 10 min in running tap water.

6. Stain nuclei for 10 min in Mayer's haematoxylin. Do not use Ehrlich's haematoxylin, which will also stain some PAS-positive tissue components.

7. Dehydrate for 2 min each in ascending concentrations of alcohol.

8. Clear in xylene 1–2 min each.

9. Mount in a synthetic resin, such as DPX.

- PAS-positive substances stain red or magenta, nuclei blue

For some experiments it is necessary to exclude glycogen. This can be done by incubating the slides with diastase before staining them; *Protocol 27* is slightly modified accordingly.

Protocol 28. Modified PAS staining

1. Bring sections to distilled water.

2. Incubate them for 12 min with a 1% solution of malt diastase.

3. Rinse in distilled water.

4. Wash in running tap water for 5 min.

Protocol 28. *Continued*

5. Transfer to 1% periodic acid for 15 min. The remainder follows steps 3–9 of *Protocol 27*.
   - nuclei, blue to blue-black
   - chromosomes, blue-black
   - nucleoli, purplish
   - cartilage, light to dark blue
   - basement membrane, pink
   - cytoplasm, various shades of pink
   - muscle fibres, thyroid colloid, elastic fibres, deep pink
   - collagen and osteoid tissue, light pink

Azan substitute staining provides a very clear differentiation between epithelium, stroma, and cartilage, with a well-defined basement membrane between epithelium and underlying stroma.

Protocol 29. Azan substitute staining

1. De-wax sections and bring down to distilled water.
2. Place in carmalum for 20 min.
3. Wash in distilled water for 1 min.
4. Transfer to 1% phosphomolybdic acid for 5 min.
5. Wash in distilled water for 1 min.
6. Transfer to aniline blue for 10 min.
7. Wash in distilled water for 1 min.
8. Differentiate in 96% alcohol for 5 min.
9. Transfer to: absolute alcohol, 10 dips; absolute ethanol, 2 min; xylene, 5 min; xylene, 5 min.
10. Mount in a synthetic resin such as DPX.
    - nuclei red
    - chromosomes, red
    - epithelial cytoplasm, pink
    - basement membrane, brilliant blue
    - collagen, blue
    - cartilage, blue

11. Autoradiography

Measurements of isotope incorporation of labelled tissues by extraction and counting give the total radioactivity, but its distribution between different tissue components or cells is uncertain. Autoradiography provides this information and pinpoints the incorporation in different tissue components or individual cells, their cytoplasm or nucleus. This label can be recognized in suitably treated sections by light microscopy.

Autoradiographs were initially prepared by the method of Doniach and Pelc (75) using Kodak AR 10 stripping film, but more recently a dipping method has been used routinely. The latter method is faster and, because very thin films can be produced, results in better resolution. The dipping method is described here.

Protocol 30. Labelling of tissues

Remove explants from the strips of lens paper on Millipore filter and immerse in 2 ml of fresh medium warmed to 37°C.

- *DNA synthesis.* Add [^{3}H]thymidine 37 KBq/ml (1 μCi/ml). Mix well and incubate cultures for 5 h.
- *RNA synthesis.* Add [^{3}H]uridine 370 KBq/ml to (10 μCi/ml) medium. Mix well and incubate for 40 min. With this short labelling time the uptake is confined mainly to rapidly labelling RNA in the nucleus. Longer labelling periods will show progressive cytoplasmic labelling from rRNA and transferred mRNA.
- *Protein synthesis.* Add [^{3}H]leucine 370 KBq/ml (10 μCi/ml). Mix well and incubate for 40 min.

At the end of incubation, lift explants from medium and wash them thoroughly with HBSS to remove unincorporated radioactivity. For fixation, immerse them in acetic alcohol (1 part of glacial acetic acid, 3 parts of 80% alcohol) for 25 min. Transfer them directly to 10% formal saline, two changes, 1 h each or overnight.

For dehydration and embedding, follow the procedure described in *Protocol 25*, except that tissue sections are cut at 4 μm.

Protocol 31. Dipping procedure

Apparatus and materials

- Ilford K5 nuclear research emulsion in gel form. Store in a refrigerator at 4°C. When in use it lasts for about 6 weeks; if unopened it lasts 3 months or longer.

Protocol 31. *Continued*

- 2 small beakers
- glass or polypropylene cell (Raymond Lamb) and spoon
- forceps
- test slides
- pegs
- drying rack
- black plastic box (Cole Parmer)
- silica gel
- glass-distilled water

Preparation of emulsion

1. Open emulsion in a darkroom illuminated with a Kodak red filter.
2. Using a plastic or glass spoon, spoon enough emulsion for immediate requirements into a chemically clean beaker.
3. Melt the emulsion in a water bath maintained at 40–42°C.
4. When melted, fill 2/3 of glass cell with emulsion, then fill almost to the top with glass-distilled water warmed to 40–42°C.
5. Close the cell and invert it once or twice.
6. Replace the cell in the water bath and allow it to sit for 1–2 min before dipping.

Application of emulsion

1. Before dipping the slides, clear in xylene then transfer to 50:50 xylene-ethanol, 100, 90, 70, 50, 30% ethanol, and distilled water, 2 min each.
2. Acid-soluble precursors of DNA, RNA, or protein may be removed in ice-cold 10% trichloroacetic acid (2 × 10 min) followed by three rinses in ice-cold distilled water.
3. Bring the slides in distilled water to the water bath, rinse once or twice with glass-distilled water, drain in an upright position but do not allow them to dry.
4. First dip test slides into emulsion to make sure it is thoroughly melted and homogeneous. Slides should be uniformly coated with a thin grey film when held up to safe light.
5. Dip the experimental slides into emulsion. On withdrawal drain them against the edge of the cell for 5 s, then turn them the right way up and place the labelled end in the groove of the drying rack (i.e. slides should be in a vertical position).
6. Slides will dry in 1 h if left alone, alternatively they can be dried with a fan for 5–10 min.

7. When dry, place them in light-tight black plastic boxes containing tubes of silica gel and place the boxes in a refrigerator at 4°C.

8. It is recommended to develop test slides at different intervals to find the most effective duration of exposure. The following times have been found effective and the results could be used as a guide: [³H]thymidine, 6–7 days; [³H]uridine, 8–10 days; [³H], 8–10 days.

Development

1. Use Kodak D19 developer.

2. Filter solution before use.

3. Develop for 4–5 min at 18–21°C.

4. Rinse once with distilled water.

5. Fix in Amfix (May & Baker) (1:10) for 2–3 min or until film is clear.

6. Rinse and place in hypoclearing agent (Kodak) for 2 min.

7. Wash for 5 min in running tap water, then with three changes of deionized or distilled water.

Slides can be dried, but better results are obtained if they are taken directly from distilled water to stain. Stain slides routinely with haemotoxylin–eosin or carmalum.

On microscopic examination, the label appear as dark grains over the nuclei ([³H]thymidine, [³H]uridine) or over the cytoplasm ([³H]leucine). If the incubation with [³H]uridine is prolonged beyond 40 min, the label will have moved partially to the cytoplasm.

As the film is very thin and homogeneous, the histological structure of the tissues and changes of their growth pattern can be recognized and related to the localization and degree of labelling.

The incorporation can be assessed quantitatively by counting the number of labelled and unlabelled cells and expressing the result as a percentage of total cell number and its standard deviation (labelling index). The uptake per cell is estimated by counting the number of grains over individual cells and expressed as average grain number per cell and its standard deviation (grain count). In general, a higher isotope concentration, and resultant grain density, should be used for labelling index measurements, so that the nucleus is totally black and easier to score, and a lower grain density (20–50 grains/cells) for grain counts.

11.1 Steroid autoradiography

The autoradiographic technique used for measuring the uptake of tritiated thymidine, uridine, and leucine is not suitable for studying the uptake of steroid hormones and their receptor sites. Steroid hormones are alcohol soluble, and contact with alcohol during the fixation procedure would remove them from the tissue. Instead, dry-mount or thaw-mount techniques introduced by Stumpf and

STEROID AUTORADIOGRAPHIC TECHNIQUE
(THAW – MOUNT)

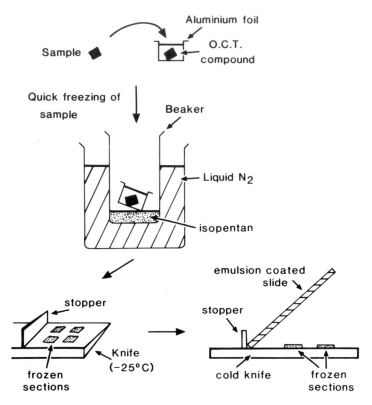

Figure 17. Details of steroid autoradiographic technique (77).

Sar (76) are more appropriate. A comparison of these techniques showed that the uptake of steroids was identical with both methods, but that the thaw-mount technique avoids any chemical artefact seen after freeze drying. It is also simpler and more time saving. The application of this technique to prostatic tissues in organ culture as developed by Takeda *et al.* (77) is described in *Protocol 32*. (*Figure 17*).

Protocol 32. Thaw-mount autoradiography in organ culture

Labelling medium: medium 199 with Earle salts supplemented with 5% fetal bovine serum containing 2.5 µCi (92.5 kBq)/ml ^{3}H-testosterone (94 Ci/mmol, 3.48 GBq/µmol) (Amersham) or 2.5 µCi (92.5 kBq)/ml ^{3}H-dihydrotesto-sterone (146 Ci/mmol, 5.40 GBq/µmol) (Amersham).

1. Incubate explants for 5 h.

2. To remove non-specific binding of ^{3}H-labelled hormones, wash explants in 30 ml HBSS, supplemented with 3% newborn calf serum or bovine serum. Agitate them gently.

3. After washing, place them into an OCT compound (Tissue Tek II, Lab-Tek Division, Miles Laboratories) and freeze them rapidly in isopentane chilled by liquid nitrogen.

4. Place frozen samples in liquid nitrogen until sectioning.

5. Preparation of emulsion and dipping procedure are the same as in conventional autoradiography (see *Protocol 31*), but coat slides with emulsion and store them in air- and light-tight boxes unit frozen tissue is sectioned.

6. Section frozen samples at 5 μm with a cryostat kept at −25°C in the dark or with safe light. Keep sections on knife and place over them the emulsion-coated slide to which they adhere.

7. Transfer slide with sections to an air- and light-tight box containing silica gel, and store the box in a refrigerator at 4°C throughout the exposure time of 2, 3, 4, or 6 weeks.

8. At these times, remove slides from boxes and breathe once or twice on the area of the sections to improve their adherence to the slide.

9. Fix with 70% alcohol and wash briefly with deionized water. Develop with Kodak D19 developer for 4 min at 15°C.

10. Fix for 10 min, then rinse in tap water (see p. 255, steps 5–7), and stain with haematoxylin–eosin. The hormones appear as dark grains under the nuclei. Cells possessing more than 10 grains under their nuclei are considered labelled.

11. Count them in at least 500 cells in selected areas and express binding as percentage of labelled cells.

---

12. Guide to quantitation of results

Changes in morphology induced by chemical or physical means can be quantitated, and alterations of morphology related to changes in cell proliferation or DNA, RNA, or protein synthesis. The average number of explants used for each point should be six.

To eliminate possible variations of effect in different areas of the treated explants it is recommended to use serial sections and count either all sections or every alternate one.

For example, to quantitate the incidence of hyperplasia induced by a carcinogen or mitogen, hyperplastic and normal structures are counted and the result expressed as the percentage of hyperplastic structures and its standard error.

To determine the incidence of mitosis, dividing and resting cells are counted in serial sections of explants incubated with colcemid (2 μg/ml) or another metaphase-arresting agent, for 5 h prior to fixation. The results are expressed as the percentage of dividing cells over total cell number and its standard error.

DNA, RNA, or protein synthesis can be determined in autoradiographs by counting labelled and unlabelled cells in serial sections of explants labelled with the appropriate isotopes. The result is expressed as the percentage of labelled cells over total cell number and its standard error.

This method has provided reliable data which were reproducible in replicate experiments. One or two replicates sufficed to confirm the results.

13. Advantages and limitations of organ culture

In organ culture the various tissue components, their spatial relationship, and functional activity are, under suitable conditions, preserved *in vitro*. The preservation of the stroma is of particular importance, as it seems indispensable for the growth and differentiation of epithelium. These features make tissues in organ culture a more physiological experimental model than cell cultures. Contrary to some beliefs, the setting up of organ cultures does not require more effort than the establishment of cell cultures.

Fetal organs or organ rudiments develop *in vitro* in a similar manner as *in vivo*. Epithelia, in the presence of their stroma, grow and differentiate in the appropriate manner and manufacture specific secretory products. Hormone-dependent organs such as prostate and mammary glands, vagina, and uterus remain hormone-sensitive and responsive *in vitro*. Endocrine organs, such as ovaries, testes, adrenals, and pituitary, continue to secrete specific hormones *in vitro*.

Thus, tissues in organ culture can provide important basic information on normal development, growth, and differentiation and the influence of extrinsic factors on these parameters. The results can usually be obtained more quickly *in vitro* than *in vivo* and, if necessary, be easily quantitated. This quantitation has proved reliable and reproducible in replicate experiments.

The point has to be considered whether the use of organ culture can fully replace animal experimentation. *In vivo*, the interpretation of experimental results is complicated by the presence of systemic factors; their elimination *in vitro* reduces the problem to be investigated to its essentials. However, it is not certain whether the results obtained *in vitro* can always be extrapolated to the situation *in vivo*. The very presence of systemic factors may have a controlling influence and modify the response seen in culture. Studies on drug action sometimes produce contradictory results *in vitro* and *in vivo*, because some drugs are metabolized *in vivo* but not *in vitro*.

The time factor poses another problem. The period of cultivation in organ culture is not unlimited and, so far, has not exceeded a few months. This period may be too short where prolonged exposure to extrinsic factors is necessary. In such cases, a combined *in vitro–in vivo* procedure may be of benefit, in which tissues treated *in vitro* are implanted into suitable host animals. The use of nude mice facilitates such procedures.

In summary, experiments in organ culture do not always replace animal experiments but the results seen *in vitro* serve as a valuable guide to the events taking place *in vivo* and thus considerably reduce the number of animal experiments necessary for studying any particular problem.

References

1. Loeb, B. (1897). *Über die Entstehung von Bindegewebe, Leucocyten und roten Blutkorperchen aus Epithel und eine Methode, isolierte Gewebsteile zu züchten.* M. Stern & Co., Chicago.
2. Harrison, R. G. (1907). *Proc. Soc. Exp. Biol. Med.,* **4,** 140.
3. Loeb, L. and Fleischer, M. S. (1919). *J. Med. Res.,* **40,** 509.
4. Parker, R. C. (1936). *Science,* **83,** 379.
5. Medawar, P. B. (1948). *Quart. J. Microsc. Sci.,* **89,** 187.
6. Fell, H. B. and Robison, R. (1929). *Biochem. J.,* **23,** 767.
7. Lasnitzki, I. (1963). *Natl. Cancer Inst. Monogr.,* **12,** 281.
8. Rudnick, D. (1983). *J. Exp. Zool.,* **78,** 369.
9. Gaillard, P. J. (1951). In *Methods in Medical Research* (ed. M. B. Visscher), Vol. 4. Year Book Publishers, Chicago.
10. Spratt, N. T., Jr. (1947). *Science,* **106,** 542.
11. Spratt, N. T., Jr. (1948). *J. Exp. Zool.,* **107,** 39.
12. Wolff, E. T. and Haffen, K. (1952). *Tex. Rep. Biol. Med.,* **10,** 463.
13. Chen, J. M. (1954). *Exp. Cell Res.,* **5,** 10.
14. Richter, K. M. (1958). *J. Oklahoma State Med. Assoc.,* **51,** 252.
15. Lash, J., Holtzer, S., and Holtzer, H. (1957). *Exp. Cell. Res.,* **13,** 292.
16. Shaffer, B. B. (1956). *Exp. Cell Res.,* **11,** 244.
17. Trowell, O. A. (1954). *Exp. Cell Res.,* **6,** 246.
18. Trowell, O. A. (1959). *Exp. Cell Res.,* **16,** 118.
19. Trowell, O. A. (1961). *Coll. Intern. CNRS,* **101,** 237.
20. Barret, L. A., McDowell, E. M., Frank, A. L., Harris, C. C., and Trump, B. F. (1976). *Cancer Res.,* **36,** 1003.
21. Lasnitzki, I. (1951). *Br. J. Cancer,* **5,** 345.
22. Fell, H. B. and Canti, R. G. (1934). *Proc. R. Soc. Lond.,* B **119,** 470.
23. Fell, H. B. (1939). *Trans. R. Soc. Lond.,* B **229,** 407.
24. Fell, H. B. and Mellanby, E. (1953). *J. Physiol.,* **119,** 470.
25. Lasnitzki, I. (1963). *J. Exp. Med.,* **118,** 1.
26. Lasnitzki, I. (1955a). *J. Endocrinol.,* **12,** 236.
27. Lasnitzki, I. (1955b). *Br. J. Cancer,* **9,** 435.
28. Lasnitzki, I. (1958). *Br. J. Cancer,* **12,** 547.
29. Lasnitzki, I. (1956). *Br. J. Cancer,* **10,** 510.
30. Murray, M. R. and Stout, A. O. (1947). *Am. J. Anat.,* **80,** 225.
31. Hardy, M. H. (1978). *Annual meeting of Am. Tissue Culture Assoc.*
32. Hardy, M. H. (1967). *Exp. Cell Res.,* **46,** 377.

33. Biggers, J. D., Claringbold, P. J., and Hardy, M. H. (1956). *J. Physiol.,* **131,** 497.
34. Moon, Y. S. and Hardy, M. H. (1973). *Am. J. Anat.,* **138,** 253.
35. Wolff, Et., Haffen, K., Kieny, M., and Wolff, Em. (1953). *J. Embryol. Exp. Morphol.,* **1,** 55.
36. Wolff, Et. (1960). *C.R. Acad. Sci. Paris,* **250,** 3881.
37. Wolff, Et., Smith, J., and Wolff, Em. (1975). In *Organ Culture in Biomedical Research* (ed. M. Balls and M. A. Monnickendamm), p. 405. Cambridge University Press.
38. Lasnitzki, I. and Mizuno, T. (1977). *J. Endocrinol.,* **74,** 47.
39. Lasnitzki, I. and Mizuno, T. (1979). *J. Endocrinol.,* **82,** 171.
40. Bronckers, A. L. J. J. (1983). *J. Biol. Buccale,* **11,** 195.
41. Navia, J. M., Snider, C., Punya, J., and Harris, S. (1984). *Arch. Oral Biol.,* **29,** 11.
42. Wigglesworth, D. J. and Hayward, A. F. (1973). *Z. für Zellforschung,* **138,** 171.
43. Sakakura, Y. (1986). *Calci. Tissue Int.,* **39,** 271.
44. Sakakura, Y. (1989). *In Vitro Cell. Dev. Biol.,* **25,** 959.
45. New, D. A. T., Coppola, P. T., and Cockcroft, D. L. (1976). *J. Reprod. Fertil.,* **48,** 219.
46. New, D. A. T. (1983). In *Methods for Assessing the Effects of Chemicals on Reproductive Functions* (ed. V. B. Vouk and F. J. Sheehan), p. 277. Wiley, New York.
47. Cockcroft, D. L. (1980). *Acta Morphol. Acad. Sci. Hung.,* **28,** 117.
48. Fainstat, T. (1968). *Fertil. Steril.,* **19,** 317.
49. Lasnitzki, I. (1976). In *Organ Culture in Biomedical Research* (ed. M. Balls and M. E. Monnickendamm), p. 241. Cambridge University Press.
50. McMahon, M. J. and Thomas, G. H. (1973). *Br. J. Cancer,* **27,** 323.
51. Lasnitzki, I., Whitaker, R. H., and Withycombe, J. F. R. (1975). *Br. J. Cancer,* **32,** 168.
52. Lasnitzki, I., Takeda, H., and Mizumo, T. (1989). *J. Endocrinol.,* **120,** 167.
53. Hodges, G. M., Hicks, R. M., and Spacey, G. D. (1977). *Cancer Res.,* **37,** 3720.
54. Knowles, M. A., Finesilver, A., Harvey, A. E., Berry, R. Z., and Hicks, R. M. (1983). *Cancer Res.,* **43,** 374.
55. Lasnitzki, I. (1968). *Cancer Res.,* **28,** 510.
56. Lasnitzki, I. and Goodman, D. S. (1974). *Cancer Res.,* **34,** 1564.
57. Lasnitzki, I. (1976). *Br. J. Cancer,* **34,** 239.
58. Lasnitzki, I. and Bollag, W. (1982). *Cancer Treat. Rep.,* **66,** 1375.
59. Steinswaag, S. K. and Laerum, O. D. (1985). *In Vitro,* **41,** 1517.
60. Hart, A., Matteyse, F. J., and Balinsky, J. B. (1983). *In Vitro,* **19,** 841.
61. Mans, R. J. and Novelli, G. D. (1961). *Arch. Biochem. Biophysics,* **94,** 48.
62. Pederson, D. C. and Bowyer, D. E. (1985). *Am. J. Pathol.,* **119,** 264.
63. Johnson, D. C. and Tschumper, R. C. (1987). *Invest. Ophthalm. Vis. Sci.,* **28,** 945.
64. Howdle, P. D. (1984). *Postgraduate Med. J.,* **60,** 645.
65. Scott, B. B. and Losowski, M. S. (1977). *J. Roy. Coll. Physicians,* **11,** 405.
66. Falchuck, Z. M., Gebhard, R. L., Sessons, C., and Strober, W. (1974). *J. Clin. Invest.,* **53,** 487.
67. Serafini, E. P., Kirk, A. P., and Chambers, T. J. (1977). *Gut,* **22,** 648.
68. Rachmilewitz, D., Ligumsky, M., Sharon, P., Karmeli, F., and Zor, U. (1978). *Gastroenterology,* **75,** 638.

69. Hawkey, C. J. and Truelove (1981). *Gut,* **22,** 190.
70. Trump, B. F. (1976). *Cancer Res.,* **36,** 1003.
71. Hillman, E. A., Valerio, M. G., Halter, S. A., Barret-Boone, L. A., and Trump, B. F. (1983). *Cancer Res.,* **43,** 245.
72. Hillman, E. A., Vocci, M. J., Church, W., Harris, C. C., and Trump, B. F. (1980). In *Methods in Cell Biology,* Vol. 21 (ed. C. C. Harris, B. G. Trump, and G. D. Stoner), p. 331. Academic Press, New York.
73. Schurch, W., McDowell, E. M., and Trump, B. F. (1978). *Cancer Res.,* **38,** 3723.
74. Paul, J. (1975). *Cell and Tissue Culture,* 5th edn. Churchill Livingstone, Edinburgh.
75. Doniach, S. R. and Pelc, T. (1950). *Br. J. Radiol.,* **23,** 184.
76. Stumpf, W. E. and Sar, M. (1975). *Meth. Enzymol.,* **36,** 135.
77. Takeda, H., Mizuno, T., and Lasnitzki, I. (1985). *J. Endocrinol.,* **104,** 87.

8

Cytotoxicity and viability assays

ANNE P. WILSON

1. Introduction

Drug development programmes for the identification of new cancer chemo-therapeutic agents involve extensive preclinical evaluation of vast numbers of chemicals for detection of antineoplastic activity. The safety evaluation of compounds such as drugs, cosmetics, food additives, pesticides, and industrial chemicals necessitates the screening of even greater numbers of chemicals. Animal models have always played an important role in both contexts, and although cell culture systems have figured largely in the field of cancer chemotherapy, where the potential value of such systems for cytotoxicity and viability testing is now widely accepted, there is increasing pressure for a more comprehensive adoption of *in vitro* testing in both spheres of application. The impetus for change originates partly from financial considerations, since *in vitro* testing has considerable economic advantages over *in vivo* testing. There is also an increasing realization of the limitations of animal models in relation to human metabolism, as increasing numbers of metabolic differences between species come to be identified. Finally, there is the moral pressure to reduce animal experimentation.

The safety evaluation of chemicals involves an extensive range of studies on mutagenicity, carcinogenicity, and chronic toxicity. The former two aspects have been covered in other books (1, 2, 3) and are outside the scope of this text. Whilst the major application of *in vitro* cultures is currently with analysing acute toxicity, the existence of adequate culture systems would also improve the prospects of chronic toxicity testing. Toxicology and cancer chemotherapy therefore share the aim of determining the acute toxicity of a range of chemicals against a variety of cell types. In both areas there may be several ultimate goals:

- identification of potentially active compounds
- identification of the mechanism by which the compound exerts its toxic effect
- prediction of effective cytotoxic drug for treatment of cancer patients
- screening to identify the range of activity of a compound

- identification of a potential target cell population
- identification of the toxic concentration range
- relationship of concentration to exposure time ($C \times T$)

The fundamental requirements for both applications are similar. Firstly, the assay system should give a reproducible dose–response curve with low inherent variability over a concentration range which includes the *in vivo* exposure dose. Secondly, the selected response criterion should show a linear relationship with cell number, and thirdly, the information obtained from the dose–response curve should relate predictively to the *in vivo* effect of the same drug.

2. Background

Use of *in vitro* assay systems for the screening of potential anticancer agents has been common practice almost since the beginnings of clinical cancer chemotherapy in 1946, following the discovery of the antineoplastic activity of nitrogen mustard. A number of reviews describe the historical development of these techniques and their application (4, 5) and more recent publications bring the story up to date (6–8).

Early cytotoxicity studies were largely qualitative, in that explant cultures growing in poorly defined media were used for the study of drug effect, which could be 'quantitated' by assessment either of morphological damage or of inhibition of the zone of outgrowth. The development of better-defined growth media, together with techniques for growing dispersed cells as a monolayer on glass, allowed the screening of identical replicate cell samples in reproducible growth conditions, which therefore meant that drug effects could be quantitated in a meaningful way. Using these techniques in conjunction with measurement of protein content of treated and untreated cells, Eagle and Foley found a correlation between the *in vitro* and *in vivo* activities of neoplastic agents (9) demonstrating the validity of the method. The assay system was subsequently included in the Cancer Chemotherapy National Service Centre (CCNSC) screening programme, and currently plays an essential role in pre-screening during the stepwise purification of natural fermentation beers, because its rapidity, minimal requirement for test compound, and economy offer advantages which cannot be found in the *in vivo* screening systems (6).

Accumulated experience, both clinically and experimentally, demonstrated that heterogeneity of chemosensitivity existed between tumours, even those of identical histology. The successful development of the *in vitro* agar plate assay for antibiotic screening precipitated interest in the development of an analogous technique for 'tailoring' chemotherapy to suit the individual tumour and patient, thus removing the undesirable combination of ineffective chemotherapy in the presence of non-specific toxicity. The idea was first applied experimentally by Wright *et al.* (10) using explants of human tumour tissue, and this report has been succeeded by numerous others which aim to investigate the correlation between *in vitro* and *in vivo* results in humans. The methodology which has been used is varied, and represents the multiplicity of factors which must be considered in devising these assays. In spite of this diversity the consensus of the majority of reports is that more than 90% positive correlation can be expected between *in vitro* resistance and clinical resistance, and approximately 60% positive correlation between *in vitro* sensitivity and clinical response. This is true of most retrospective studies, and also of the one prospective study, in which patients were treated on the basis of *in vitro* results (11). There is also an indication that the frequency of responding cultures *in vitro* is similar to the clinical response rates for different drugs in a particular tumour type (12, 13).

The relationship between *in vitro* drug sensitivity exhibited by primary cultures of human tumours and their *in vivo* counterparts argues for their use in drug evaluation programmes, since they provide a closer approximation to the human clinical situation than do the limited number of cell lines which are currently used. The role of the 'Human Tumour Stem Cell Assay' is under investigation in this context (14) and more recently it has been used for analysing dose responses, with a view to designing effective regimes for high-dose or regional chemotherapy (15).

The developments in the field of safety evaluation of drugs has been reviewed by several authors in recent years (16, 17). There are a number of advantages of *in vitro* testing for safety evaluation which include analysis of species specificity, feasibility of using only small amounts of test substances, and the facility to do mechanistic studies. Reference 17 also highlights the progress which has been made in developing suitable models for different target organs.

3. Specific techniques

Making a decision on the final choice of assay is a function of the context in which the assay is to be used, the origin of the target cells, and the nature of the test compounds. Parameters which vary between different assays include:

- culture method (Section 3.1)
- duration of drug exposure and drug concentrations (Section 3.2)
- duration of recovery period after drug exposure (Section 3.3)
- endpoint used to quantitate drug effect (Section 4)

3.1 Culture methods

The choice of culture method depends on the origin of the target cells and the duration of the assay and, to some extent, dictates the endpoint.

3.1.1 Organ culture

The advantages offered by organ culture (see Chapter 7) relate to the main-tenance of tissue integrity and cell–cell relationships *in vitro*, thereby giving a closer analogy to the *in vivo* situation than the majority of other culture methods available. However, reliable quantitation of drug effect is impaired by difficulties associated with variation between replicates in size and cellular heterogeneity. Although the method has been used extensively to study the hormone sensitivity of potentially responsive target tissues, the number of studies relating to drug sensitivity is limited. The topic has been covered in a review article (18) (see also Chapter 7).

3.1.2 Spheroids

Spheroids result from the spontaneous aggregation of cells into small spheri-cal masses which grow by proliferation of the component cells. Their struc-ture is analogous to that of a small tumour nodule or micrometastasis, and the use of spheroids for drug sensitivity testing therefore permits an *in vitro* analysis of the effects of three-dimensional relationships on drug sensitivity, without the disadvantages previously mentioned for organ culture. Specific parameters which can be studied are:

- drug penetration barriers in avascular areas
- the effects of metabolic gradients (e.g. pO_2, pCO_2)
- the effects of proliferation gradients

The majority of studies have been carried out using spheroids derived from cell lines, but primary human tumours also have the capacity to form spheroids in approximately 50% of cases; the spheroid-forming capacity of normal cells is limited in comparison with tumour cells, so stromal elements may be excluded during reaggregation of human tumour biopsy material. Culture times in excess of 2 weeks are usually necessary for drug sensitivity testing, and the method is therefore not suitable in a situation where results are required quickly.

3.1.3 Suspension cultures

i. Short-term cultures (4–24 h)

The short-term maintenance of cells in suspension for assay of drug sensitivity is applicable to all cell sources. When the cells are derived from human tumour biopsy material the assay system has several theoretical advantages in that the ability to grow is not a limiting factor since no growth is required, stromal cell overgrowth and clonal selection are minimized, and results can be obtained rapidly which is important if the assay is to be used in a clinical context. The method has been used extensively in Germany for chemo-sensitivity studies on a variety of tumour types (19, 20). A modified method using either tissue fragments or cells has been described by Silvestrini *et al.*

(13), again with a variety of tumour types. Both groups used the incorporation of tritiated nucleotides into DNA or RNA as an endpoint. Limitations of
the method relate mainly to the short time period of the assay, which pre-
cludes long drug exposures over one or more cell cycles and also takes no
account of either the reversibility of the drug's effect or of delayed cytotoxicity.

ii. Intermediate duration (4–7 days)

Suspension cultures of intermediate duration are particularly suited to chemosensitivity studies on haematological malignancies, and have been described in
several reports (21–23).

3.1.4 Monolayer culture

The technique of growing cells as a monolayer has been most frequently
applied to the cytotoxicity testing of cancer cell lines, but the method has also
been used with some success for studies on the chemosensitivities of biopsies
from a variety of different tumour types. In the case of human biopsy material
the greatest problems associated with the method are, firstly, that the success
rate is limited because adherence and proliferation of tumour cells is not
always obtained and, secondly, that contamination of tumour cell cultures by
stromal cells (fibroblastic or mesothelial) occurs all too frequently. The problem is greater with some tumour types than others, and some method of cell
identification is therefore an essential part of such assays. The technique has
also found wide acceptance with toxicologists since appropriate cell lines may
be available for the development of models for specific organ systems (17).

The culture method probably offers the greatest flexibility in terms of
possible drug exposure and recovery conditions, and also in methods of
quantitation of drug effect. Of all methods described, the growth of cells in
monolayers requires the lowest cell numbers, and it is therefore amenable to
microscale methodology which permits multiple drug screening over a wide
concentration range, and also facilitates automation. There have been major
developments in this area in recent years. When cell numbers are really
sparse it is often feasible to culture the cells until sufficient numbers are
available for assay, although reports on changes in chemosensitivity after
subculture are conflicting. Two to three subcultures are probably acceptable
and, indeed, subculture has been recommended because variability between
replicates is reduced (24). When stromal cell contamination is unacceptably
high, subculturing also offers the possibility of 'purifying' tumour cell cultures
by differential enzyme treatment or physical cell separation (see Chapters 5
and 6).

3.1.5 Clonogenic growth in soft agar

Although monolayer cloning can be applied to cells cultured directly from the
tumour, the majority of reports in recent years have used suspension cloning
to minimize growth of anchorage-dependent stromal cells. Clonogenic assays

have the theoretical advantage that the response is measured in cells with a high capacity for self-renewal (potentially the stem cells of the tumour), and cells with limited proliferative capacity, which make up much of the bulk of the tumour, are not assayed. However, this is only true if colonies are truly clones (i.e. were initiated from one cell and not from a clump) and are scored only after many cell generations in clonal growth. Regrettably this is often not the case. Cloning efficiencies of 0.01–0.1% are often quoted where it may be difficult to exclude the possibility of clumps, and while 10 generations (about 1000 cells) may be readily obtained in monolayer cloning, suspension colonies are often scored after 4–6 cell generations (16–64 cells). Given that some of these colonies started out as clumps of 3–4 cells, the generation number may be as low as 2 and their capacity for self-renewal still in some considerable doubt.

Nevertheless, growth in soft agar is undoubtedly a useful assay system for cell lines which have a comparatively high plating efficiency. However, a number of technical problems have been encountered using solid tumours and effusions from patients, which unfortunately influence interpretation of results. These include difficulties in obtaining a pure single-cell suspension from epithelial tumours (which is an essential requisite for the definition of clonogenic growth), very low plating efficiencies (≪1%), the formation of colonies from anchorage-dependent cells under certain growth conditions, requirements for large cell numbers, and finally the somewhat subjective nature of colony quantitation. These represent a failure of present technology rather than a failure of the assay method, and the importance of optimizing methods of disaggregation and selective growth media for different tumour types has been emphasized (25).

Critical assessment of the technical difficulties which have been encountered with the 'Human Tumour Stem Cell Assay' can be found in several reports (26–28). Results obtained using the double agar method developed by Hamburger *et al.* (29) have been described in a review publication (30). An alternative methodology, developed by Courtnay and others (31), gives higher plating efficiencies, and has been compared with the 'Hamburger–Salmon' system by Tveit *et al.* (32), and it was apparent that the methodology used influenced the chemosensitivity profile obtained (*Figure 1*).

In recent years comparatively few publications have dealt with the applied use of the assay, but there have been several reports describing methods of improving plating efficiencies in soft agar (33–35).

3.2 Duration of drug exposure and drug concentrations

The choice of drug concentrations should be dictated by consideration of the therapeutic levels which can be achieved with clinically used drug dosages. When the compound is undergoing preclinical screening for potential anti-tumour activity this is not possible and, in the face of accumulated evidence

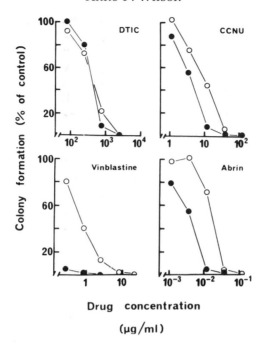

Figure 1. Dose–response curves of a melanoma xenograft (V. N.) cultured in soft agar using either (a) the 'Courtnay' method (●) or (b) the 'Hamburger-Salmon' method (●). Cells were exposed to the drugs for 1 h, plated at 3×10^4 cells per tube or dish and scored for colonies after 14 days incubation. Control cultures showed about 400 colonies using both methods. Method (b) showed greater sensitivity with three of the four drugs tested. (Reproduced with permission of the publisher from ref. 32.)

on effective *in vitro* levels of compounds with known *in vivo* activity, an upper limit of 100 μg/ml is recommended. Pharmacokinetic data are available for many of the clinically used anticancer drugs, and parameters which are relevant to *in vitro* assays include the peak plasma concentration and the plasma clearance curves (*Figure 2*). A detailed description of pharmacological considerations may be found elsewhere (36). Pharmacokinetic data on most cancer chemotherapeutic agents, which includes peak plasma concentration, the $C \times T$ parameter (where C is concentration and T is time in hours), and the terminal half-life ($t_{1/2}$) of the drug in plasma, have been summarized (37). When no pharmacokinetic data are available, an approximation of the plasma levels can be obtained by calculation of the theoretical concentration obtained when the administered dose is evenly distributed throughout the total body fluid compartment. It is axiomatic that the concentration range adopted should give a dose–response curve, and the range selected by different groups reflects the different sensitivity levels of the various assays. The same provisos apply when deciding on appropriate drug concentrations for

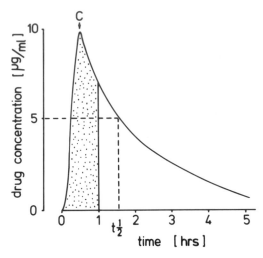

Figure 2. Typical plasma clearance curve for intravenously administered drug. C, peak plasma concentration; $t_{1/2}$, terminal half-life of drug in plasma   is the area under curve for $T = 1$ h (μg ml^{-1} h^{-1}).

the assessment of acute toxicity, though upper concentration levels may be in excess of 100 μg/ml.

Pharmacokinetic data show that maximum exposure to drug occurs in the first hour after intravenous injection and, for this reason, an exposure period of 1 h has been chosen by many investigators. Whilst this may be adequate for cycle-specific drugs, such as the alkylating agents, longer exposure times over several cell cycles are necessary for phase-specific drugs. Prolonged drug exposure using a variety of cancer chemotherapeutic agents has been shown to result in gradually decreasing ID_{50} values, as exposure times increase (38) (*Figure 3*). In some cases these reach a stable minimum plateau value, but not always.

Rate of penetration of the drug may also be a limiting factor when short exposure times are used. The level of cell kill achieved with a short exposure time is also related to the method of assay; whilst high levels of cell kill can be obtained with a 1 h exposure to an alkylating agent using primary suspension cloning, a similar duration is insufficient to show cytotoxicity in monolayer.

Ultimately the question of duration of drug exposure becomes one of practicality. If a significant effect is achievable in 1 h, then this should be used. Many drugs may bind irreversibly to intracellular constituents and the actual exposure may therefore be in excess of 1 h due to drug retention. Others, principally the antimetabolites and antitubulins, are more likely to be reversible if not present at the sensitive phase of the cell cycle, and prolonged exposure spanning one or more cell cycles may be required. Resistance of the

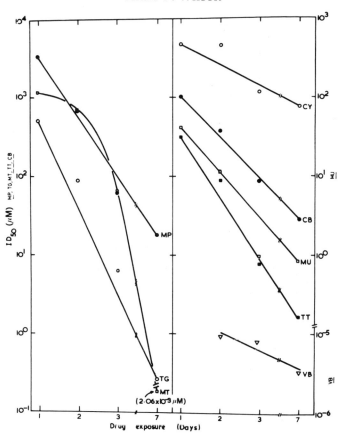

Figure 3. Effect of prolonged drug exposure on ID_{50} values for 6-mercaptopurine (MP), thioguanine (TG), methotrexate (MT), cyclophosphamide (CY), chlorambucil (CB), mustine (MU), thiotepa (TT), and vinblastine (VB), tested against HeLa cells using a microtitration plate assay. Drugs were replaced at 24 h intervals for the 48 h and 72 h exposures, and at 24, 48, and 72 h for the 7-day exposure. A break in the time axis is therefore shown between 3 and 7 days. (Reproduced with permission of the publishers from ref. 38.)

surviving fraction when short exposures are used may be due to an inappropriate phase of the cell cycle during drug exposure, but truly resistant cells (i.e. with resistance even at the appropriate phase of the cell cycle) can only be demonstrated unequivocally after a prolonged exposure.

When prolonged drug exposure times are used, it should be remembered that the theoretical $C \times T$ value is only equal to the actual $C \times T$ value when the drug retains full activity at 37°C over the entire exposure period and the response is linear with time. Data on the stability of some anticancer drug solutions at 37°C have been reported (38). The effective concentration of drug may also be reduced by binding of drug to the surface of the incubation

271

vessel. Considerations such as those described above are also relevant for assessment of the acute toxicity of a compound.

3.3 Recovery period

The inclusion of a recovery period following drug exposure is important for three reasons:

(a) When metabolic inhibition is used as an index of drug effect, it allows recovery of metabolic perturbations which are unrelated to cell death.

(b) Sublethal damage can be repaired.

(c) Delayed cytotoxicity may occur with some drugs, which is not expressed for 1–2 cell cycles.

Depending upon the nature of the drug and the endpoint of the assay, absence of a recovery period can either underestimate or overestimate the level of cell kill achieved. However, it is equally important that the recovery period is not too long, because cell kill can then be masked by overgrowth of a resistant population. In monolayer assays which monitor cell counts or precursor incorporation, the cells must remain in the log phase of growth throughout the exposure and recovery period. In clonogenic assays the recovery period is the period of clonal growth; the time taken to form measurable colonies is a minimum of five or six cell generations (32–64 cells/colony) in suspension assays and usually much greater than this when monolayer cloning is in use.

4. Endpoints

4.1 Cytotoxicity, viability, and survival

Interpretation of the significance of assay results is dependent upon distinguishing between assays which measure cytotoxicity and assays which measure cell survival. Cytotoxicity assays measure drug-induced alterations in metabolic pathways or structural integrity which may or may not be related directly to cell death, whereas survival assays measure the end result of such metabolic perturbations which may be either cell recovery or cell death. Theoretically, the only reliable index of survival in proliferating cells is the demonstration of reproductive integrity, as evidenced by clonogenicity. Metabolic parameters may also be used as a measure of survival when the cell population has been allowed time for metabolic recovery following drug exposure.

When the test compound exhibits toxicity which is not specifically related to proliferative potential and results in loss of one or more specific and essential cell functions rather than loss of reproductive capacity, a cytotoxicity test may be more appropriate.

4.2 Cytotoxicity and viability

Some cytotoxicity assays offer instantaneous interpretation, such as the up-take of a dye by dead cells, or release of ^{51}Cr or fluorescein from pre-labelled cells. These have been termed tests of viability, and are intended to predict survival rather than measure it directly. On the whole these tests are good at identifying dead cells but may overestimate long-term survival. Most imply a breakdown in membrane integrity and irreversible cell death.

Other aspects of cytotoxicity, measuring metabolic events, may be more accurately quantified and are very sensitive, but prediction of survival is less certain as many forms of metabolic inhibition may be reversible. In these cases impairment of survival can only be inferred if depressed rates of pre-cursor incorporation into DNA, RNA, or protein are maintained after the equivalent of several cell population doubling times has elapsed.

4.2.1 Membrane integrity

This is the commonest measurement of cell viability at the time of assay. It will give an estimate of instantaneous damage (e.g. by cell freezing and thawing), or progressive damage over a few hours. Beyond this, quantitation may be difficult due to loss of dead cells by detachment and autolysis. These assays are of particular importance for those toxic agents which exert their primary effect on membrane integrity.

i. ^{51}Cr release

Labelling cells with ^{51}Cr results in covalent binding of chromate to basic amino acids of intracellular proteins. These labelled proteins leak out of the cell when the membrane is damaged, at a rate which is proportional to the amount of damage incurred. The method is used extensively in immunological studies for determining cytotoxic T-cell activity against tumour target cells. Natural leakage of ^{51}Cr from undamaged cells may be high, and therefore the time period over which the assay can be used is restricted to approximately 4 h. Additionally, the target cells must be pre-labelled prior to incubation with drug or effector cells, and the final preparation for counting as well as counting itself is time consuming. In one comparative study which evaluated ^{51}Cr release as an endpoint for anticancer drug cytotoxicity testing the method was found to be of no value (39).

ii. Enzyme-release assays

Enzyme-release assays for measuring membrane integrity have been developed which overcome some of the problems described for the ^{51}Cr-release assay. Different enzymes have been used, though LDH has been found to be generally useful, since it is released by a range of cell types (40). This original paper describes the assay in the context of determining cytotoxicity of antibody-dependent and T-cell mediated reactions. The assay also has application in the wider context of toxicity testing, and its use has been described as part of a panel of assays for investigating hepatotoxicity (41).

iii. Dye exclusion

Viability dyes which have been used to determine membrane integrity include trypan blue, eosin Y, naphthalene black, nigrosin (green), erythrosin B, and fast green. Staining for viability assessment is more suited to suspension cultures than to monolayers, because dead cells detach from the monolayer and are therefore lost from the assay. A major disadvantage may be the failure of reproductively dead cells to take up dye, as was demonstrated when cells with impaired clonogenicity showed 100% viability according to dye exclusion (39). The method has been renovated, however, and technical innovations introduced which attempt to circumvent some of the problems commonly associated with such assays. In the methodology developed by Weisenthal *et al.* (42) a 4-day assay period is used to permit reproductively dead cells to lose their membrane integrity and the inaccuracies produced by either overgrowth of viable cells or lysis of dead cells is compensated for by incorporation of fixed duck erythrocytes as an internal standard. Comparison of cell counts versus percentage viability versus viable cell/duck cell ratio demonstrated the increased sensitivity of the latter method (*Figure 4*). The

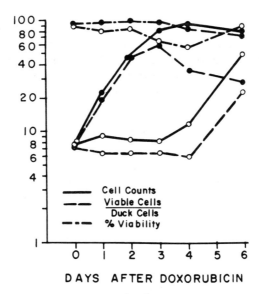

DAYS AFTER DOXORUBICIN

Figure 4. Effect of doxorubicin (0.12 μ ml^{-1} h^{-1}) on MDAY-D2 cells as assessed by three different techniques: Coulter counter particle counts (⎯) (*y*-axis units are cells/ml × 10^{-4}); ratio of living tumour cells to duck red blood cells, normalized to the same scale as the Coulter counts (⎯ ⎯ ⎯), and percentage viability (⎯ . ⎯) *y*-axis units are percentage viability (living cells/living and dead cells × 100). ●, control cultures; ○, doxorubicin-treated cultures. (Reproduced with permission of the publishers from ref. 42.)

method has been applied with equal success to solid tumours, effusions, and haematological malignancies.

4.2.2 Respiration and glycolysis

Drug-induced changes in respiration (oxygen utilization) and glycolysis (carbon dioxide production) have been measured using Warburg manometry, both parameters showing dose-related depression (43, 44). Other authors have determined dehydrogenase activity by incorporating methylene blue into agar containing drug-treated cells, cell death being indicated by non-reduction of the dye (45). The latter method has the disadvantage of being non-quantitative, whilst the former, although quantitative, has not been widely adopted because the technical manipulations involved are extensive and unsuited to multiple screening. The more direct approach of monitoring pH changes in cultures containing an appropriate pH indicator has also been described (46).

4.2.3 Radioisotope incorporation

Measurement of the incorporation of radiolabelled metabolites is a frequently used endpoint for cytotoxicity assays of intermediate and short-term duration.

i. Nucleotides

Measurement of [^{3}H]-thymidine incorporation into DNA and [^{3}H]-uridine incorporation into RNA are two of the most commonly used methods of quantitation of drug cytotoxicity (13, 19, 20, 22). In short-term assays, which do not include a recovery period, there are a number of disadvantages, all of which relate to a failure of [^{3}H]-thymidine incorporation to reflect the true DNA synthetic capacity of the cell. These are:

(a) Changes may relate to changes in size of the intracellular nucleotide pools rather than changes in DNA synthesis.

(b) Some drugs such as 5-fluorouracil and methotrexate which inhibit pyrimidine biosynthesis (*de novo* pathway) cause increased uptake of exogenous [^{3}H]-thymidine due to a transfer to the 'salvage' pathway, which utilizes pre-formed pyrimidines.

(c) Continuation of DNA synthesis in the absence of [^{3}H]-thymidine incorporation can occur (47).

The choice of isotope appears, to some extent, to be dependent on the drug and should probably be evaluated in pilot studies. The low labelling index of human tumours with resultant low levels of nucleotide incorporation in short-term assays necessitates the use of high cell densities, which can restrict the number of drugs and range of concentrations tested when cell numbers are limited. Two different 'hybrid' techniques have recently been reported, which combine the 'stromal cell inhibition' advantage offered by the soft agar culture system with the facilitated quantitation offered by the use of radio-isotopes. Both assays are of intermediate duration (about 4 days) and use [³H]-thymidine incorporation into DNA as an endpoint; in one method the cells are grown in liquid suspension over soft agar (48), whilst in the other the cells are incorporated in the soft agar (49).

Given that a homogeneous cell population is available, [³H]-nucleotide incorporation can be used after an appropriate recovery period to measure survival or, in the presence of drug, to measure an antimetabolic effect, but with the reservations expressed above.

ii. [¹²⁵I]-Iododeoxyuridine ([¹²⁵I]-Udr)

[¹²⁵I]-Udr is a specific, stable label for newly synthesized DNA which is minimally re-utilized and can therefore be used over a 24 h period to measure the rate of DNA synthesis (50); quantitation is facilitated because the isotope is a gamma emitter. Disadvantages include its variable toxicity to different cell populations, which therefore means that more cells are required because [¹²⁵I]-Udr must be used in low concentrations.

iii. [³²P]-Phosphate (³²P)

The rate of release of ³²P into the medium from pre-labelled cells is a function of the cell type and is increased in damaged cells. This has been used as a measure of drug efficacy (51). The incorporation of ³²P into nucleotides has also been used as an index of drug cytotoxicity (52). Neither method has been routinely adopted.

iv. [¹⁴C]-Glucose

Glucose incorporation is used as a cytotoxicity endpoint because it is a precursor which is common to a number of biochemical pathways (53). The method has not been widely used.

v. [³H]-Amino acids

Protein synthesis may be considered as an essential metabolic process without which the cell will not survive, and incorporation of amino acids into proteins has been used successfully as an index of cytotoxicity. The most extensive studies have utilized monolayers of cells growing in microtitration plates, using either incorporation of [³H]-leucine (38) measured by liquid scintillation counting, or [³⁵S]-methionine incorporation, measured using autofluoro-graphy (54).

vi. ⁴⁵Calcium (⁴⁵Ca)

Unrelated compounds may produce alterations in the permeability of cell membranes to calcium, such that increased calcium uptake results. Measurements of ⁴⁵Ca uptake can therefore be used (55).

Anne P. Wilson

4.2.4 Total protein content

Protein content determination is a relatively simple method for estimating cell number. It is particularly suited to monolayer cultures, and has the advantage that washed, fixed samples can be stored refrigerated for some time before analysis without impairment of results, facilitating large-scale screening. Overestimation of cell number may arise with some drugs which inhibit replication without inhibiting protein synthesis (e.g. BrUdr, methotrexate). Assessment of cytotoxicity by this method requires the demonstration of an alteration in the accumulation of protein per culture against time, preferably taken at several points, or at one point after prolonged drug exposure and recovery, as described above.

4.2.5 Colorimetric assays

The advent of sophisticated microplate readers which allow rapid quantitation of colorimetric assays on a microscale has paralleled the development of a variety of assays which use some form of colour development as an endpoint for quantitating cell number. These include methods which reflect:

- protein content (methylene blue, Coomassie blue, Kenacid blue, sulphorhodamine B, Bichinoninic acid (BCA))
- DNA content (Hoechst 33342) or DNA synthesis (BrdU uptake)
- lysosome and Golgi body activity (neutral red)
- enzyme activity (hexosaminidase, mitochondrial succinate dehydrogenase)

Linear relationships between endpoint and cell number have been demonstrated for all the methods described. Discrimination between live and dead cells in monolayer assays is not relevant, since dead cells will usually detach provided the assay time is long enough for this to occur. In suspension cultures this aspect is relevant, however, and needs to be considered when choosing an appropriate assay.

i. Protein content

Several methods are available for measuring the protein content of cell monolayers. These include the use of the Folin–Ciocalteau reagent according to the method of Lowry (56), and amido black (6). Several new techniques for colorimetric determination of protein content are also available. Methylene blue binds to basophilic proteins and nucleic acids of fixed cells at pH 8.4 and has been used for microtitration plate assays (57), its main disadvantage being the need for extensive washing to remove unfixed dye. In an evaluation of 20 histological dyes for suitability in determining cellular protein or biomass, sulphorhodamine B was found to give best results over both high and low cell densities (58), and the method has been found to compare favourably with the MTT assay (59). Kenacid blue has also been used to measure protein content in cell monolayers (60), and this method is the standard cytotoxicity test used by FRAME (Fund for the Replacement of Animals in Medical Experiments, 34 Storey St., Nottingham).

Protein determination using Coomassie blue G-250 was first described in 1976 (61). Binding of protein to dye causes a shift in the absorption maximum of the dye from 465 nm to 595 nm, which can be monitored. The assay has gained widespread acceptance for protein determinations in solution because it has several advantages when compared with other assays (see ref. 61). It has been used successfully on cell monolayers without solubilization of protein (61), though its accuracy in this context is dependent upon even cell distribution since OD is only determined in one position on each well. The BCA method is a development of the biuret reaction in which protein in alkaline solution containing Cu^{2+} results in the formation of a coloured complex due to bonding between protein and Cu^{2+}. This is also the basis of the Lowry method, but the Folin–Ciocalteau reagent is replaced with BCA which is a sensitive, stable and highly specific reagent for Cu^{2+} and overcomes the problems associated with the use of the Folin–Ciocalteau reagent (62). The dye binding assay and the BCA assay were found to give comparable results using several different protein solutions (63, 64), the main differences being in the type of chemical which interferes with the reaction. The BCA assay does not appear to have been used for cell monolayers, although it seems to have potential since protein can be solubilized using either sodium hydroxide or detergent, neither of which interferes with the assay (63). Because the reactions are dependent upon a shift in absorption maxima they also have the theoretical advantage that destaining is not necessary.

ii. DNA content

DNA content may be measured in microtitration plates by determining the DNA-associated fluorescence obtained after staining with dyes whose fluorescence is enhanced by intercalation at A–T specific sites on the chromatin. Dyes which have been investigated include Hoechst 33342 and 2-diamidinophenylindole (DAPI) (65). DNA synthesis may also be determined using bromodeoxyuridine (BrUdr). The amount of BrUdr incorporated is detected immunohistochemically using a monoclonal antibody to BrUdr, and the binding may be quantitated using appropriate conjugates and chromogenic substrates (66).

iii. Lysosomal and Golgi body activity

The uptake of neutral red by lysosomes and Golgi bodies has been used to quantitate cell numbers (67, 68). The stain appears to be specific for viable cells, but the main limitation of the method is the difference in uptake between cell types. Thus, some cell types, such as activated macrophages and fibroblasts, take up large amounts very rapidly whereas others, such as lymphocytes, show negligible staining.

iv. Enzyme activity

N-acetyl-β-D-hexosaminadase is a lysosomal enzyme with a widespread distribution. Its presence can be detected colorimetrically using the substrate p-nitrophenol-N-acetyl-β-D-glucosaminide, in the absence of serum, which contains large amounts of the enzyme (69).

v. Tetrazolium dye reduction

The most widely used technique involves the use of a tetrazolium salt (MTT, or 3,(4,5-dimethylthiazol-2-yl)-2,5-diphenyl tetrazolium bromide) which is metabolized to a coloured formazan salt by mitochondrial enzyme activity in living cells. The method was first described in 1983 (69) as a rapid colorimetric method for immunological studies, and modifications for this application have been described (70). The technique is particularly useful for assaying cell suspensions because of its specificity for living cells; this consideration is not as important for monolayers since dead cells become non-adherent. One disadvantage of the technique is the need to use unfixed cells, which may impose time restraints. The potential of the technique for drug sensitivity testing of human tumours was recognized, and early reports of its use are promising (71–75). More recent publications describe some technical problems which affect interpretation of the assay results and need to be considered when planning protocols. These include the possibility of increased mitochondrial enzyme activity in drug-treated cells and the effect of medium-conditioning by cells on formazan production (76). The absorption spectrum of MTT-formazan has also been shown to be dependent upon pH and cell density (77), and a method has been described which overcomes these problems and increases linearity between MTT-formazan and cell density especially at high cell numbers (77).

4.3 Survival (reproductive integrity)

Survival assays give a direct measure of reproductive cell death by measuring plating efficiency either in monolayer or in soft agar. The endpoints which have been described in the previous section can also be used as an index of reproductive integrity providing the design of the assay incorporates a recovery period. Increases in total protein cell number or protein synthetic capacity have been taken to imply proliferative ability, although interpretation is more difficult due to differential responses in elements of a heterogeneous cell population.

4.3.1 Cloning in monolayer

Cells generally have a higher cloning efficiency in monolayer than they do in soft agar, and the method is frequently used for cell lines. Normal cells and tumour cells will form colonies in monolayer, and the method is not therefore applicable to tumour biopsy material, which commonly shows high levels of

stromal cell contamination, unless criteria are available to discriminate between tumour and stroma. Feeder layers of irradiated or mitomycin C-treated
cells can be used to increase plating efficiencies, and indeed small drug-
resistant fractions are more likely to be detected in the improved culture
conditions existing when feeder cells are used (78).

4.3.2 Cloning in soft agar

The advantages and disadvantages of this method have been discussed in
Section 3.1.5.

4.3.3 Spheroids

Various methods can be used to quantitate the effect of drugs on spheroidal
growth. These are:

- relative changes in volume of treated and untreated spheroids
- cloning efficiency in soft agar of disaggregated spheroids
- cell proliferation from spheroids adherent to culture surfaces

The first method is rather insensitive because spheroidal growth tends to
plateau, and the second may be affected by difficulties with disaggregation to
a single cell suspension and low plating efficiencies. The choice of endpoint is
largely a function of the individual characteristics of the particular spheroids
under study.

4.3.4 Cell proliferation

An increase in cell number in a proliferating cell line can be regarded as an
index of normal behaviour. Growth curves may be determined and the
doubling time during exponential growth derived. An increase in doubling
time is taken as an indication of cytotoxicity, but it must be stressed that this is
a kinetic measurement averaged over the whole population and cannot distinguish between a reduced growth rate of all cells and an increase in cell loss
at each cell generation. Estimates of cytotoxicity based on cell growth in mass
culture must utilize the whole growth curve, or they may be open to misinterpretation. If 50% of cells die at the start of the experiment, the growth rate of
the residue, determined in log phase, may be the same, but will show a delay.
In practice it is very difficult to distinguish between early cell loss and a
prolonged lag period where cells are simply adapting. For these reasons, cell
growth rates must be taken as only a rough guide to cytotoxicity, and accurate
measurements of cell survival and cell proliferation should be made by
colony-forming efficiency (survival) and colony size (proliferation).

5. Assay comparisons

In spite of the diversity of methodologies used for cytotoxicity and viability
testing, the same levels of correlation between *in vitro* sensitivity and *in vivo*
response have been reported when the methods have been applied to human

tumour biopsy material. A number of comparative studies have been under-taken, which generally indicate that appropriately designed cytotoxicity assays give results which are comparable to clonogenic assays (23, 24, 32, 39, 48, 72, 78, 79), and the relative merits of clonogenic and non-clonogenic assays have been discussed and reviewed (80).

6. Technical protocols

6.1 Drugs and drug solutions

6.1.1 Drug sources

Pharmaceutical preparations for intravenous administration frequently con-tain various additives which may themselves be cytotoxic. Such preparations are therefore not suitable for aliquotting by weight and, if they are used as a stock solution, the cytotoxicity of the additional components should be deter-mined in the assay system used. The problems can be avoided by obtaining pure compounds from the drug manufacturers.

6.1.2 Storage

It is recommended that dry compounds be stored at −20°C to −70°C over desiccant; this is especially important with compounds which are unstable in aqueous solution. It is routine practice to make up stock drug solutions which are then aliquotted and stored at −70°C. Storage at higher temperature is not recommended, and some drugs (nitrosoureas) are unstable even under these conditions (81). As some drugs may bind to conventional cellulose nitrate or acetate filters, sterile filtration must be carried out under controlled conditions to check for binding of drug. Use the maximum drug concentration and maximum volume to saturate the filter within the first few ml of filtrate, which may then be discarded. Alternatively, use non-absorbent filters, e.g. nylon. In either case the filtrate should be assayed to make sure no activity has been lost. In practice it has been found that handling of non-pharmaceutical prepa-rations of pure drugs under aseptic conditions (i.e. use of sterile blade for weighing out into sterile container) is sufficient to prevent contamination, even in experiments of prolonged duration.

6.1.3 Diluents

i. Solvents

Solvents commonly used for compounds which are not soluble in aqueous solution include ethanol, methanol, and DMSO. Different cell populations exhibit different sensitivities to these organic solvents, and appropriate sol-vent controls should therefore always be included. A minimum dilution factor of 1000 is recommended to avoid solvent toxicity.

ii. Medium components

Certain drugs bind avidly to serum proteins (e.g. *cis*-platinum), and the presence of serum in the incubation medium may therefore reduce the amount of available drug. Components of some media can protect cells against the cytotoxic effects of antimetabolites (e.g. thymidine and hypoxanthine protect against methotrexate, thymidine protects against 5-bromodeoxyuridine and 5-fluorodeoxyuridine). The presence of amino acids may also affect uptake of drugs which use the amino acid transport system. A comprehensive list has been detailed elsewhere (6).

6.1.4 Drug activation

Many components which are not themselves cytotoxic are converted to cytotoxic metabolites by the P450 mixed oxidase system of the liver. Active metabolites for drugs known to require metabolic conversion may be available, e.g. phosphoramide mustard and 4-hydroperoxycyclophosphamide for cyclophosphamide (82) but for drugs with less clearly characterized pathways of conversion some method of *in vitro* activation is necessary. Methods include the use of S9 microsome fractions, cultured rat or human hepatocytes, or liver biopsy material (83–86). The use of intact cells, in which the level of cofactors resemble those *in vivo* and are high enough to sustain the associated reactions, provides a closer approximation to the *in vivo* situation, and may give results which are different to those obtained using liver homogenates. Species differences exist in the complement of cytochrome P450 system found in the liver, and for this reason human hepatocytes may be preferred. The major problem with human hepatocyte culture is loss of cytochrome P450 activity with increasing time in culture, though appropriate conditions can reduce this (87).

6.2 Drug incubation

It is common procedure to incubate cells with drug solutions immediately after enzyme disaggregation of solid tissue, or harvesting of cell monolayers by trypsinization. There is evidence to suggest that susceptibility of cells to drug is altered by enzyme treatment and does not return to 'untrypsinized' levels until approximately 12 h after enzyme exposure (88). It may therefore be expedient to include a pre-incubation recovery period for freshly disaggregated cells to allow for this.

Maintenance of pH at 7.4 is essential during the incubation period since alterations in pH will alter cell growth, and alkaline pH particularly will markedly reduce cell viability.

Comparison of static versus non-static incubations of cell suspensions revealed marked differences in dose–response profiles (89), and cells should therefore be kept in continuous suspension to allow equal drug distribution. If the surface area:depth ratio of the incubation vessel is small, incubation in a water-shaker bath will not keep the cells in suspension, and intermittent shaking by hand is recommended.

6.3 Assay by survival and proliferative capacity

6.3.1 Clonogenicity

One of the most generally accepted methods of assaying for survival is the measurement of the ability of cells to form colonies in isolation. This is usually achieved by simple dilution of a single-cell suspension, and determination of survival by counting the colonies which form. A lower threshold must be set in line with the doubling time of the cells being studied and the total duration of the assay, usually five or six doublings (32 or 64 cells/colony).

As some drugs may have an effect on cell proliferation as well as, or instead of, survival *per se*, it may be necessary to do a colony size analysis as well. This may be done by counting the number of cells per colony (very tedious and only possible in small colonies), by measuring the diameter (prone to error if cell size or degree of piling up changes) or by measuring absorbance of colonies stained with 1% crystal violet.

Protocol 1. Monolayer cloning

Adherent cells are plated on to a flat surface of glass or tissue culture-treated plastic, allowed to grow and form colonies, stained, and counted. Drug treatment is best performed before subculture for plating, where highly toxic substances are being tested. For low-grade toxins, where chronic application is required, they may be applied 24–48 h after plating, and retained throughout the clonal growth period, provided they are stable.

1. Prepare replicate 50 ml (25 cm^2) flasks, two for each intended concentration of drug, and two for controls.

2. When cultures are at the required stage of growth (usually mid log phase but, in special circumstances, the plateau phase of the growth cycle) add the drug to the test and solvent to the control for 1 h at 37°C.

3. Remove the drug, rinse the monolayer with PBS, and prepare a single-cell suspension by conventional trypsinization (see Chapter 1).

4. Count the cells in each suspension and dilute to the appropriate cell concentration to give 100–200 colonies per Petri dish (5 or 6 cm). This figure depends on the plating efficiency of the cells and the effect of the drug (e.g. control plates with no drug from a cell culture of a known plating efficiency of 20% will require 500–1000 cells per dish; at the ID$_{50}$ of the drug they will require 1000–2000 cells per dish, and at the ID$_{90}$, 5000–10 000 cells per dish). A trial plating should be done first, to determine the plating efficiency of the cells, and to determine approximately the ID$_{50}$ and ID$_{90}$ of the drug. In practice it is more usual to set up dishes at two cell concentrations, one to give a satisfactory number of colonies at low drug concentrations and controls, and one for higher drug concentrations, with

Protocol 1. *Continued*

some overlap in the middle range. Experience will usually determine where this is likely to fall.

5. Plate out the appropriate number of cells per dish and place in a humid incubator at 37°C with 5% carbon dioxide.

6. Grow until colonies form. For rapidly growing cells (15–24 h doubling time) this will take 7–10 days; for slower growing cells (36–48 h doubling time) 2–3 weeks are required. In general, for a survival assay the colonies should grow to 1000 cells or more on average (10 generations). As the colonies increase in size the growth rate (particularly of normal cells) will slow down as the colonies tend to grow from the edge, and the slower growing colonies will tend to catch up. Hence if cell proliferation (colony size) is the main parameter, clonal growth should be determined at shorter incubation times giving smaller colonies with a wider size distribution.

7. Rinse dishes with PBS, fix in methanol or gluteraldehyde and stain with 1% crystal violet, rinse in running tap water, and dry.

8. Count colonies above threshold and calculate as a fraction of control. Plot on a log scale against drug concentration.

Protocol 2. Clonogenicity in soft agar using a double layer agar system

The following procedure utilizes a 1 h drug exposure period, and gives four replicates per test condition.

1. Prepare 35 mm Petri dishes with a 1 ml base layer of 0.5% agar in growth medium by mixing 1% agar (melted by autoclaving or boiling) at 45°C with an equal volume of double-strength medium at 45°C and dispensing 1 ml aliquots in a pre-heated pipette. Note that autoclaved agar may be toxic (89).

2. Prepare a single cell suspension of the target cell population (see Chapters 1 and 6), adjust the cell concentration in growth medium to give 20 times the final concentration desired at plating and store at 4°C until ready to use.

3. Prepare drug dilutions in growth medium and aliquot out 900 µl of each concentration into duplicate tissue-culture-grade tubes, including control tubes containing growth medium only and appropriate solvent controls. Because of the time-span involved when plating out multiple drugs and concentrations, additional controls are recommended for plating out at the beginning, middle, and end of setting up the experiment. The controls thus incorporate variability due to the time involved in setting up the assay.

4. Check that the stock cell suspensions still comprises single cells, and add 100 μl to each of the prepared tubes.

5. Incubate the tubes at 37°C for 1 h (see Section 8.2).

6. At the end of the incubation period centrifuge the tubes (2 min, 100 g) and wash the cells in 5 ml of saline. Repeat once more, and resuspend the cell pellet in 2 ml of growth medium. Keep the cells on ice to maintain cell viability whilst plating out replicates.

7. Centrifuge the duplicate set of tubes for one drug concentration; remove medium and add 2 ml of warmed growth medium containing 0.3% agar to each tube. Needle gently to disperse cell aggregates and plate out 1 ml aliquots on to each of four prepared bases. The final plating out is most easily accomplished using a 1 ml micropipette (Finn pipette or equivalent); if about 2 mm is cut off the end of the tips this prevents problems due to blockage of the small aperture by solidified agar.

8. Put the dishes to solidify on a cooled, horizontal surface.

9. Repeat for all test conditions, including controls at appropriate intervals throughout.

10. Incubate the plates in a humidified atmosphere of 95% air/5% carbon dioxide.

11. Score the plates for colonies when control colonies have reached a pre-determined size. This is usually more than 50 cells for cell lines, but a size of more than 20 or 30 cells has been used when the population has a slow growth rate and a low plating efficiency, as with human tumour biopsies.

6.3.2 Modifications

i. The 'Courtnay' method

The 'Courtnay' method for suspension cloning (89) utilizes rat red blood cells as feeder cells, and a 5% oxygen tension. The procedure outlined above may be used, with appropriate modification of final plating conditions.

ii. Feeder cells

A linear relationship between plating efficiency and plated cell numbers may not occur from low to high cell numbers. When cell lysis has occurred due to drug treatment, the reduction in cell numbers can reduce the plating efficiency disproportionately in relation to seeding density rather than reflecting the clonogenicity of the remaining cell population. This problem can be circumvented by incorporating homologous feeder cells in the assay which have either been lethally irradiated with a ^{60}Co or ^{137}Cs source, or treated for approximately 12 h with 2 μg/10^6 cells of mitomycin C. The radiation dose needs to be established for each cell type, e.g. 20–30 Gy for lymphocytes, 60 Gy for lymphoid cell lines.

iii. Use of 2-(p-iodophenyl)-3-(p-nitrophenyl)-5 phenyl tetrazolium chloride (INT)

Viable cells reduce colourless tetrazolium salts to a water-insoluble coloured formazan product, and the reaction has been used to distinguish viable colonies from degenerate clumps when scoring finally. It should be noted, however, that viable colonies may become degenerate due to nutrient deficiency which can cause misleading results. The stain is made by dissolving INT violet in buffered saline to a final concentration of 0.5 mg/ml; dissolution is slow and the stain needs to be prepared 24 h prior to use. Add 0.5–1 ml to each 35 mm Petri dish and incubate overnight at 37°C. Viable colonies then stain a reddish-brown colour.

iv. 20% oxygen versus 5% oxygen

The lower oxygen tension, which more closely resembles physiological levels of oxygen, is recommended for clonogenic assays because it results in higher cloning efficiencies. There is evidence to suggest that it also modifies the chemosensitivity profile of the cell population (91), producing enhanced cytotoxicity.

6.3.3 Spheroids

The experimental protocols outlined below are based on techniques described elsewhere (92). Three endpoints may be used to determine the cytotoxic effect of drugs on spheroids; these are:

- volume growth delay
- clonogenic growth
- outgrowth as a cell monolayer

Pre-selection of similarly sized spheroids and drug incubation is a common starting point.

Protocol 3. Spheroid incubation

1. Pre-select spheroids in the chosen range. Sizes in the range 150–250 μm are just visible to the naked eye and can be selected using a Pasteur pipette.
2. The method of incubation with drug depends to some extent on the exposure period to be used. For 1 h exposures, which have been most commonly utilized, incubate spheroids in agar-coated (0.5–1%) Petri dishes or in glass universal containers. For longer exposure times, incubate the spheroids in a spinner vessel which will keep them in continuous suspension and prevent adherence to the vessel walls.
3. At the end of incubation rinse the spheroids in two to three 5-min washes of drug-free medium.

Treated spheroids can be grown as a mass culture and the volume of individual randomly chosen spheroids determined at set time intervals. However, from a statistical viewpoint it is recommended that successive measurements are made on individually isolated spheroids placed in either 24-well or 96-well multidishes with agar-coated bases. The smaller wells can be used for spheroids up to 600 μm in diameter.

Protocol 4. Volume growth delay

1. Plate out 12–24 spheroids per drug concentration and 24 controls into individual wells. For a 24-well plate use 0.5 ml of 1% agar for the base and add the spheroid in 1 ml of growth medium.

2. At 2–3 day intervals measure two diameters (x, y) at right angles to each other, using an eyepiece graticule in an inverted microscope. Calculate the size of the spheroid as 'mean diameter' $(\sqrt{x, y})$ or as 'volume'

$$\frac{4}{3}\pi\left(\frac{\sqrt{xy}}{2}\right)^3$$

and use to construct growth curves for drug-treated and control spheroids. Results are usually normalized to pre-treatment spheroid size and expressed as V_t/V_0 where V_t is the volume at time t and V_0 is the initial volume. An example is shown in *Figure 5* for spheroids derived from the xenograft of an adenocarcinoma of the lung (93). The method assumes that the spheroid is symmetrical, but flattening of spheroids to a dome shape has been observed for some cell lines, and this will lead to over-estimation of volume.

Protocol 5. Clonogenic growth

1. Disaggregate spheroids to a single-cell suspension using appropriate conventional treatment with 0.25% trypsin, or 0.125% trypsin plus 1000 units/ml type I CLS grade collagenase (Worthington).

2. Adjust the cell number to 20–200/ml for cell lines, or up to 10^3/ml for primary cloning. If cloning in suspension use 10^4/ml for cell lines and up to 5×10^5/ml for primary cloning.

3. Score for colonies when the designated size criterion in controls has been attained.

Obtaining a pure single-cell suspension may be virtually impossible, and plating efficiencies may be very low. The following method described in *Protocol 6* circumvents these problems.

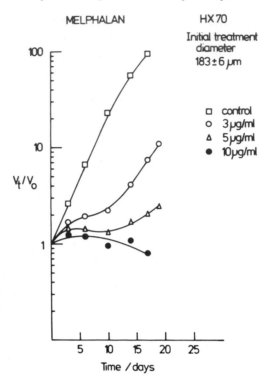

Figure 5. The effect of increasing concentrations of a 1 h exposure of melphalan on the growth of adenocarcinoma lung-derived spheroids. V_t, volume at time interval shown; V_o, volume at start of experiment. Significant differences ($p < 0.001$; Student's t test) were observed at day 14 between control and 3.0 µg/ml, and 3.0 µg/ml and 10.0 µg/ml. The diameter shown on the figure is the mean diameter ± s.e. of the plated spheroids. (Reproduced with permission of the publishers from ref. 94.)

Protocol 6. Outgrowth as cell monolayer

1. Allow individual spheroids to attach to individual wells of tissue-culture plastic microtitration plates.

2. After 2–3 weeks remove the central spheroid and determine the number of cells in the monolayer which has proliferated from the spheroid. This may be done directly using cell counts of the detached cell population. Alternatively, the methodology described for quantitation of drug effect on cell monolayers in Protocols 8 and 9 could be adopted.

6.4 Cytotoxicity assays

The following methods have all been developed as automated cytotoxicity assays in microtitre plates. The same procedure for drug exposure of cells can be used as a starting point for all the methods described.

Protocol 7. Preparation of plates

1. Add 200 μl of cells in growth medium to each well of a microtitration plate. Evaporation from outer wells may be excessive (edge effect) resulting in poor cell growth, and it is therefore recommended that a 10 × 6 matrix is used for the cytotoxicity assay. Edge effects due to evaporation can be reduced by using plate sealers (Mylar, Flow Labs). Some drugs are volatile, or give off volatile metabolites (e.g. formamides release formaldehyde) and this can cause variable non-specific cytotoxicity in adjacent wells. The use of sealers minimizes this effect. Outer wells may be filled with either saline or cell suspension. The appropriate cell concentration depends on the cells used, and varies as a function of growth at low density, cell size, rate of growth, and saturation density. The plating conditions need to be optimized to ensure that cells are in exponential growth phase over the period of the assay. That is, cells are exposed to drugs after lag phase, and the assay is terminated before, or as, controls reach confluence.

2. When the cells are in exponential growth, remove medium and add drugs serially across the plate using replicates of 3–6 wells per concentration and including appropriate solvent controls. Add fresh growth medium to control wells.

3. Incubate plates for selected drug exposure times at 37°C in 5% carbon dioxide, re-feeding with fresh medium or drugs at appropriate intervals.

4. Remove medium and wash wells three times with 200 μl of saline to remove drugs. Add 200 μl of fresh growth medium to each well.

5. Incubate plates for selected recovery period, changing medium if necessary.

6. Proceed to selected endpoint. Note that it is important to avoid damaging the cell monolayer during wash procedures. This is best done by tilting the plate to 45° and inserting the pipette tip in the angle between the base and side of the well.

Protocol 8. Protein determination of cell monolayers using methylene blue

This protocol is based on the procedure described by Pelletier *et al.* (57).

1. Remove medium from wells and wash twice with 200 μl of PBS including wells without cells for reagent blank.

2. Add 200 μl of 10% v/v formalin in PBS and fix cells at room temperature for 10 min.

Protocol 8. *Continued*

3. Remove formalin, and wash cells twice with 200 μl of borate buffer (0.01 M, pH 8.4, 3.8 g/litre in distilled water).

4. Add 100 μl of methylene blue per well (1% w/v in borate buffer) and stain for 10 min at room temperature.

5. Remove methylene blue and wash extensively with borate buffer, blotting the plate surface between washes. A minimum of five washes is needed to remove excess dye.

6. Leave washed plates to dry at room temperature for 2–3 h.

7. Add 200 μl of 0.1 M HCl per well and shake plate for 15 min at room temperature using a plate shaker (Dynatech) to solubilize the stain.

8. Read OD in microplate reader (Dynatech, Molecular Devices) at 660 nm (maximum absorbance).

Protocol 9. Protein determination of cell monolayers using sulphorhodamine B (SRB)

This protocol follows that developed by Skehan (58) and described recently by Fricker (59).

1. Remove medium from the cells and fix in 10% trichloracetic acid for 30 min at 4°C.

2. Wash cells five times with tap water.

3. Add 100 μl of SRB (0.05%–0.1%) in 1% acetic acid and stain for 15 min at room temperature. (The concentration of SRB needs to be optimized for each batch of dye.)

4. Wash cells four times in 1% acetic acid.

5. Air dry plates and dissolve stain in 200 μl 10 mM unbuffered Tris.

6. Measure absorbance at 540 nm, using a reference filter of 630 nm.

Protocol 10. Protein determination of cell monolayers using Kenacid Blue (FRAME cytotoxicity test)

This protocol is based on procedures described by Knox *et al.* (60) and modified as described in INVITTOX Protocol number 3 (93).

1. Wash cells three times with PBS and fix in 150 μl glacial acetic acid/ ethanol/water (1:50:49 v/v/v) for 20 min at room temperature.

2. Remove fixative and stain with 150 μl Kenacid Blue solution. (The stain is made as follows: dissolve 0.4 g Kenacid Blue (BDH) in 250 ml ethanol and 630 ml water (stock); immediately before use add 12 ml of glacial acetic

acid to 88 ml of stock solution and filter.) Stain for 20 min at room temperature with gentle agitation.

3. Remove unbound stain using three washes with ethanol/glacial acetic acid/water (10:5:85, v/v/v). The final wash should be done for 20 min with agitation.

4. Add 150 μl of destain solution per well (1 M potassium acetate in 70% ethanol), and agitate rapidly for 20 min until homogeneous colour distribution is obtained.

5. Measure absorbance at 577 nm against a reference filter of 404 nm.

Protocol 11. Determination of amino acid incorporation into cell monolayers

This protocol is based on a cytotoxocity assay first described by Freshney *et al.* (38).

1. At the end of the assay period add 50 μl of 5–20 μCi (185–740 kBq)/ml [^{3}H]-leucine (L-4,5 [^{3}H]-leucine, Amersham plc) in growth medium to each well and incubate for 3 h at 37°C.

2. Remove the isotope and wash the cell monolayer three times in PBS.

3. Add 100 μl of PBS and 100 μl of methanol to each well, remove and add 100 μl of methanol per well. Fix the cells for 30 min at room temperature (this prevents detachment during subsequent processing).

4. Remove methanol and air-dry plates. Fixed plates can be stored for at least 48 h at 4°C before further processing.

5. Put plates on to ice and wash the monolayers in three 5 min washes of ice-cold 10% trichloracetic acid (TCA). Following fixation, wash solutions can be removed by inversion of the plate and shaking sharply.

6. Wash off TCA with methanol, using three washes from a wash bottle and air-dry the monolayer.

7. Add 100 μl of 1 M NaOH to each well; leave overnight at room temperature to solubilize protein.

8. Transfer NaOH to minivials and add 2.4 ml of scintillant (e.g. Fisofluor, Fisons) to each vial followed by 100 μl of 1.1 M HCl to acidify the contents.

9. Cap the vials, mix to homogenize and clarify the contents, and count for 5–10 min on a β-counter.

The method can also be used for non-adherent cell suspensions provided the plates are centrifuged at 1000 g. for 3–5 min prior to medium removal.

The previous methodology using liquid scintillation counting is labour-intensive during processing, and can cause problems when many plates are processed, particularly when β-counting facilities are restricted. Minivials tend to lose scintillant through their walls after some days storage and, although this can be reduced by storing at 4°C prior to counting, it does mean that samples should be counted as soon as possible after preparation. An alternative method (*Protocol 12*), which is well suited to automation, has been developed by Freshney *et al.* (54).

Protocol 12. Measurement of amino acid incorporation into protein using autofluorography

1. Perform the procedures described in *Protocol 11*, steps 1–4, using 5 μCi (185 kBq)/ml of [^{35}S]methionine (42 Ci/mol (1.55 GBq/mmol)) to label the cells in step 1; [^{3}H]leucine can be used, but longer development times are required.

2. Add 50 μl of scintillant to each well and centrifuge the open plates in microtitration plate carriers at 500 g. for 1 h at room temperature. Even evaporation of scintillant is obtained using this method which gives a flat layer of scintillant on the base of the well, and therefore improves resolution of spots. Sodium salicylate may be substituted for conventional scintillant, avoiding solubilization of the plate with some solvent-based scintillants, and evolution of potentially toxic toluene vapour during evaporation. The original method used toluene-based scintillation fluid, which is potentially toxic on evaporation. Recent developments have produced non toxic, solvent-based scintillants, such as Ecoscint, which should work as well.

3. Place a sheet of X-ray film under the plate in a dark room, and secure it with a layer of polyurethane sponge and a pressure plate (metal or glass) using adhesive tape or bulldog clips. Expose the plates in a light-proof box at −70°C with desiccant. An exposure time of 5 days gives an absorbance of 0.92 when 3000 cells per well are exposed to 10 μCi (370 kBq)/ml of [^{35}S]methionine (84 μCi (3.1 mBq)/mol).

4. Remove the film and develop for 5 min in Kodak D19 at 20°C. Fix in Ilfofix for 4 min, wash in Hypo clearing agent for 2 min, and tap water for 5 min. Dry the film.

5. The results are quantitated by scanning the images on a scanning densitometer (e.g. Chromoscan, Joyce Loebl) with a thin-layer attachment using an 11 mm circular aperture and a blue (465 nm) filter. ID$_{50}$ values can be obtained directly from the OD readings.

Protocol 13. Incorporation of [³H]-nucleotides into DNA/RNA

This procedure can be used for cell monolayers which have been harvested using enzymes or for cell suspensions.

1. Add isotopes to each well (2.5 μCi (92.5 KBq)/ml final concentration; e.g. 6-[³H]-uridine (22 Ci (814 GBq)/mmol)); methyl-[³H]-thymidine (5 Ci (185 GBq)/mmol); 6-[³H]-deoxyuridine (25 Ci (925 GBq)/mmol) and incubate for 1–3 h at 37°C.

2. Remove isotope and wash cells in three washes of PBS to remove unincorporated radioactivity. When using cell suspensions, plates must be centrifuged prior to medium removal as in *Protocol 11*.

3. For cell monolayers, add 100 μl of appropriate enzyme solution and incubate cells at 37°C to detach cells. Shake plate on a plate shaker for a few minutes to ensure cell detachment is complete.

4. Harvest the cells on to filter paper using cell harvester (e.g. Titertek, Dynatech, Inotech), and wash filters three times with 10% TCA.

5. Wash filters with methanol to remove water, and air dry.

6. Transfer filters to minivials and add about 3 ml of an emulsifier cocktail scintillant (e.g. Beckman Ready Solv HP). A system for counting dry filters is now available (Inotech).

7. Leave the samples for at least 1 h in the dark before counting, to remove chemiluminescence, which produces artefactually high counts.

8. Count the samples for 5–10 min or as long as is required to reduce counting error to ±5%.

Problems associated with the use of the heterogeneous counting systems as described in these protocols have been detailed in an excellent technical review (95).

The modified procedure in *Protocol 14* has been described for measuring [³H]-thymidine incorporation into DNA in cells growing in liquid suspension over soft agar in Petri dishes (48).

Protocol 14. Measurement of [³H]-thymidine incorporation

1. Add 25 μl of methyl-[³H]-thymidine (6.7 Ci (248 GBq)/mM diluted to 75 μCi (2.78 MBq)/ml in PBS) to cells for 24 h.

2. Harvest the cells and wash in PBS.

3. Add 5% TCA at 4°C to the washed cells for 30 min.

4. Centrifuge and repeat step 3 for 10 min.

5. Wash in methanol.

Protocol 14. *Continued*

6. Resuspend the pellet in 0.5 ml of 10× hyamine hydroxide, and heat at 60°C for 1 h.

7. Place in 12 ml of Hydrofluor scintillation fluid and count.

Protocol 15. Total DNA synthesis measured by [^{125}I]UdR uptake

The methodology is based on descriptions in the literature (50).

1. Add 0.06 µCi (2.22 KBq)/ml [^{125}I]UdR to growth medium of monolayer cultures for 24 h.

2. Wash the culture three times in saline and add fresh growth medium.

3. After 6 h, repeat the wash procedure to remove unbound label and label released from lysed dead cells, and count bound ^{125}I activity in a gamma counter.

As for the microtitration assay, confluence should be avoided in control tubes because overcrowding leads to reduced [^{125}I]UdR incorporation, and therefore underestimation of cell kill. When cells have been grown as monolayers in inclined test tubes the upper limit of the assay was found to be about 2×10^5 cells, and the lower limit about 2×10^4 cells.

Protocol 16. Measurement of viable cell number using neutral red

This protocol is based on a procedure described by Fiennes *et al.* (68).

1. Make up an aqueous stock solution of 0.1% w/v neutral red, autoclave and acidify with 2 drops glacial acetic acid/100 ml. Store in the dark at 4°C for up to 3 weeks. Immediately before use dilute the stock 1 in 10 using warm PBSA and check that precipitation of neutral red does not occur by adding a drop to a cold glass slide.

2. Add 200 µl of neutral red to each well of the microtitration plate and incubate for 90 min at 37°C.

3. Remove dye and wash cells with two washes of warm PBS.

4. Extract dye by adding 100 µl of absolute ethanol: 0.1 M citrate buffer, pH 4.2 (21.01 gM citric acid + 200 ml of 1 M NaOH per litre (A); 60 ml A + 40 ml 0.1 M HCl), mixed 1:1 v/v.

5. Agitate for 20 min at room temperature and read OD at 540 nm on a standard Elisa plate reader.

Protocol 17. Measurement of viable cell number using MTT

The protocol is based on a method described by Plumb *et al.* (77), which eliminates the effect of pH on the results of the assay and gives a linear relationship between cell number and formazan production at low and high cell densities.

1. Remove medium and add 200 μl of fresh medium per well, containing 10 mM Hepes (pH 7.4).

2. Immediately add 50 μl MTT (Sigma) in PBS at a concentration from 2–5 mg/ml. The optimal concentration may need determining for individual cell lines.

3. Wrap the plates in foil and incubate for 4 h at 37°C.

4. Remove the medium and MTT and immediately add 200 μl of DMSO per well, followed by 25 μl of Sorensen's glycine buffer (0.1 M glycine plus 0.1 M NaCl equilibrated to pH 10.5 with 0.1 M NaOH).

5. Read the OD of the plates at 570 nm on an Elisa plate reader.

7. Interpretation of results

7.1 Relationship between cell number and cytotoxicity index

The validity of a cytotoxicity index is dependent upon the degree of linearity between cell number and the chosen cytotoxicity parameter, and this should be established for any cytotoxicity assay. In clonogenic assays a linear relationship may not occur at low cell numbers due to the dependence of clonogenic growth on conditioning factors, whilst at high cell densities linearity is lost due to nutritional deficiencies. In cytotoxicity assays linearity may be lost at the upper end due to density-dependent inhibition of the relevant metabolic pathway, whilst the sensitivity limit of the assay may affect linearity at the lower end. This would cause apparent stimulation at low drug levels and an overestimation of cell kill at higher concentrations. Control cell numbers at the end of the assay must therefore fall on the linear portion of the curve. The accuracy at higher levels of cell kill is dependent upon the range over which linearity extends, and influences the number of decades of cell kill which can be measured. If results are plotted on a log scale this may imply that the assay is accurate down to 3–4 decades of cell kill, which should be confirmed before expressing results in this way. This is particularly important in *in vitro* drug combination studies when synergism or additivity is often observed beyond the second decade of cell kill.

7.2 Dose–response curves

Results are commonly plotted as dose–response curves using a linear scale for percentage inhibition (of isotope incorporation, for example) and a log scale for surviving fraction on clonogenic tests. Assay variation for replicate points is routinely depicted as mean ± standard deviation; a minimum of three replicates is therefore required for each test point. Some method is required for defining the sensitivity of a cell population in relation to other cell populations, or different test conditions; several parameters are available and are shown in *Figure 6*.

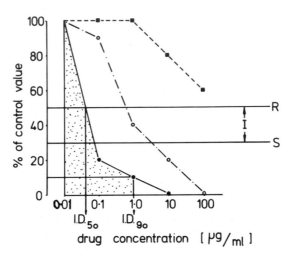

Figure 6. Interpretation of results using linear plot of response against drug concentration. ID_{50}, and ID_{90} are the concentration required to reduce cytotoxicity index to 50% or 90% of control values. is the area under curve between 0 and 1 µg/ml (for peak plasma concentration of 10 µg/ml). *S*, cut-off boundary for sensitivity; *R* cut-off boundary for resistance; *I*, intermediate zone. ●—●, sensitive population; ○—○, intermediate population; ■—■, resistant population.

7.2.1 Area-under-curve method

The use of this method acknowledges the probability that the shape of the dose–response curve may be instrumental in influencing the outcome to drug exposure, rather than cell kill at any one concentration. It is calculated using the trapezoidal method which adds the area of rectangles and triangles under the survival curve. The method has been applied most extensively in the 'Human Tumour Stem Cell Assay' (96).

7.2.2 Cut-off points for definition of sensitivity and resistance

If plots of dose–response curves from comparisons of cell lines show that they maintain their relative sensitivity rankings at different concentrations (i.e.

crossing-over of dose–response curves is minimal), then information on the relative sensitivities of different cell lines can be obtained by defining sensitivity at one concentration, and this is the most commonly used method for *in vitro* predictive testing. When retrospective correlations between *in vitro* data are made for defining these cut-off points, an intermediate zone is found where tumours cannot be defined as sensitive or resistant, and there is no clear-cut correlation between *in vitro* results and clinical response. The size of the intermediate zone will be at least partly related to the inherent variability of the assay, larger zones being associated with higher standard deviations. Although sensitivity may be defined at one concentration it is recommended that more than one concentration is tested, particularly in the developmental stage of the assay.

7.2.3 ID_{50} and ID_{90} values

Cell sensitivity may also be defined by the ID_{50} and ID_{90} values (i.e. drug concentration required to inhibit viability/cytotoxicity by 50% or 90%).

7.2.4 Correlation between *in vitro* and *in vivo* results

Criteria for defining tumours as sensitive or resistant are based on retrospective correlations between *in vitro* results and clinical responses, using a training set of data. Even when a laboratory is using an established method for tumour sensitivity testing, 'own laboratory' sets of training data should be obtained to allow for inter-laboratory variation. The response of patients with tumours of intermediate sensitivity may be influenced by prognostic factors other than tumour sensitivity (e.g. tumour burden at onset of chemotherapy, stage of disease, histology, tumour cell doubling time, previous chemotherapy, and performance status). When analysing results for correlations some attempt to stratify patients according to these parameters may assist in providing more meaningful data. Quantitative assessment of tumour response is also of paramount importance. It is pointed out that *in vitro* chemosensitivity can be expected only to indicate that *some* degree of cell kill will be achieved *in vivo*, not that the patient will achieve a complete response to treatment, the latter being under the influence of other factors also. The true positive correlation rate of an assay is defined as

$$\frac{S/S + S/R}{S/S} \times 100\%$$

where the numerator of each fraction is the *in vitro* response and the denominator is the *in vivo* response. The true negative rate is defined as

$$\frac{R/R + R/S}{R/R} \times 100\%$$

In assessing the significance of the correlation rates obtained, these should be compared with the correlation rates which would be obtained were the *in vitro* results randomly distributed (97). For example, a drug gives a 50% response rate *in vivo*, and 50% of tumours show *in vitro* sensitivity to this drug. If the *in vitro* results are randomly distributed between sensitivity and resistance, then the chances of obtaining a positive correlation between *in vitro* sensitivity and *in vivo* response are 50% of 50% (i.e. 25%), and also of obtaining a positive correlation between *in vitro* resistance and *in vivo* resistance. The overall apparent positive correlation rate is therefore 50%. *In vitro* versus *in vivo* correlations are also complicated by the use of combination regimes to treat patients. Strictly speaking, correlations should be made only when *in vitro* data is available for all drugs used. Whether or not they are tested in combination depends on the treatment protocol, since some drugs are administered sequentially. Also, if the assay can only measure two decades of cell kill it may be too insensitive to detect additive or synergistic effects.

8. Pitfalls and trouble shooting

Problems which may be encountered with these assays include:

- large standard deviations
- variability between assays done on the same cell population
- stimulation to above control levels

8.1 Large standard deviations

Possible reasons for large standard deviations include:

(a) Faults in aliquotting cell suspension, which are most likely to be made due to inadequate mixing of cell suspension during dispensing leading to uneven distribution of cells between replicates.

(b) The presence of large cell aggregates in the original cell suspension, leading to uneven distribution of cells between replicates.

(c) Non-specificity of cytotoxicity endpoint (e.g. due to measurement of non-specific binding of radioactivity—see *Protocol 10*).

8.2 Variation between assays

Replicate assays on different days cannot be performed on human tumour biopsy material to check day to day reproducibility, but this can be evaluated using cell lines. It is a recognized problem that cell lines which show consistent sensitivity profiles may show 'deviant' results occasionally, for reasons which cannot be identified. Specific reasons for failure to obtain reproducible results may include:

(a) Failure to harvest the cell population at an identical time point (e.g. exponential growth versus early confluence versus late confluence).

(b) Deterioration of stock drug solutions (see Section 5).

(c) When drug solutions have a short half-life they must be used immediately after diluting to ensure consistency in the drug levels available to cells in each assay.

(d) Failure to standardize incubation conditions (Section 6.2).

The assay system must be checked for reproducibility before applying it to human biopsy material.

8.3 Stimulation to above control levels

Stimulation can be a true measure of cellular events but may be due to technical artefacts. These include:

(a) Non-specific binding of radioactivity.

(b) Density-dependent inhibition of metabolic pathways in controls which is not evident in test situations where some cell kill has been achieved.

(c) Stimulation of uptake of metabolic precursors by anti-metabolites (e.g. thymidine by 5-fluorouracil and methotrexate).

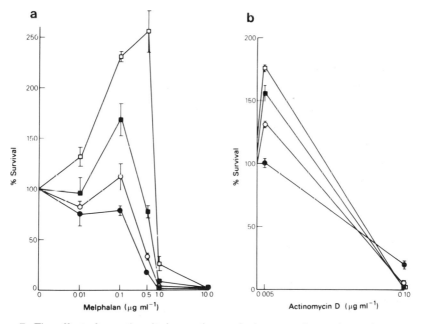

Figure 7. The effect of growth unit size on the survival curves of a murine melanoma cell line (CCL) to melphalan (a) and of a human melanoma biopsy to actinomycin D (b). Growth unit size and frequency was measured using FAS II automated image analysis system. ○, ⩾60 μm; ○, ⩾104 μm; ■, ⩾124 μm; □, ⩾140 μm. Mean ± s.e. shown. (Reproduced with permission of the publishers from ref. 96.)

Plated cell density influences the distribution in size of growth units in clonogenic assays, with large units decreasing as plated cell numbers increase (98). The effect of this on drug sensitivity profiles was examined and, as expected, the dose–response curve was strongly influenced by the size criterion used for colony-scoring, with stimulation to above control levels occurring when large colonies were scored (*Figure 7*).

9. Automation and future developments

Technology now exists for automating all stages of cytotoxicity assays which are performed in microtitration plates. It is hard to envisage further developments in this area. Areas for future development which were mentioned previously have been researched extensively in recent years with promising results, most particularly with the discovery of the multidrug resistant phenotype in tumour cells and the development of monoclonal antibodies which recognize this epitope. The use of these monoclonal antibodies opens up exciting possibilities for *in vitro* studies on drug sensitivity, for example in defining heterogeneity within a population, and comparing *in vitro* cytotoxicity testing results with multidrug resistance expression, and for identifying compounds which overcome the multidrug resistance phenotype.

References

1. Balmain, A. and Brown, K. (1988). *Adv. Cancer Res.*, **51**, 147.
2. Kilbey, B. J., Legator, M., Nichols, W., and Ranel, C. (ed.) (1984). *Handbook of Mutagenicity Testing Procedures*. Elsevier, Amsterdam.
3. Veritt, S. and Parry, J. M. (ed.) (1984). *Mutagenicity Testing: A Practical Approach*. IRL, Oxford.
4. Gellhorn, A. and Hirschberg, E. (1955). *Cancer Res., 15*, Suppl. 3, 1.
5. Foley, G. E. and Epstein, S. S. (1964). *Adv. Chemother., 1*, 175.
6. Hakala, M. I. and Rustrum, Y. M. (1979). In *Methods in Cancer Research: Cancer Drug Development*. Part A (ed. V. T. DeVita and H. Busch), p. 247. Academic Press, New York.
7. Dendy, P. P. (ed.) (1976). *Human Tumours in Short-term Culture: Techniques and Clinical Application*. Academic Press, New York.
8. Dendy, P. P. and Hill, B. T. (ed.) (1983). *Human Tumour Drug Sensitivity Testing in Vitro: Techniques and Clinical Applications*. Academic Press, New York.
9. Eagle, H. and Foley, G. E. (1956). *Am. J. Med., 21*, 739.
10. Wright, J. C., Cobb, J. P., Gumport, S. L., Golomb, F. M., and Safadi, D. (1957). *New Engl. J. Med., 257*, 1207.
11. Von Hoff, D. D. *et al.* (1983). *Cancer Res., 43*, 1926.
12. Wilson, A. P. and Neal, F. E. (1981). *Br. J. Cancer, 44*, 189.
13. Silvestrini, R., Sanfilippo, O., and Daidone, M. G. (1983). In *Human Tumour Drug Sensitivity Testing in Vitro* (ed. P. P. Dendy and B. T. Hill), p. 281. Academic Press, New York.

14. Salmon, S. E. (1980). In *Cloning of Human Tumour Cells. Progress in Clinical and Biological Research*, Vol. **48** (ed. S. E. Salmon), p. 281. Alan R. Liss, New York.
15. Von Hoff, D. D. *et al.* (1986). *J. Clin. Oncol.*, **4**, 1827.
16. Turner, P. (ed.) (1983). *Animals in Scientific Research: An Effective Substitute for Man?* Macmillan, London.
17. Atterwill, C. K. and Steele, C. E. (ed.) (1987). *In Vitro Methods in Toxicology.* Cambridge University Press.
18. Masters, J. R. W. (1983). In *Human Tumour Drug Sensitivity Testing in Vitro* (ed. P. P. Dendy and B. T. Hill), p. 163. Academic Press, New York.
19. Volm, M., Wayss, K., Kaufmann, M., and Mattern, J. (1979). *Eur. J. Cancer*, **15**, 983.
20. KSST. Group for Sensitivity Testing of Tumours (1981). *Cancer*, **48**, 2127.
21. Durkin, W. J., Ghanta, V. K., Balch, C. M., Davis, D. W., and Hiramoto, R. N. (1979). *Cancer Res.*, **39**, 402.
22. Raich, P. C. (1978). *Lancet*, **i**, 74.
23. Weisenthal, L. M. and Marsden, J. (1981). *Proc. Am. Assoc. Cancer Res.*, **22**, 155.
24. Morgan, D., Freshney, R. I., Darling, J. L., Thomas, D. G. T., and Celik, F. (1983). *Br. J. Cancer*, **47**, 205.
25. Hill, B. T. (1983). In *Human Tumour Drug Sensitivity Testing in Vitro* (ed. P. P. Dendy and B. T. Hill), p. 91. Academic Press, New York.
26. Agrez, M. W., Kovach, J. S., and Lieber, M. M. (1982). *Br. J. Cancer*, **46**, 88.
27. Bertoncello, I. *et al.* (1982). *Br. J. Cancer*, **45**, 803.
28. Rupniak, H. T. and Hill, B. T. (1980). *Cell Biol. Int. Rep.*, **4**, 479.
29. Hamburger, A. W., Salmon, S. E., Kim, M. B., Trent, J. M., Soehnlen, B., Alberts, D. S., and Schmidt, H. J. (1978). *Cancer Res.*, **38**, 3438.
30. Salmon, S. E. (ed.) (1980). *Cloning of Human Tumour Stem Cells. Progress in Clinical and Biological Research*, Vol. **48**. Alan R. Liss, New York.
31. Courtnay, V. D. and Mills, J. (1978). *Br. J. Cancer*, **37**, 261.
32. Tveit, K. M., Endersen, L., Rugstad, H. E., Fodstad, O., and Pihl, A. (1981). *Br. J. Cancer*, **44**, 539.
33. Von Hoff, D. D., Forseth, B. J., Huong, M., Buchok, J. B., and Lathan, B. (1986). *Cancer Res.*, **46**, 4012.
34. Hanauske, A.-R., Hanauske, U., and Von Hoff, D. D. (1987). *Eur. J. Cancer, Clin. Oncol.*, **23**, 603.
35. Baker, F. L., Ajani, J., Spitzer, G., Tomasovic, B. J., Williams, M., Finders, M., and Brock, N. A. (1988). *Int. J. Cell Cloning*, **6**, 95.
36. Alberts, D. S., George Chen, H-S., and Salmon, S. E. (1980). In *Cloning of Human Tumour Stem Cells* (ed. S. E. Salmon), p. 197. Alan R. Liss, New York.
37. Alberts, D. S. and Chen, H-S. G. (1980). In *Cloning of Human Tumour Stem Cells* (ed. S. E. Salmon), Appendix 4. Alan R. Liss, New York.
38. Freshney, R. I., Paul, J., and Kane, I. M. (1975). *Br. J. Cancer*, **31**, 89.
39. Roper, P. R. and Drewinko, B. (1976). *Cancer Res.*, **36**, 2182.
40. Korzeniewski, C. and Callewaert, D. M. (1983). *J. Immunol. Meth.*, **64**, 313.
41. Chenery, R. J. (1987). In *In Vitro Methods in Toxicology* (ed. C. K. Atterwill and C. E. Steele), p. 211. Cambridge University Press.

42. Weisenthal, L. M., Dill, P. L., Kurnick, N. B., and Lippman, M. E. (1983). *Cancer Res.*, **43**, 258.
43. Bickis, I. J., Henderson, I. W. D., and Quastel, J. H. (1966). *Cancer*, **19**, 103.
44. Dickson, J. A. and Suzanger, M. (1976). In *Human Tumours in Short-term Cultures: Techniques and Clinical Applications* (ed. P. P. Dendy), p. 107. Academic Press, New York.
45. Buskirk, H. H., Crim, J. A., Van Giessen, G. J., and Petering, H. G. (1973). *J. Natl Cancer Inst.*, **51**, 135.
46. Cosma, G. N. and Wenzel, D. G. (1984). *J. Tissue Culture Meth.*, **9**, 29.
47. Grunicke, H., Hirsch, F., Wolf, H., Bauer, V., and Kiefer, G. (1975). *Exp. Cell Res.*, **90**, 357.
48. Friedman, H. M. and Glaubiger, D. L. (1983). *Cancer Res.*, **42**, 4683.
49. Sondak, V. K. *et al.* (1984). *Cancer Res.*, **44**, 1725.
50. Dendy, P. P., Dawson, M. P. A., Warner, D. M. A., and Honess, D. J. (1976). In *Human Tumours in Short-term Culture: Techniques and Clinical Applications* (ed. P. P. Dendy), p. 139. Academic Press, New York.
51. Forbes, I. J. (1963). *Aust. J. Exp. Biol.*, **41**, 255.
52. Izsak, F. Ch., Gotlieb-Stematsky, T., Eylan, E., and Gazith, A. (1968). *Eur. J. Cancer*, **4**, 375.
53. Edwards, A. J. and Rowlands, G. F. (1968). *Br. J. Surg.*, **55**, 687.
54. Freshney, R. I. and Morgan, D. (1978). *Cell Biol. Int. Rep.*, **2**, 375.
55. Ramos, K. and Acosta, D. (1984). *J. Tissue Culture Meth.*, **9**, 3.
56. Oyama, V. I. and Eagle, H. (1956). *Proc. Soc. Exp. Biol. Med.*, **91**, 305.
57. Pelletier, B., Dhainaut, F., Pauly, A., and Zahnd, J-P. (1988). *J. Biochem. Biophys. Meth.*, **16**, 63.
58. Skehan, P., Storeng, R., Scudiero, D., Monks, A., McMahon, J., Vistica, D., Warren, J., Bokesch, H., Kenney, S., and Boyd, M. R. (1989). *Proc. Am. Assoc. Cancer Res.*, **30**, 612.
59. Fricker, S. (1990). BACR Meeting 1990, Poster 30.
60. Knox, P., Uphill, P. F., Fry, J. R., Berford, J., and Balls, M. (1986). *Fd. Chem. Toxic.*, **24**, 457.
61. Bradford, M. M. (1976). *Anal. Biochem.*, **72**, 248.
62. Laughton, C. (1984). *Anal. Biochem.*, **140**, 417.
63. Smith, P. K. *et al.* (1985). *Anal. Biochem.*, **150**, 76.
64. Lane, R. D., Federman, D., Flora, J. L., and Beck, B. L. (1986). *J. Immunol. Meth.*, **92**, 261.
65. McCaffrey, T. A., Agarwal, L. A., and Weksler, B. B. (1988). *In Vitro Cell. Dev. Biol.*, **24**, 247.
66. In *Amersham Res. News*, (1989). **3**, 13.
67. Parish, C. R. and Mullbacher, A. (1983). *J. Immunol. Meth.*, **58**, 225.
68. Fiennes, A. G. T. W., Walton, J., Winterbourne, D., McGlashar, D., and Hermon-Taylor, J. (1987). *Cell Biol. Int. Rprts.*, **11**, 373.
69. Landegren, U. (1984). *J. Immunol. Meth.*, **67**, 379.
70. Mosmann, T. (1983). *J. Immunol. Meth.*, **65**, 55.
71. Denizot, F. and Lang, R. (1986). *J. Immunol. Meth.*, **89**, 271.
72. Cole, S. P. C. (1986). *Cancer Chemother. Pharmacol.*, **17**, 259.
73. Carmichael, J., DeGraff, W. G., Gazdar, A. F., Minna, J. D., and Mitchell, J. B. (1987). *Cancer Res.*, **47**, 936.

74. Park, J-G., Kramer, B. S., Steinberg, S. M., Carmichael, J., Collins, J. M., Minna, J. D., and Gazdar, A. F. (1987). *Cancer Res., 47,* 5875.
75. Pieters, R., Huismans, D. R., Leyva, A., and Veerman, A. J. (1989). *Br. J. Cancer, 59,* 217.
76. Jabbar, S. A. B., Twentyman, P. R., and Watson, J. V. (1989). *Br. J. Cancer, 60,* 523.
77. Plumb, J. A., Milroy, R., and Kaye, S. B. (1989). *Cancer Res., 49,* 4435.
78. Freshney, R. I., Celik, F., and Morgan, D. (1982). In *The Control of Tumour Growth and its Biological Base* (ed. W. Davis, C. Maltoni, and St. Tanneberger). Fortschritte in der Onkologie, Band 10, Berlin, Akademie-Verlag.
79. Wilson, A. P., Ford, C. H. J., Newman, C. H., and Howell, A. (1984). *Br. J. Cancer, 49,* 57.
80. Weisenthal, L. M. and Lippman, M. E. (1985). *Cancer Treat. Rep., 69,* 615.
81. Bosanquet, A. G. (1984). *Br. J. Cancer, 49,* 385.
82. Powers, J. F. and Sladek, N. E. (1983). *Cancer Res., 43,* 1101.
83. Fry, J. R. (1983). In *Animals in Scientific Research: An Effective Substitute for Man?* (ed. P. Turner), p. 69. Macmillan, London.
84. Begue, J. M., Le Bigot, J. F., Guguen-Guillouzo, C., Kiechel, J. R., and Guillouzo, A. (1983). *Biochem. Pharmacol., 32,* 1643.
85. Davies, D. S. and Boobis, A. R. (1983). In *Animals in Scientific Research: An Effective Substitute for Man?* (ed. P. Turner), p. 69. Macmillan, London.
86. Guillouzo, A. and Guguen-Guillouzo, C. (ed.) (1986). *Research in Isolated and Cultured Hepatocytes.* John Libbey, Eurotext Ltd./INSERM, London.
87. Guillouzo, A., Beaune, P., Gascoin, M-N., Begue, J. M., Campion, J-P., Guengerich, P. F., and Guguen-Guillouzo, C. (1985). *Biochem. Pharmacol., 34,* 2991.
88. Barranco, S. C., Bolton, W. E., and Novak, J. K. (1980). *J. Natl Cancer Inst., 64,* 913.
89. Courtnay, V. D. and Mills, J. (1981). *Br. J. Cancer, 44,* 306.
90. Dixon, R. A., Linch, D., Baines, P., and Rosendaal, M. (1981). *Exp. Cell Res., 131,* 478.
91. Gupta, V. and Krishnan, A. (1982). *Cancer Res., 42,* 1005.
92. Nederman, T. and Twentyman, P. (1984). In *Spheroids in Cancer Research. Methods and Perspectives.* Ch. 5. (ed. H. Acker, J. Carlsson, R. Durand, and R. M. Sutherland) Springer-Verlag, Berlin.
93. INVITTOX (1990). *The FRAME Cytotoxicity Test.* INVITTOX Protocol number 3. Frame, 34 Stoney St., Nottingham.
94. Jones, A. C., Stratford, I., Wilson, P. A., and Peckham, M. J. (1982). *Br. J. Cancer, 46,* 870.
95. Kolb, A. I. (1981). *Lab. Equipment Digest, 19,* 87.
96. Moon, T. E. (1980). In *Cloning of Human Tumour Stem Cells* (ed. S. E. Salmon), p. 209. Alan R. Liss, New York.
97. Berenbaum, M. C. (1974). *Lancet, ii,* 1141.
98. Meyskens, F. L. Jr, Thomson, S. P., Hickie, R. A., and Sipes, N. J. (1983). *Br. J. Cancer, 46,* 863.

9

In situ hybridization

S. M. LANG, A. H. WYLLIE, and D. CONKIE

1. Introduction

In situ hybridization (ISH) is a technique which permits detection of DNA or RNA sequences in cell smears, tissue sections, and metaphase chromosome spreads. The method is based on the formation of double-stranded hybrid molecules, which form between a DNA or RNA target sequence and the complementary, single-stranded, labelled probe.

2. Hybridization technique

There are many common features in the great variety of methods which have been published in the past few years (1–6). A typical *in situ* hybridization protocol includes the following steps: preparation of slides, fixation and pre-treatment of materials on slides, denaturation of target DNA, hybridization *in situ*, washing, and detection.

2.1 Preparation of slides

Slides and cover slips are pre-treated to avoid non-specific adhesion of probe and at the same time ensure adhesion of cells, see *Protocol 1*. The pretreatment protocol employed can contribute to background to a variable degree, especially when bovine serum albumin (BSA) is used.

Protocol 1. Pre-treatment of slides and cover slips

Slides

A. *RNA*

1. Wash slides in 0.1 M HCl for 20 min, rinse in absolute ethanol and allow to dry.
2. Incubate in 3 × SSC, 0.02% Ficoll, and 0.02% polyvinylpyrrolidone at 65°C for 2 h.
3. Fix in ethanol/acetic acid (3:1 v/v) for 20 min, dry, and bake at 180°C for 2 h.

Protocol 1. *Continued*

B. *DNA*

1. Coat slides with TESPA (three-amino-propyl-triethoxy-silane) (7)

Cover slips

1. Immerse cover slips in 0.1 M HCl for 20 min, rinse in absolute ethanol and air dry.

2. Siliconize with dimethyldichlorosilane and bake at 180°C for 2 h.

2.2 Fixation

Optimal fixation procedures should preserve morphology and at the same time ensure the accessibility of the target RNA or DNA sequence. Useful fixatives for RNA–RNA *in situ* hybridization are 4% paraformaldehyde and PLPD (phosphate buffer, lysine, periodate, dichromate) (8). Fixation with EAA (ethanol/acetic acid 3:1 v/v) decreases the intensity of the signal and formaldehyde containing fixatives such as formalin or AFE (2% acetic acid, 4% formaldehyde, 85% absolute ethanol) are not very suitable for *in situ* hybridization, see *Figure 1* (9).

Protocol 2. Preparation of cell smears for ISH

1. Wash cell monolayers in PBS supplemented to 0.02% with EDTA and expose briefly to 0.1% trypsin in PBS. Wash the harvested cells once in PBS, count, and smear on to glass slides.

2. For RNA fix in 4% paraformaldehyde (pH 7.0) in PBS at 4°C for 60 min, rinse in PBS for 5 min, dehydrate in a series of graded alcohol, and store dry at −20°C until required.

3. For DNA fix in 4% paraformaldehyde in PBS for 16–24 h, rinse twice in 0.1 M Tris pH 7.4 for 5 min, immerse twice in 0.25% (v/v) Triton X-100, 0.25% (v/v) Nonidet P40 in 0.1 M Tris pH 7.4 for 5 min, and wash twice in 0.1 M Tris pH 7.4 for 5 min.

4. Incubate 10 min at 37°C in 100 µg/ml proteinase K in 50 mM Tris, 5 mM EDTA, wash twice in 0.1 M Tris with 2 mg/ml glycine for 5 min.

5. Immerse for 15 s in 20% (v/v) acetic acid at 40°C and wash twice for 5 min in 0.1 M Tris.

Many protocols include digestion with pronase or proteinase K to optimize the accessibility of the target sequence, but in our experience this is more important for paraffin-processed tissue sections than for frozen sections or cell smears. Immediately following this digestion, the cells may then be incubated in paraformaldehyde to cross-link the exposed target sequence to

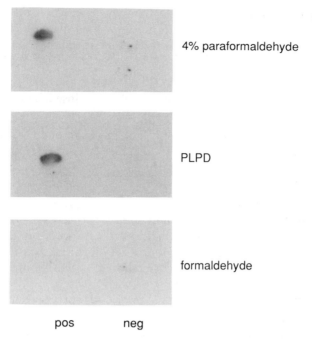

4% paraformaldehyde

PLPD

formaldehyde

pos   neg

Figure 1. Influence of fixative on RNA–RNA ISH. X-ray film exposed to pairs of smears of RFHO6N1 cells, containing Ha-ras target sequence (top) and non-expressing 208F cells (bottom) hybridized with [^{32}P]-labelled SP64-Ha-ras-3A/4 probe (negative probe on right-hand side) and [^{32}P]-labelled SP64-Ha-ras-2A/8 (positive probe on left-hand side).

the remaining cell components, with a resulting twofold increase in sensitivity (10).

For DNA–DNA *in situ* hybridization, paraformaldehyde and formalin can be employed successfully. For the digoxigenin detection method it is essential to fix with 4% paraformaldehyde for at least 16 h, to allow for a more aggressive treatment with proteinase K. This substantially reduces background.

2.3 Hybridization *in situ*

Renaturation (hybridization) of DNA or RNA duplexes depends on salt concentration, temperature, and formamide concentration. It is also influenced by pH, and the characteristics of the probe used (see ref. 4). The hybridization buffer contains substances which reduce background. Some of these factors are briefly discussed below.

The maximum rate of renaturation of DNA is at a temperature between 16 and 32 °C below the melting temperature (T_m). For DNA–DNA *in situ* hybridization a temperature of 25 °C below T_m is recommended. Due to the higher thermal stability of RNA–DNA and RNA–RNA hybrids higher temperatures can be employed for RNA *in situ* hybridization. In the pH range 5–9 the rate of renaturation is relatively independent of pH. For hybridization involving RNA it is preferable to chose mildly acidic conditions, because of the susceptibility of RNA to alkaline hydrolysis. Most mixes include agents like citrate or EDTA to complex divalent cations, which strongly stabilize duplex DNA. With decreasing salt concentrations the electrostatic repulsion between two DNA strands of duplex DNA increases and the duplex is destabilized.

Formamide, an organic solvent, reduces the thermal stability of double-stranded polynucleotides. It reduces the T_m of DNA–DNA duplexes in a linear fashion by $0.72\,°C$ for each 1% increment in formamide concentration. The relationship between T_m of RNA-containing duplexes and formamide concentration is non-linear, and optimum conditions are best found empirically. Formamide can be used to increase the stringency of the hybridization and washing steps without having to use very high temperatures, which could prove damaging to the tissues, or prejudice the adhesion of tissue sections to slides.

2.4 Washing procedure

The washing steps are intended to remove probe hybridized to sequences with incomplete homology, but to conserve the binding of probe in perfectly matched hybrids. The degree to which mismatches decrease the stability of duplexes is inversely related to the probe length. In short probes, such as synthetic oligonucleotides, the effect of single mismatches on hybrid stability is pronounced.

The stringency can be varied by alterations in salt concentration, temperature, or formamide concentration. Higher stringency of the washing procedure leads to improved signal/noise ratio, with low background. If the stringency of the wash is too high, even perfectly matched hybrids become destabilized, resulting in a decrease of both signal and background.

2.5 Detection of signals

For microscopic assessment of the signal emitted by the radioactive label, slides are dipped in liquid autoradiographic emulsion. After development they are counterstained with haematoxylin/eosin. For rapid detection of specific RNAs in small numbers of cells in smears and for optimizing methods applied to a series of replicate sections or smears, we have found it useful to compare the radiographic signal in directly overlaid X-ray film.

In one of the most sensitive non-radioactive detection methods—commercially available in kit form (Boehringer)—the DNA probe is tagged with digoxigenin, a glycoside, which is recognized by a specific antibody (raised against it as a hapten). The antibody is conjugated to alkaline phosphatase, which can be detected by an enzyme-linked colour reaction.

3. Applications of *in situ* hybridization

In situ hybridization can be used to detect DNA sequences which are part of normal or abnormal genes, to study gene expression by mRNA detection, or to identify the chromosomal location of DNA sequences (1–6). The method is particularly useful if target sequences are distributed in a non-random way in tissues, for the visualization of heterogeneity and the study of cell differentiation. The method described for RNA–RNA hybridization (11) is generally applicable to rapid screening of small numbers of cultured cells for expression of oncogene mRNA. It circumvents the problem of extracting rare RNAs in sufficient amounts for detection and the intracellular or intranuclear localization of RNA is possible. Similar methods have been used to visualize homeotic gene expression in developing larvae (12, 13). DNA–DNA ISH is also suitable for the detection of viral genomes in sections of routinely processed archival paraffin blocks of human tissues.

4. The choice of probe

In principle, probes can be RNA or double- or single-stranded DNA. Double-stranded probes have a number of disadvantages in comparison with single-stranded probes. For use, the probes must be denatured; and renaturation may occur in solution, thus reducing the concentration of probe available for hybridization. The tendency of double-stranded probes to form long concatenates in solution limits their tissue penetration. RNA probes are preferable to conventional nick-translated DNA probes because they are easy to prepare and can increase the sensitivity of the detection method. Furthermore, DNA–RNA or RNA–RNA hybrids are thermally more stable than DNA–DNA hybrids, which allows for more stringent conditions. Single-stranded RNA probes are particularly effective for *in situ* hybridization. Using SP6 or T7 cloning vectors, transcription (with the appropriate bacterial polymerase) results in the synthesis of labelled RNA (14, 15). The versatility of the vector allows transcription of RNA from almost any desired DNA sequence, which can be labelled to high specific activity. An additional RNase digestion step can be included to reduce background by removing excess probe, which may be bound non-specifically.

Hybrid formation in solution is proportional to the single-stranded probe length. For ISH the probes have to diffuse into the dense matrix of cells and chromosomes. Reduction of probe length is usually desirable and can be achieved by measured exposure to alkali conditions (RNA probes) (16). The use of short synthetic DNA probes (oligonucleotides) permits choice of conditions for highly selective binding of probe, but the overall stability of the duplexes formed is poor (17–19). The reannealing rate is also proportional to the probe concentration. On the other hand, high probe concentrations

309

increase non-specific binding and it is advisable to test the optimal concentration for each probe.

5. The choice of label

The most sensitive detection methods employ radioactive label. The autoradiographic detection of labelled probes depends on the isotope used and the specific activity, which has to be high enough to permit detection after hybridization within a reasonable exposure time and with a good signal/noise ratio. Several isotopes are available for radioactive labelling. The use of $[^{32}P]$ allows rapid detection of the signal, yet cellular localization is suboptimal because of the long path length of the $[^{32}P]$ β-rays (*Figure 2*). Autoradiographs of high quality and improved cellular localization employ ^{35}S-labelled probes. $[^{35}S]$ emits β-rays of much shorter path length. One of the major problems using $[^{35}S]$ is non-specific binding of the label to cells or sections. It has been suggested that pre-hybridization in the presence of non-labelled thio α UTP reduces this non-specific binding (20). It is also possible to label nucleic acids with tritium or by iodination with $[^{125}$iodine. Tritium-labelled probes give accurate intracellular localization, but for low abundance target molecules may require exposure of up to 100 days. This may result in high background and is unacceptably long for most purposes.

The disadvantages of the radiographic method, such as safety hazards, instability of the label, and long exposure times, can be avoided by using non-radioactive labels such as biotin (21, 22), acetylaminofluorene (23), fluorescent labels (24), mercurated (25) and sulphonated probes, or digoxigenin (26). One of the most sensitive methods employs digoxigenin as label (*Figure 3*) and is described in more detail in *Protocol 4*. Random priming with digoxigenin has the advantage of high labelling efficiency. ISH employing this method achieves high sensitivity, good intracellular localization, and low background. Non-radioactive detection also has the advantage that simultaneous detection of several DNA sequences through double ISH is possible.

6. Sensitivity

For RNA probes the reported maximum sensitivity is 20 mRNA copies per cell (1). Lynn *et al*. (4) report detection of sequences that comprise about 0.05% of the total RNA with asymmetric probes. The sensitivity of probes labelled with digoxigenin has been reported as 1–2 genomes/cell for human papilloma virus (HPV) genomes (27).

7. Negative controls

To establish the specificity of the hybridization several types of controls can be applied, such as prior tissue digestion with RNase and DNase to remove

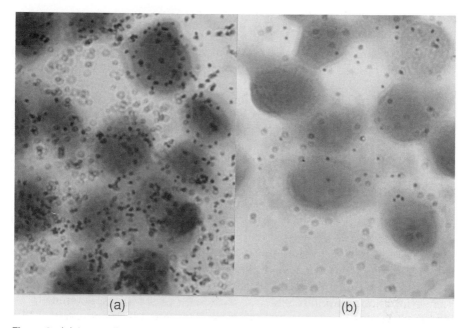

(a)   (b)

Figure 2. (a) Autoradiographic image of RFHO6N1 cells, containing Ha-ras target sequence, hybridized with [³²P]-labelled SP64-Ha-ras-2A/8 (positive probe). (b) Autoradiographic image of [²⁰⁸F] cells (negative control) hybridized with [³²P]-labelled SP64-Ha-ras-2A/8 (positive probe).

the target sequence, competition with unlabelled probes, comparison of signals obtained with adequate and inadequate probes, comparison of ISH with immunohistochemical analysis of the encoded protein, and comparison of the observed result with the known or supposed localization of the mRNA within suitable test tissues. The best biological, 'negative' controls are cells which are closely similar to the cells studied, yet known to lack the target sequence. A test system comprising positive and negative cell lines is very useful for the optimization of *in situ* hybridization; see *Figure 1*.

8. Detailed protocols for *in situ* hybridization

The following protocols provide a radioactive method for ISH of RNA target sequences, employing asymmetric RNA probes (11), and a non-radioactive method for *in situ* hybridization of DNA target sequences, employing random primed digoxigenin DNA probes (26).

Protocol 3. *In situ* hybridization with radioactive-labelled asymmetric RNA probes[a]

1. Linearize plasmid with the appropriate restriction enzyme.

Protocol 3. *Continued*

2. Generate labelled RNA probe following the supplier's protocol (Trans-probe, Pharmacia)[b]. Use 1 μg of the DNA template and 100 μCi (3.7 MBq) [α-^{32}P]UTP[c] sp. act. 800 Ci/mmol (29.6 GBq/μmol) (Amersham) or 100 μCi (3.7 MBq) [α-^{35}S]SUTP at 400 Ci/mmol (14.8 GBq/μmol) (Du Pont/NEN). Redissolve the ethanol-precipitated probes in hybridization buffer containing 50–70% deionized formamide, 2 × SSC, 0.5 × Denhardt's solution, 1 mM EDTA pH 7.0 and 200 μg/ml sonicated salmon sperm DNA. Incubate the hybridization mixture at 85°C for 5 min and count a small sample in a liquid scintillation counter.

3. Apply to each slide 8 μl hybridization mixture with 200 000–400 000 c.p.m. (total) per slide, cover with a siliconized cover slip (19 mm diameter), and seal by immersion in mineral oil. Allow hybridization to proceed for 7–18 h at a temperature suitable for the probe sequence (40–60°C).

4. Remove mineral oil by several rinses in chloroform. Allow cover-slips to float off in 2 × SSC at room temperature. Incubate slides in 50–70% deionized formamide and 0.1 × SSC for 60 min (one bath change) at hybridization temperature. Rinse in 2 × SSC at room temperature for 5 min. Remove single-stranded RNA by digestion with 50 μg/ml RNase A (Sigma) in 2 × SSC at 37°C for 60 min. Incubate slides in 50–70% de-ionized formamide and 0.1 × SSC at hybridization temperature for 15 min, and finally rinse in 0.1 × SSC for 5 min at room temperature.

5. Dehydrate slides and dip in Ilford K5 autoradiographic film emulsion diluted 1:1 with 1% glycerol. Store dry in a light-tight box containing silica gel as desiccant for 5 days at −70°C. Alternatively expose to preflashed Fuji Rx x-ray film for 24–48 h and subsequently dip in liquid emulsion. Develop and counterstain with haematoxylin/eosin in the usual way.

[a] All solutions are treated with 0.1% DEPC to destroy ribonucleases. They must then be autoclaved to inactivate DEPC and prevent carboxymethylation of RNA
[b] The components of the transcription reaction are mixed at room temperature, not on ice, because spermidine can cause DNA to precipitate at 0°C
[c] [α-^{32}P]NTP which is more than 10 days old can substantially inhibit transcription

---

Protocol 4. *In situ* hybridization with digoxigenin-labelled DNA probes

1. Linearize plasmid with the appropriate restriction enzyme.

2. Label 0.2 μg DNA with digoxigenin-11-UTP by random priming following the recommended protocol provided by the supplier (Boehringer). Re-move unincorporated nucleotides using Geneclean (Stratech).

3. Add 140 ng/ml probe to hybridization buffer containing 2 × SSC, 5%

(w/v) dextran sulphate, 50% deionized formamide and 0.2% low-fat skimmed milk. Apply 50 µl of hybridization mix per slide, cover with Gelbond (ICN), and seal edges with nail varnish. Denature target DNA and probe at 90°C for 10 min and allow hybridization to proceed for 16 h in a humidified chamber at a temperature appropriate for the sequence used as probe—in a protocol suitable for HPV detection—at 42°C.

4. Remove Gelbond and wash slides for 10 min in 2 × SSC at room temperature. Wash slides for 20 min in 2 × SSC at 60°C, for 10 min in 0.2 × SSC at room temperature, and finally for 20 min in 0.2 × SSC at 42°C.

5. In a minor modification to the supplier's protocol, process slides (at room temperature) as follows: Wash 5 min in buffer 1 (0.1 M Tris pH 7.5, 0.15 M NaCl). Immerse for 30 min in buffer 1 supplemented with 20% sheep serum. Incubate in a humidified box with buffer 1 and a 1:5000 dilution of anti-digoxigenin antibody for 20 min. Wash twice in buffer 1 for 20 min. Place in buffer 2 (0.1 M Tris pH 9.5, 0.1 M NaCl, 5 mM $MgCl_2$) for 60 min. Apply buffer 2 and 0.33 mg/ml nitro-blue tetrazolium salt and 0.17 mg/ml 5-bromo-4-chloro-3-indolyl phosphate. Stop reaction after 16 h by immersion in 20 mM Tris pH 7.5, 5 mM EDTA and mount in Aquamount (BDH).

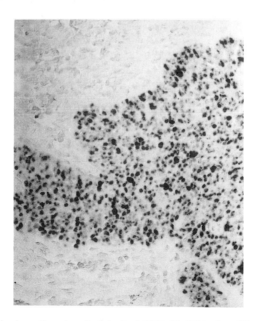

Figure 3. ISH employing digoxigenin-labelled HPV 16 DNA. Paraffin section of an anal carcinoma showing intense labelling of nuclei in the tumour and absence of label in the surrounding normal tissue.

Acknowledgements

This work is supported by the Cancer Research Campaign. We thank Robert G. Morris for assistance in preparing this manuscript.

References

1. Cox, K. H., DeLeon, D. V., Angerer, L. M., and Angerer, R. C. (1984). *Dev. Biol.*, **101,** 485.
2. Harper, M. E., Marselle, L. M., Gallo, R. C., and Wong-Staal, F. (1986). *Proc. Natl Acad. Sci. USA,* **83,** 772.
3. Lawrence, J. B. and Singer, R. H. (1985). *Nucleic Acids Res.,* **13,** 1777.
4. Lynn, D. A., Angerer, L. M., Bruskin, A. M., Klein, W. H., and Angerer, R. C. (1983). *Proc. Natl Acad. Sci. USA,* **80,** 2656.
5. Haase, A. T., Walker, D., Stowring, L., Ventura, P., Geballe, A., Blum, H., Brahic, M., Goldberg, R., and O'Brien, K. (1985). *Science,* **227,** 189.
6. Stoner, G. D., You, M., Skouv, J., Budd, G. C., Pansky, B., and Wang, Y. (1987). *Ann. Clin. Sci.,* **17,** 74.
7. Van Prooijen-Knegt, A. C., Raap, A. K., Van der Burg, M. J. M., Vrolijk, J., and Van der Ploeg, M. (1982). *Histochem. J.,* **14,** 333.
8. Going, J. J., Williams, A. R. W., Wyllie, A. H., Anderson, T. J., and Piris, J. (1988). *J. Path.,* **155,** 185.
9. Tournier, I., Bernuau, D., Poliard, A., Schoevaert, D., and Feldmann, G. J. (1987). *Histochem. Cytochem.,* **35,** 453.
10. Ostrow, R. S., Manias, D. A., Clark, A. B., Okagaki, T., Twiggs, L. B., and Faras, A. J. (1987). *Cancer Res.,* **47,** 649.
11. Lang, S. M., Morris, R. G., Rose, K. A., and Wyllie, A. H. (1989). *Anticancer Res.,* **9,** 805.
12. Akam, M. E. (1983). *EMBO J.,* **2,** 2075.
13. Hafen, E., Levine, M., Garber, R. L., and Gehring, W. J. (1983). *EMBO J.,* **2,** 617.
14. Brysch, W., Hagendorff, G., and Schlingensiepen, K.-H. (1988). *Nucleic Acids Res.,* **16,** 2333.
15. Melton, D. A., Krieg, P. A., Rebagliati, M. R., Maniatis, T., Zinn, K., and Green, M. R. (1984). *Nucleic Acids Res.,* **12,** 7035.
16. Bakkus, M. H. C., Brakel-van Peer, K. M. J., Adraansen, H. J., Wierenga-Wolf, A. F., Van den Akker, T. W., Dicke-Evinger, M. J., and Benner, R. (1989). *Oncogene* **4,** 1255.
17. Buvoli, M., Biamonti, G., Riva, S., and Morandi, C. (1987). *Nucleic Acids Res.,* **15,** 9091.
18. Uhl, G. R., Zingg, H. H., and Habener, J. F. (1985). *Proc. Natl Acad. Sci. USA,* **82,** 5555.
19. Jirikowski, G. F., Ramalho-Ortigao, J. F., and Seliger, H. (1988). *Mol. Cell Probes,* **2,** 59.
20. Bandtlow, C. E., Heumann, R., Schwab, M. E., and Thoenen, H. (1987). *EMBO J.,* **6,** 891.

21. Brigati, D. J., Myerson, D., Leary, J. J., Spalholz, B., Travis, S. Z., Fong, C. K. Y., Hsiung, G. D., and Ward, D. C. (1983). *Virology, 126,* 32.
22. Manuelidis, L. (1985). *Human Genet., 71,* 288.
23. Raap, A. K., Geelen, J. L., Van der Meer, J. W. M., Van de Rijke, F. M., Van den Boogaart, P., and Van der Ploeg, M. (1988). *Histochemistry, 88,* 367.
24. Bauman, J. G. J., Wiegant, J., Borst, P., and Van Duijn, P. (1980). *Exp. Cell Res., 138,* 485.
25. Hopman, A. H. N., Wiegant, J., Tesser, G. I., and Van Duijn, P. (1986). *Nucleic Acids Res., 14,* 6471.
26. Morris, R. G., Arends, M. J., Bishop, P., Sizer, K., Duvall, E., and Bird, C. C. (1990). *J. Clin. Path., 43,* 800.
27. Heiles, H. B. J., Genersch, E., Kessler, C., Neumann, R., and Eggers, H. J. (1988). *Biotechniques, 6,* 978.

General reagents

Hanks' balanced salt solution (HBSS) (modified)

- $CaCl_2$ 0.14 g   (1.26 mM)
- KCl   0.40 g   (5. 36 mM)
- KH_2PO_4 0.06 g   (0.44 mM)
- $MgCl_2.6H_2O$ 0.10 g   (0.49 mM)
- $MgSO_4.7H_2O$ 0.10 g   (0.41 mM)
- NaCl   8.00 g   (0.137 M)
- $Na_2HPO_4.7H_2O$ 0.09 g   (0.34 mM)
- Phenol Red   0.01 g
- water to   1000 ml

Autoclave below pH 6.5, 100 kPa (15 p.s.i., 1 bar), 20 min. Before use, pH may be adjusted to 7.4 with sterile NaOH and sterile 1 M Hepes may be added to 20 mM as a buffer.

Glucose, if required, should be autoclaved separately at 100 g/l and diluted 1:100 to give 1.0 g/l (5.55 mM).

This recipe is provided as a general handling, washing, and dissection solution. If used as a base for growth medium, glucose (see above) and 0.35 g/l $NaHCO_3$ (4 mM) must be added. It is then suitable for culture in a sealed container with air as a gas phase.

Dulbecco's phosphate-buffered saline, solution A (PBSA, Ca^{2+} and Mg^{2+} free)

- KCl   0.20 g   (2.68 mM)
- KH_2PO_4 0.20 g   (1.47 mM)
- NaCl   8.00 g   (0.137 M)
- $Na_2HPO_4.7H_2O$ 2.16 g   (8.06 mM)
- water to   1000 ml

Sterilize by autoclaving 100 kPa (15 p.s.i., 1 bar) 20 min.

Used as a washing solution before disaggregation, and as a diluent for trypsin.

Dissection BSS (DBSS)

- HBSS or PBSA with 250 u/ml penicillin, 250 μg/ml streptomycin, 100 μg/ml kanamycin, or 50 μg/ml gentamycin, and 2.5 g/ml fungizone

PBSA/EDTA

- $Na_2EDTA.2H_2O$ 0.372 g   (1 mM)
- PBSA to   1000 ml

Sterilize by autoclaving 100 kPa (15 p.s.i., 1 bar) 20 min.

Trypsin

- 2.5 g (crude Difco 1/250) or 0.1 g (3 × recrystallized, Sigma).
- HBSS or PBSA or PBSA/EDTA to 1000 ml, as required to obtain complete disaggregation

Suppliers of specialist items

Amersham International, White Lion Road, Amersham, Bucks HP7 9LL, UK

Amicon Ltd, Amicon House, 2 Kingsway, Woking, Surrey, GU21 1UR, UK and Upper Mill, Stonehouse, Gloucestershire GL10 2BJ, UK

Amicon Division, W. R. Grace & Co., 72 Cherry Hill Drive, Beverly, MA 01915, USA

APV Co. Ltd, Crawley, Sussex, UK

Arbrook, Ltd, Livingstone, Lothian, UK

Associated Bag Co., Milwaukee, WI, USA

Baird and Tatlock (London) Ltd, PO Box 1, Romford, Essex RM1 1HA, UK

BDH Chemicals Ltd, Poole, Dorset BH12 4NN, UK

Beckman-RIIC Ltd, Analytical Instruments Sales and Service Operation, Cressex Industrial Estate, Turnpike Road, High Wycombe, Bucks, UK

Becton Dickinson (UK) Ltd, Between Towns Road, Cowley, Oxford OX4 3LY, UK

Bellco Glass Inc., 340 Edrudo Road, Vineland, NJ 08360, USA (see A. & R. Horwell Ltd for UK)

Bibby, see J. Bibby

Boehringer Mannheim GmbH, Biochemica, Postfach 310120, D-6800 Mannheim 31, Germany

BRL, Bethesda Research Laboratories GmbH, Offenbacher Strasse 113, D-6078 Neu-Isenburg 1, Germany

B. Braun AG, Melsungen, Germany

BS & S (Scotland) Ltd, 5/7 West Telferton, Portobello Industrial Estate, Edinburgh EH8 6UL, UK

Buchler Instrument Inc., Fort Lee, NJ, USA

Burroughs Wellcome, Research Triangle Park, NC, USA

Camon Labor Service, PO Box 129505, D–6200 Wiesbaden, Federal Republic of Germany

Cellco Advanced Bioreactors Inc., Kensington, MD 20895, USA

Cell Enterprises Inc., 15719 Crabbs Brancy Way, Derwood, MD 20855, USA

Cell-Pharm, CD Medical Inc., Miami Lakes, FL 33014, USA

Charles River (UK) Ltd, Manson Road, Margate, Kent CT9 4LT, UK

Chemap, A C Biotechnics Inc., 230 Crossways Park Drive, Woodbury, NY 11746, USA (Holzliwisenstrasse 5, CH-8604 Volketswil, Switzerland)

CIBA Corning Diagnostics Ltd, Colchester Road, Halstead, Essex CO9 2DX, UK

Clonetics Corporation, 9620 Chesapeake Drive, San Diego, CA 92123, USA

Cole-Parmer Instrument Co., 7425 N. Oak Park Ave, Chicago, IL 60648, USA (PO Box 22, Bishop's Stortford, Herts CM23 3DX, UK)

Corning Ltd, Stone, Staffs ST15 0BG, UK (MP-21-5-8, Corning, NY 14831, USA)

Coulter Electronics Ltd, Northwell Drive, Luton, Beds LU3 3RH, UK

Collaborative Research Inc., Research Products Division, 128 Spring St, Lexington, MA 02173, USA and 1365 Main St, Waltham, MA 02154, USA

Cryo-Med, 49659 Leona Drive, Mt. Clements, MI 48043, USA

Diessel, D-3200 Hildesheim, Germany

Difco Laboratories Ltd, PO Box 14B, Central Avenue, East Molesey, Surrey, KT8 0SE, UK

Dow Chemical Co., Heathrow House, Bath Road, Hounslow TWS 9QY, UK

Du Pont UK Ltd, 2 New Road, Southampton SO2 0AA, UK

Dynatech Laboratories Ltd, Daux Road, Billingshurst, Sussex, UK

Endotronics Inc., 8500 Evergreen Blvd, Coon Rapids, Mn 55433, USA

Expanded Metal Co., Hartlepool, UK

Fisons Scientific Apparatus, Bishop Meadow Road, Loughborough LE11 0RG, UK

Flow Laboratories Ltd, PO Box 17, Second Ave Ind. Estate, Irvine, Ayrshire KA12 8NB, UK

Fluka Chemicals Ltd, Peakdale Road, Glossop, Derbyshire SK13 9XE, UK

Fluka Chemika-Biochemika, Industriestrasse 25, CH-9470 Buchs, Switzerland

Fotodyne Inc., New Berlin, WI, USA

Gelman Sciences Inc., 600 S. Wagner Road, Ann Arbor, MI 48106, USA (see also Flow Laboratories for UK)

Gen-Probe Inc., 9880 Campus Point Drive, San Diego, CA 92121, USA

Genzyme Corporation, 75 Kneeland Street, Boston, MA 02111, USA

Gibco Europe Ltd, PO Box 35, Trident House, Renfrew Road, Paisley PA3 4EF, UK

Glaxo Laboratories Ltd, Greenford, Middlesex, UK

Gurr (see BDH)

Harvard Apparatus Co, South Natick, MA, USA

Hopkins and Williams, PO Box 1, Romford, Essex RM1 1HA, UK

A. & R. Horwell Ltd, 73 Maygrove Road, North Hampstead, London NW6 2BP

Hyclone Laboratories Inc., 1725 So State Highway 89–91, Logan, Utah 84321, USA

ICN Biochemicals, PO Box 28050, Cleveland, OH 44128, USA

Ingold Electrodes Ltd, 261 Ballardvale Street, Wilmington, MA 01887, USA

Innovative Chemistry Inc., PO Box 90, Marshfield, MA 02050, USA

Irvine Scientific, 2511 Daimler Street, Santa Ana, CA 92705, USA

J Bibby Science Products Ltd, Tilling Drive, Stone, Staffordshire ST15 0SA, UK

Joyce Loebl, Dukesway, Team Valley, Gateshead NE11 0PZ, UK

JRH Biosciences, 13804 W 107th Street, Lenexa, KS 66215, USA

Kinetek, St Louis, MO 63146, USA

Kodak Ltd, PO Box 66, Kodak House, Station Road, Hemel Hempstead, Herts HP 1JU, UK LH Fermentation, Bells Hill, Stoke Poges, Slough, Bucks SL2 4EJ, UK

LH Fermentation Ltd, Porton House, Vanwall Road, Maidenhead, Berks SL6 4UB, UK 3492 Trust Way, Hayward, CA 94545, USA

Leica Instruments GmbH, Heidelberger Strasse 17–19, D-6907 Nussloch, Germany (Davy Ave., Knowhill, Milton Keynes MK5 8LB, UK; 111 Deer Lake Rd, Deerfield, IL 60015, USA)

Raymond Lamb, Sunbeam Road, London NW10, UK

Life Technologies Inc. (BRL), 8400 Helgerman Court, Gaithersburg, MD 20877, USA

Markem Co., Keene, NH, USA

May & Baker Ltd, Liverpool Road, Eccles, Manchester M30 7RT, UK

MBR Bioreactor AG, CH-8620 Wetzikon, Switzerland

Media Cult a/s, Symbion Science Park, Haraldsgade 68, DK 2100, Copenhagen, Denmark

Melles Griot B.V., Post Bus 272, NL-6900 AG Zevenaar, The Netherlands (1770 Kettering Street, Irvine, CA 92714, USA)

Microgon, Laguna Hills, CA 92653, USA

Miles Laboratories Inc, Research Products Division, PO Box 2000, 1127 Myrtle St, Elkhart, IN 46514, USA

Millipore (UK) Ltd, Millipore House, 11–15 Peterborough Rd, Harrow, Middlesex HA1 2BR, UK

Molecular Devices Corporation, Menlo Oaks Corporate Center, 4700 Bohannon Drive, Menlo Park CA 94025, USA

Morgan Sheet Metal Co., Sarasota, FL, USA

National Diagnostics, 1013–1017 Kennedy Blvd, Manville, NJ 08835, USA

NEN Research Products, 549 Albany St, Boston, MA 02118, USA

New Brunswick Scientific, 26–34 Emerald St, London WC1N 3QA, UK

New England Biolabs, Beverley, MA, USA

Northumbria Biologicals, S. Nelson Industrial Estate, Cranlington, Northumberland NE23 9HL, UK

Nuclepore Corporation, 7035 Commerce Drive, Pleasanton, CA 94566–3294, USA

Nunc (see Gibco Europe Ltd)

Nycomed Pharma AS Diagnostics, Lillogaten 3, Postbox 4284, Torshov, N-0401 Oslo 4, Norway

Ortho Pharmaceuticals Ltd, Enterprise House, Station Road, Loudwater, High Wycombe, Bucks, UK (410 University Avenue, Westwood, MA 02090, USA)

Pall Process Filtration Ltd, Europa House, PO Box 62, Portsmouth, Hants PO1 3PD, UK

Pentapharm Ltd, Engelgasse 109, CH–4002 Basel, Switzerland

Pharmacia (Great Britain) Ltd, Prince Regent Road, Hounslow, Middlesex TW3 1NE, UK (800 Centennial Ave., Piscataway, NJ 08854, USA)

Planer Products Ltd, Windmill Road, Sunbury on Thames, Middlesex TW16 7HD, UK

Polysciences Ltd, 24 Low Farm Place, Moulton Park, Northampton NN3 1HY, UK

Protein Polymer Technology Inc., 10655 Sorrenio Valley, San Diego, CA 92121, USA

Russell pH Ltd, Station Road, Auchtermuchty, Fife KY14 7DP, UK

Sartorius GmbH, PO Box 19, D-3440 Gottingen, Germany (Scotlab Ltd, Unit 15, Earn Ave., Righead Industrial Estate, Bellshill, ML4 3JQ, UK; Sartorius Instruments, 1430 Waukegan Road, McGaw Park, IL 60085, USA)

Schering Chemicals Ltd, Pharmaceutical Division, Burgess Hill, Sussex RH15 9NE, UK

Schott Glaswerke, PO Box 2480, D-500 Mainz 1, Germany

Sigma Chemical Co. Ltd, Fancy Road, Poole, Dorset BH17 7NH, UK (PO Box 14508, St Louis, MO 63178, USA)

Squibb and Sons Ltd, Regal House, Twickenham, Middlesex TW1 3QT, UK

Starna Cells Inc, PO Box 1919, Atascadero, CA 93423, USA

Sterilin Ltd, 43–45 Broad St, Teddington, Middlesex TW11 8QZ, UK

Stratech Scientific Ltd, 50 Newington Green, London N16 9PX, UK

Techne (Cambridge) Ltd, Duxford, Cambridge CB2 4PZ, UK (3700 Brunswick Pike, Princeton, NJ 08540, USA)

Union Carbide UK Ltd, Cryogenics Division, Redworth Way, Aycliffe Industrial Estate, Aycliffe, Co. Durham, UK M1L 2HS

Ventrex Laboratories Inc., 217 Read Street, Portland, ME 04103, USA

VWR, Bridgeport, NJ, USA

Watson-Marlowe Ltd, Falmouth, Cornwall TR11 4RU, UK

Whatman Biosystems Ltd, Maidstone, Kent ME14 2LE, UK

Whittaker Bioproducts Inc., 8830 Biggs Ford Road, Walkersville, MD 21793–0127, USA

Worthington Biochemical Corp., Halls Mill Road, Freehold, NJ 00728, USA

Zeiss (Carl) (Oberkochen) Ltd, Degenhardt House, 31–36 Foley St., London W1P 8AP, UK

Zeiss (Carl) Inc., 444 5th Avenue, New York, NY 10018, USA

Suppliers of growth factors and serum-free media

Some companies have specialized in the production and delivery of (1) serum-additives, (2) serum-substitutes, (3) serum-free, and (4) serum- and protein-free (i.e. totally chemically defined) media for mammalian cell cultures.

Moreover, some of these companies also offer distinct growth factors for various cell lines. The compilation of the following table—although incomplete—may assist the reader in his (her) first search for commercial sources; catalogues of these companies are worth consulting.

The full addresses are given in the main list of suppliers.

Commercial sources of serum free media and supplements

Company	City	Country	Product and Remarks
Amicon Co.	Beverly MA	USA	WRC 935
Boehringer Mannheim GmbH	Mannheim 31	FRG	Nutridoma media for hybridomas, media for various cells, growth factors
Camon Labor Service GmbH	Wiesbaden	FRG	CG medium for tumor cells, fibroblasts, macrophages, T- and B-cells
Cell Enterprises	Harrisonburg VA	USA	ABC medium (F12 modified, protein free)
Clonetics Co.	San Diego CA	USA	Various media for keratinocytes, endothelial and mammary cells etc.
Collaborative Research Inc.	Bedford MA	USA	Growth factors
Endotronics	Coon Rapids MN	USA	HL-1, CHO-1, PC-1 (for hybridomas and other transformed cells)
Genzyme Co.	Cambridge MA	USA	Growth factors
Gibco BRL Life Technol. Inc.	Grand Island NY	USA	PFHM-II (protein free media for hybridomas), SFM (serum free media) for keratinocytes etc.
Hyclone Laboratories Inc.	Logan UT	USA	HYQ-CCM-1 (for hybridomas)
ICN Biochemicals	Costa Mesa CA	USA	Hybri-Clone SF (for hybridomas)
Irvine Scientific	Santa Ana CA	USA	HB 101 (for murine and human hybridomas), HB 102, HB 104 (for hybridomas, T, B, LAK cells)
JRH Biosciences	Lenexa	USA	Ex-Cell, special media for particular cells
Media Cult a/s	Copenhagen	Denmark	SSR 1,2,3 (synthetic serum replacements)
Miles Inc. Diagnostic Div.	Kankakee IL	USA	Pentex Ex-Cyte (water-soluble mammalian lipoprotein additive)
Pentapharm AG	Basel	Switzerland	Prionex (albumin substitute, collagen fraction)
Protein Polymer Technol. Inc.	San Diego CA	USA	Pronectin F (attachment factor substitute)
Sigma Chemicals Co.	St. Louis MO	USA	Various media, SFPF (serum free, protein free) media, growth factors
Ventrex Laboratories Inc.	Portland ME	USA	HL-1, CHO-1, PC-1 (for hybridomas and other transformed cells)
Whittaker Bio-products Inc.	Walkersville MD	USA	Ultradoma media (for hybridomas)

Index

Index

ORDER OTHER TITLES OF INTEREST TODAY

138. Plasmids (2/e) Hardy, K.G. (Ed)
...... Spiralbound hardback 0-19-963445-9 £30.00
...... Paperback 0-19-963444-0 £19.50
136. RNA Processing: Vol. II Higgins, S.J. & Hames, B.D. (Eds)
...... Spiralbound hardback 0-19-963471-8 £30.00
...... Paperback 0-19-963470-X £19.50
135. RNA Processing: Vol. I Higgins, S.J. & Hames, B.D. (Eds)
...... Spiralbound hardback 0-19-963344-4 £30.00
...... Paperback 0-19-963343-6 £19.50
134. NMR of Macromolecules Roberts, G.C.K. (Ed)
...... Spiralbound hardback 0-19-963225-1 £32.50
...... Paperback 0-19-963224-3 £22.50
133. Gas Chromatography Baugh, P. (Ed)
...... Spiralbound hardback 0-19-963272-3 £40.00
...... Paperback 0-19-963271-5 £27.50
132. Essential Developmental Biology Stern, C.D. & Holland, P.W.H. (Eds)
...... Spiralbound hardback 0-19-963423-8 £30.00
...... Paperback 0-19-963422-X £19.50
131. Cellular Interactions in Development Hartley, D.A. (Ed)
...... Spiralbound hardback 0-19-963391-6 £30.00
...... Paperback 0-19-963390-8 £18.50
129 Behavioural Neuroscience: Volume II Sahgal, A. (Ed)
...... Spiralbound hardback 0-19-963458-0 £32.50
...... Paperback 0-19-963457-2 £22.50
128 Behavioural Neuroscience: Volume I Sahgal, A. (Ed)
...... Spiralbound hardback 0-19-963368-1 £32.50
...... Paperback 0-19-963367-3 £22.50
127. Molecular Virology Davison, A.J. & Elliott, R.M. (Eds)
...... Spiralbound hardback 0-19-963350-4 £35.00
...... Paperback 0-19-963357-6 £25.00
126. Gene Targeting Joyner, A.L. (Ed)
...... Spiralbound hardback 0-19-963407-6 £30.00
...... Paperback 0-19-9634036-8 19.50
125. Glycobiology Fukuda, M. & Kobata, A. (Eds)
...... Spiralbound hardback 0-19-963372-X £32.50
...... Paperback 0-19-963371-1 £22.50
124. Human Genetic Disease Analysis (2/e) Davies, K.E. (Ed)
...... Spiralbound hardback 0-19-963309-6 £30.00
...... Paperback 0-19-963308-8 £18.50
122. Immunocytochemistry Beesley, J. (Ed)
...... Spiralbound hardback 0-19-963270-7 £35.00
...... Paperback 0-19-963269-3 £22.50
123. Protein Phosphorylation Hardie, D.G. (Ed)
...... Spiralbound hardback 0-19-963306-1 £32.50
...... Paperback 0-19-963305-3 £22.50
121. Tumour Immunobiology Gallagher, G., Rees, R.C. & others (Eds)
...... Spiralbound hardback 0-19-963370-3 £40.00
...... Paperback 0-19-963369-X £27.50
120. Transcription Factors Latchman, D.S. (Ed)
...... Spiralbound hardback 0-19-963342-8 £30.00
...... Paperback 0-19-963341-X £19.50
119. Growth Factors McKay, I. & Leigh, I. (Eds)
...... Spiralbound hardback 0-19-963360-6 £30.00
...... Paperback 0-19-963359-2 £19.50
118. Histocompatibility Testing Dyer, P. & Middleton, D. (Eds)
...... Spiralbound hardback 0-19-963364-9 £32.50
...... Paperback 0-19-963363-0 £22.50

117. Gene Transcription Hames, B.D. & Higgins, S.J. (Eds)
...... Spiralbound hardback 0-19-963292-8 £35.00
...... Paperback 0-19-963291-X £25.00
116. Electrophysiology Wallis, D.I. (Ed)
...... Spiralbound hardback 0-19-963348-7 £32.50
...... Paperback 0-19-963347-9 £22.50
115. Biological Data Analysis Fry, J.C. (Ed)
...... Spiralbound hardback 0-19-963340-1 £50.00
...... Paperback 0-19-963339-8 £27.50
114. Experimental Neuroanatomy Bolam, J.P. (Ed)
...... Spiralbound hardback 0-19-963326-6 £32.50
...... Paperback 0-19-963325-8 £22.50
113. Preparative Centrifugation Rickwood, D. (Ed)
...... Spiralbound hardback 0-19-963208-1 £45.00
...... Paperback 0-19-963211-1 £25.00
...... Paperback 0-19-963099-2 £25.00
112. Lipid Analysis Hamilton, R.J. & Hamilton, Shiela (Eds)
...... Spiralbound hardback 0-19-963098-4 £35.00
...... Paperback 0-19-963097-6 £25.00
111. Haemopoiesis Testa, N.G. & Molineux, G. (Eds)
...... Spiralbound hardback 0-19-963366-5 £32.50
...... Paperback 0-19-963365-7 £22.50
110. Pollination Ecology Dafni, A.
...... Spiralbound hardback 0-19-963299-5 £32.50
...... Paperback 0-19-963298-7 £22.50
109. In Situ Hybridization Wilkinson, D.G. (Ed)
...... Spiralbound hardback 0-19-963328-2 £30.00
...... Paperback 0-19-963327-4 £18.50
108. Protein Engineering Rees, A.R., Sternberg, M.J.E. & others (Eds)
...... Spiralbound hardback 0-19-963139-5 £35.00
...... Paperback 0-19-963138-7 £25.00
107. Cell-Cell Interactions Stevenson, B.R., Gallin, W.J. & others (Eds)
...... Spiralbound hardback 0-19-963319-3 £32.50
...... Paperback 0-19-963318-5 £22.50
106. Diagnostic Molecular Pathology: Volume I Herrington, C.S. & McGee, J. O'D. (Eds)
...... Spiralbound hardback 0-19-963237-5 £30.00
...... Paperback 0-19-963236-7 £19.50
105. Biomechanics-Materials Vincent, J.F.V. (Ed)
...... Spiralbound hardback 0-19-963223-5 £35.00
...... Paperback 0-19-963222-7 £25.00
104. Animal Cell Culture (2/e) Freshney, R.I. (Ed)
...... Spiralbound hardback 0-19-963212-X £30.00
...... Paperback 0-19-963213-8 £19.50
103. Molecular Plant Pathology: Volume II Gurr, S.J., McPherson, M.J. & others (Eds)
...... Spiralbound hardback 0-19-963352-5 £32.50
...... Paperback 0-19-963351-7 £22.50
102 Signal Transduction Milligan, G. (Ed)
...... Spiralbound hardback 0-19-963296-0 £30.00
...... Paperback 0-19-963295-2 £18.50
101. Protein Targeting Magee, A.I. & Wileman, T. (Eds)
...... Spiralbound hardback 0-19-963206-5 £32.50
...... Paperback 0-19-963210-3 £22.50
100. Diagnostic Molecular Pathology: Volume II: Cell and Tissue Genotyping Herrington, C.S. & McGee, J.O'D. (Eds)
...... Spiralbound hardback 0-19-963239-1 £30.00
...... Paperback 0-19-963238-3 £19.50
99. Neuronal Cell Lines Wood, J.N. (Ed)
...... Spiralbound hardback 0-19-963346-0 £32.50
...... Paperback 0-19-963345-2 £22.50

48. **Protein Sequencing** Findlay, J.B.C. & Geisow, M.J. (Eds)
...... Spiralbound hardback 0-19-963012-7 **£15.00**
...... Paperback 0-19-963013-5 **£12.50**
47. **Cell Growth and Division** Baserga, R. (Ed)
...... Spiralbound hardback 0-19-963026-7 **£15.00**
...... Paperback 0-19-963027-5 **£12.50**
46. **Protein Function** Creighton, T.E. (Ed)
...... Spiralbound hardback 0-19-963006-2 **£32.50**
...... Paperback 0-19-963007-0 **£22.50**
45. **Protein Structure** Creighton, T.E. (Ed)
...... Spiralbound hardback 0-19-963000-3 **£32.50**
...... Paperback 0-19-963001-1 **£22.50**
44. **Antibodies: Volume II** Catty, D. (Ed)
...... Spiralbound hardback 0-19-963018-6 **£30.00**
...... Paperback 0-19-963019-4 **£19.50**
43. **HPLC of Macromolecules** Oliver, R.W.A. (Ed)
...... Spiralbound hardback 0-19-963020-8 **£30.00**
...... Paperback 0-19-963021-6 **£19.50**
42. **Light Microscopy in Biology** Lacey, A.J. (Ed)
...... Spiralbound hardback 0-19-963036-4 **£30.00**
...... Paperback 0-19-963037-2 **£19.50**
41. **Plant Molecular Biology** Shaw, C.H. (Ed)
...... Paperback 1-85221-056-7 **£12.50**
40. **Microcomputers in Physiology** Fraser, P.J. (Ed)
...... Spiralbound hardback 1-85221-129-6 **£15.00**
...... Paperback 1-85221-130-X **£12.50**
39. **Genome Analysis** Davies, K.E. (Ed)
...... Spiralbound hardback 1-85221-109-1 **£30.00**
...... Paperback 1-85221-110-5 **£18.50**
38. **Antibodies: Volume I** Catty, D. (Ed)
...... Paperback 0-947946-85-3 **£19.50**
37. **Yeast** Campbell, I. & Duffus, J.H. (Eds)
...... Paperback 0-947946-79-9 **£12.50**
36. **Mammalian Development** Monk, M. (Ed)
...... Hardback 1-85221-030-3 **£15.00**
...... Paperback 1-85221-029-X **£12.50**
35. **Lymphocytes** Klaus, G.G.B. (Ed)
...... Hardback 1-85221-018-4 **£30.00**
34. **Lymphokines and Interferons** Clemens, M.J., Morris, A.G. & others (Eds)
...... Paperback 1-85221-035-4 **£12.50**
33. **Mitochondria** Darley-Usmar, V.M., Rickwood, D. & others (Eds)
...... Hardback 1-85221-034-6 **£32.50**
...... Paperback 1-85221-033-8 **£22.50**
32. **Prostaglandins and Related Substances** Benedetto, C., McDonald-Gibson, R.G. & others (Eds)
...... Hardback 1-85221-032-X **£15.00**
...... Paperback 1-85221-031-1 **£12.50**
31. **DNA Cloning: Volume III** Glover, D.M. (Ed)
...... Hardback 1-85221-049-4 **£15.00**
...... Paperback 1-85221-048-6 **£12.50**
30. **Steroid Hormones** Green, B. & Leake, R.E. (Eds)
...... Paperback 0-947946-53-9 **£19.50**
29. **Neurochemistry** Turner, A.J. & Bachelard, H.S. (Eds)
...... Hardback 1-85221-028-1 **£15.00**
...... Paperback 1-85221-027-3 **£12.50**
28. **Biological Membranes** Findlay, J.B.C. & Evans, W.H. (Eds)
...... Hardback 0-947946-84-5 **£15.00**
...... Paperback 0-947946-83-7 **£12.50**
27. **Nucleic Acid and Protein Sequence Analysis** Bishop, M.J. & Rawlings, C.J. (Eds)
...... Hardback 1-85221-007-9 **£35.00**
...... Paperback 1-85221-006-0 **£25.00**
26. **Electron Microscopy in Molecular Biology** Sommerville, J. & Scheer, U. (Eds)
...... Hardback 0-947946-64-0 **£15.00**
...... Paperback 0-947946-54-3 **£12.50**
25. **Teratocarcinomas and Embryonic Stem Cells** Robertson, E.J. (Ed)
...... Paperback 1-85221-004-4 **£19.50**
24. **Spectrophotometry and Spectrofluorimetry** Harris, D.A. & Bashford, C.L. (Eds)
...... Hardback 0-947946-69-1 **£15.00**
...... Paperback 0-947946-46-2 **£12.50**
23. **Plasmids** Hardy, K.G. (Ed)
...... Paperback 0-947946-81-0 **£12.50**
22. **Biochemical Toxicology** Snell, K. & Mullock, B. (Eds)
...... Paperback 0-947946-52-7 **£12.50**
19. **Drosophila** Roberts, D.B. (Ed)
...... Hardback 0-947946-66-7 **£32.50**
...... Paperback 0-947946-45-4 **£22.50**

17. **Photosynthesis: Energy Transduction** Hipkins, M.F. & Baker, N.R. (Eds)
...... Hardback 0-947946-63-2 **£15.00**
...... Paperback 0-947946-51-9 **£12.50**
16. **Human Genetic Diseases** Davies, K.E. (Ed)
...... Hardback 0-947946-76-4 **£15.00**
...... Paperback 0-947946-75-6 **£12.50**
14. **Nucleic Acid Hybridisation** Hames, B.D. & Higgins, S.J. (Eds)
...... Hardback 0-947946-61-6 **£15.00**
...... Paperback 0-947946-23-3 **£12.50**
13. **Immobilised Cells and Enzymes** Woodward, J. (Ed)
...... Hardback 0-947946-60-8 **£15.00**
12. **Plant Cell Culture** Dixon, R.A. (Ed)
...... Paperback 0-947946-22-5 **£19.50**
11a. **DNA Cloning: Volume I** Glover, D.M. (Ed)
...... Paperback 0-947946-18-7 **£12.50**
11b. **DNA Cloning: Volume II** Glover, D.M. (Ed)
...... Paperback 0-947946-19-5 **£12.50**
10. **Virology** Mahy, B.W.J. (Ed)
...... Paperback 0-904147-78-9 **£19.50**
9. **Affinity Chromatography** Dean, P.D.G., Johnson, W.S. & others (Eds)
...... Paperback 0-904147-71-1 **£19.50**
7. **Microcomputers in Biology** Ireland, C.R. & Long, S.P. (Eds)
...... Paperback 0-904147-57-6 **£18.00**
6. **Oligonucleotide Synthesis** Gait, M.J. (Ed)
...... Paperback 0-904147-74-6 **£18.50**
5. **Transcription and Translation** Hames, B.D. & Higgins, S.J. (Eds)
...... Paperback 0-904147-52-5 **£12.50**
3. **Iodinated Density Gradient Media** Rickwood, D. (Ed)
...... Paperback 0-904147-51-7 **£12.50**

Sets

Essential Molecular Biology: 2 vol set Brown, T.A. (Ed)
...... Spiralbound hardback 0-19-963114-X **£58.00**
...... Paperback 0-19-963115-8 **£40.00**
Antibodies: 2 vol set Catty, D. (Ed)
...... Paperback 0-19-963063-1 **£33.00**
Cellular and Molecular Neurobiology: 2 vol set Chad, J. & Wheal, H. (Eds)
...... Spiralbound hardback 0-19-963255-3 **£56.00**
...... Paperback 0-19-963254-5 **£38.00**
Protein Structure and Protein Function: 2 vol set Creighton, T.E. (Ed)
...... Spiralbound hardback 0-19-963064-X **£55.00**
...... Paperback 0-19-963065-8 **£38.00**
DNA Cloning: 2 vol set Glover, D.M. (Ed)
...... Paperback 1-85221-069-9 **£30.00**
Molecular Plant Pathology: 2 vol set Gurr, S.J., McPherson, M.J. & others (Eds)
...... Spiralbound hardback 0-19-963354-1 **£56.00**
...... Paperback 0-19-963353-3 **£37.00**
Protein Purification Methods, and Protein Purification Applications: 2 vol set Harris, E.L.V. & Angal, S. (Eds)
...... Spiralbound hardback 0-19-963048-8 **£48.00**
...... Paperback 0-19-963049-6 **£32.00**
Diagnostic Molecular Pathology: 2 vol set Herrington, C.S. & McGee, J. O'D. (Eds)
...... Spiralbound hardback 0-19-963241-3 **£54.00**
...... Paperback 0-19-963240-5 **£35.00**
RNA Processing: 2 vol set Higgins, S.J. & Hames, B.D. (Eds)
...... Spiralbound hardback 0-19-963473-4 **£54.00**
...... Paperback 0-19-963472-6 **£35.00**
Receptor Biochemistry; Receptor-Effector Coupling; Receptor-Ligand Interactions: 3 vol set Hulme, E.C. (Ed)
...... Paperback 0-19-963097-6 **£62.50**
Human Cytogenetics: 2 vol set (2/e) Rooney, D.E. & Czepulkowski, B.H. (Eds)
...... Hardback 0-19-963314-2 **£58.50**
...... Paperback 0-19-963313-4 **£40.50**
Behavioural Neuroscience: 2 vol set Sahgal, A. (Ed)
...... Spiralbound hardback 0-19-963460-2 **£58.00**
...... Paperback 0-19-963459-9 **£40.00**
Peptide Hormone Secretion/Peptide Hormone Action: 2 vol set Siddle, K. & Hutton, J.C. (Eds)
...... Spiralbound hardback 0-19-963072-0 **£55.00**
...... Paperback 0-19-963073-9 **£38.00**

ORDER FORM for UK, Europe and Rest of World

(Excluding USA and Canada)

Qty	ISBN	Author	Title	Amount
			P&P	
			*VAT	
			TOTAL	

Please add postage and packing: £1.75 for UK orders under £20; £2.75 for UK orders over £20; overseas orders add 10% of total.

* EC customers please note that VAT must be added (excludes UK customers)

Name ....

Address ....

....

.... Post code ....

[ ] Please charge £ .... to my credit card
Access/VISA/Eurocard/AMEX/Diners Club (circle appropriate card)

Card No .... Expiry date ....

Signature ....

Credit card account address if different from above:

....

.... Postcode ....

[ ] I enclose a cheque for £......................

Please return this form to: OUP Distribution Services, Saxon Way West, Corby, Northants NN18 9ES, UK

OR ORDER BY CREDIT CARD HOTLINE: Tel +44-(0)536-741519 or
Fax +44-(0)536-746337

ORDER OTHER TITLES OF INTEREST TODAY

Price list for: USA and Canada

128. **Behavioural Neuroscience: Volume I** Sahgal, A. (Ed)
- Spiralbound hardback   0-19-963368-1   **$57.00**
- Paperback   0-19-963367-3   **$37.00**

127. **Molecular Virology** Davison, A.J. & Elliott, R.M. (Eds)
- Spiralbound hardback   0-19-963358-4   **$49.00**
- Paperback   0-19-963357-6   **$32.00**

126. **Gene Targeting** Joyner, A.L. (Ed)
- Spiralbound hardback   0-19-963407-6   **$49.00**
- Paperback   0-19-9634036-8   **$34.00**

124. **Human Genetic Disease Analysis (2/e)** Davies, K.E. (Ed)
- Spiralbound hardback   0-19-963309-6   **$54.00**
- Paperback   0-19-963308-8   **$33.00**

123. **Protein Phosphorylation** Hardie, D.G. (Ed)
- Spiralbound hardback   0-19-963306-1   **$65.00**
- Paperback   0-19-963305-3   **$45.00**

122. **Immunocytochemistry** Beesley, J. (Ed)
- Spiralbound hardback   0-19-963270-7   **$62.00**
- Paperback   0-19-963269-3   **$42.00**

121. **Tumour Immunobiology** Gallagher, G., Rees, R.C. & others (Eds)
- Spiralbound hardback   0-19-963370-3   **$72.00**
- Paperback   0-19-963369-X   **$50.00**

120. **Transcription Factors** Latchman, D.S. (Ed)
- Spiralbound hardback   0-19-963342-8   **$48.00**
- Paperback   0-19-963341 X   **$31.00**

119. **Growth Factors** McKay, I. & Leigh, I. (Eds)
- Spiralbound hardback   0-19-963360-6   **$48.00**
- Paperback   0-19-963359-2   **$31.00**

118. **Histocompatibility Testing** Dyer, P. & Middleton, D. (Eds)
- Spiralbound hardback   0-19-963364-9   **$60.00**
- Paperback   0-19-963363-0   **$41.00**

117. **Gene Transcription** Hames, B.D. & Higgins, S.J. (Eds)
- Spiralbound hardback   0-19-963292-8   **$72.00**
- Paperback   0-19-963291-X   **$50.00**

116. **Electrophysiology** Wallis, D.I. (Ed)
- Spiralbound hardback   0-19-963348-7   **$56.00**
- Paperback   0-19-963347-9   **$39.00**

115. **Biological Data Analysis** Fry, J.C. (Ed)
- Spiralbound hardback   0-19-963340-1   **$80.00**
- Paperback   0-19-963339-8   **$60.00**

114. **Experimental Neuroanatomy** Bolam, J.P. (Ed)
- Spiralbound hardback   0-19-963326-6   **$59.00**
- Paperback   0-19-963325-8   **$39.00**

113. **Preparative Centrifugation** Rickwood, D. (Ed)
- Spiralbound hardback   0-19-963208-1   **$78.00**
- Paperback   0-19-963211-1   **$44.00**

111. **Haemopoiesis** Testa, N.G. & Molineux, G. (Eds)
- Spiralbound hardback   0-19-963366-5   **$59.00**
- Paperback   0-19-963365-7   **$39.00**

110. **Pollination Ecology** Dafni, A.
- Spiralbound hardback   0-19-963299-5   **$56.95**
- Paperback   0-19-963298-7   **$39.95**

109. **In Situ Hybridization** Wilkinson, D.G. (Ed)
- Spiralbound hardback   0-19-963328-2   **$58.00**
- Paperback   0-19-963327-4   **$36.00**

108. **Protein Engineering** Rees, A.R., Sternberg, M.J.E. & others (Eds)
- Spiralbound hardback   0-19-963139-5   **$64.00**
- Paperback   0-19-963138-7   **$44.00**

107. **Cell-Cell Interactions** Stevenson, B.R., Gallin, W.J. & others (Eds)
- Spiralbound hardback   0-19-963319-3   **$55.00**
- Paperback   0-19-963318-5   **$38.00**

106. **Diagnostic Molecular Pathology: Volume I** Herrington, C.S. & McGee, J. O'D. (Eds)
- Spiralbound hardback   0-19-963237-5   **$50.00**
- Paperback   0-19-963236-7   **$33.00**

105. **Biomechanics-Materials** Vincent, J.F.V. (Ed)
- Spiralbound hardback   0-19-963223-5   **$70.00**
- Paperback   0-19-963222-7   **$50.00**

104. **Animal Cell Culture (2/e)** Freshney, R.I. (Ed)
- Spiralbound hardback   0-19-963212-X   **$55.00**
- Paperback   0-19-963213-8   **$35.00**

103. **Molecular Plant Pathology: Volume II** Gurr, S.J., McPherson, M.J. & others (Eds)
- Spiralbound hardback   0-19-963352-5   **$65.00**
- Paperback   0-19-963351-7   **$45.00**

102. **Signal Transduction** Milligan, G. (Ed)
- Spiralbound hardback   0-19-963296-0   **$60.00**
- Paperback   0-19-963295-2   **$38.00**

101. **Protein Targeting** Magee, A.I. & Wileman, T. (Eds)
- Spiralbound hardback   0-19-963206-5   **$75.00**
- Paperback   0-19-963210-3   **$50.00**

100. **Diagnostic Molecular Pathology: Volume II: Cell and Tissue Genotyping** Herrington, C.S. & McGee, J.O'D. (Eds)
- Spiralbound hardback   0-19-963239-1   **$60.00**
- Paperback   0-19-963238-3   **$39.00**

99. **Neuronal Cell Lines** Wood, J.N. (Ed)
- Spiralbound hardback   0-19-963346-0   **$68.00**
- Paperback   0-19-963345-2   **$48.00**

98. **Neural Transplantation** Dunnett, S.B. & Björklund, A. (Eds)
- Spiralbound hardback   0 19 063206-3   **$69.00**
- Paperback   0-19-963285-5   **$42.00**

97. **Human Cytogenetics: Volume II: Malignancy and Acquired Abnormalities (2/e)** Rooney, D.E. & Czepulkowski, B.H. (Eds)
- Spiralbound hardback   0-19-963290-1   **$75.00**
- Paperback   0-19-963289-8   **$50.00**

96. **Human Cytogenetics: Volume I: Constitutional Analysis (2/e)** Rooney, D.E. & Czepulkowski, B.H. (Eds)
- Spiralbound hardback   0-19-963288-X   **$75.00**
- Paperback   0-19-963287-1   **$50.00**

95. **Lipid Modification of Proteins** Hooper, N.M. & Turner, A.J. (Eds)
- Spiralbound hardback   0-19-963274-X   **$75.00**
- Paperback   0-19-963273-1   **$50.00**

94. **Biomechanics-Structures and Systems** Biewener, A.A. (Ed)
- Spiralbound hardback   0-19-963268-5   **$85.00**
- Paperback   0-19-963267-7   **$50.00**

93. **Lipoprotein Analysis** Converse, C.A. & Skinner, E.R. (Eds)
- Spiralbound hardback   0-19-963192-1   **$65.00**
- Paperback   0-19-963231-6   **$42.00**

92. **Receptor-Ligand Interactions** Hulme, E.C. (Ed)
- Spiralbound hardback   0-19-963090-9   **$75.00**
- Paperback   0-19-963091-7   **$50.00**

91. **Molecular Genetic Analysis of Populations** Hoelzel, A.R. (Ed)
- Spiralbound hardback   0-19-963278-2   **$65.00**
- Paperback   0-19-963277-4   **$45.00**

90. **Enzyme Assays** Eisenthal, R. & Danson, M.J. (Eds)
...... Spiralbound hardback 0-19-963142-5 **$68.00**
...... Paperback 0-19-963143-3 **$48.00**
89. **Microcomputers in Biochemistry** Bryce, C.F.A. (Ed)
...... Spiralbound hardback 0-19-963253-7 **$60.00**
...... Paperback 0-19-963252-9 **$40.00**
88. **The Cytoskeleton** Carraway, K.L. & Carraway, C.A.C. (Eds)
...... Spiralbound hardback 0-19-963257-X **$60.00**
...... Paperback 0-19-963256-1 **$40.00**
87. **Monitoring Neuronal Activity** Stamford, J.A. (Ed)
...... Spiralbound hardback 0-19-963244-8 **$60.00**
...... Paperback 0-19-963243-X **$40.00**
86. **Crystallization of Nucleic Acids and Proteins** Ducruix, A. & Giegé, R. (Eds)
...... Spiralbound hardback 0-19-963245-6 **$60.00**
...... Paperback 0-19-963246-4 **$50.00**
85. **Molecular Plant Pathology: Volume I** Gurr, S.J., McPherson, M.J. & others (Eds)
...... Spiralbound hardback 0-19-963103-4 **$60.00**
...... Paperback 0-19-963102-6 **$40.00**
84. **Anaerobic Microbiology** Levett, P.N. (Ed)
...... Spiralbound hardback 0-19-963204-9 **$75.00**
...... Paperback 0-19-963262-6 **$45.00**
83. **Oligonucleotides and Analogues** Eckstein, F. (Ed)
...... Spiralbound hardback 0-19-963280-4 **$65.00**
...... Paperback 0-19-963279-0 **$45.00**
82. **Electron Microscopy in Biology** Harris, R. (Ed)
...... Spiralbound hardback 0-19-963219-7 **$65.00**
...... Paperback 0-19-963215-4 **$45.00**
81. **Essential Molecular Biology: Volume II** Brown, T.A. (Ed)
...... Spiralbound hardback 0-19-963112-3 **$65.00**
...... Paperback 0-19-963113-1 **$45.00**
80. **Cellular Calcium** McCormack, J.G. & Cobbold, P.H. (Eds)
...... Spiralbound hardback 0-19-963131-X **$75.00**
...... Paperback 0-19-963130-1 **$50.00**
79. **Protein Architecture** Lesk, A.M.
...... Spiralbound hardback 0-19-963054-2 **$65.00**
...... Paperback 0-19-963055-0 **$45.00**
78. **Cellular Neurobiology** Chad, J. & Wheal, H. (Eds)
...... Spiralbound hardback 0-19-963106-9 **$73.00**
...... Paperback 0-19-963107-7 **$43.00**
77. **PCR** McPherson, M.J., Quirke, P. & others (Eds)
...... Spiralbound hardback 0-19-963226-X **$55.00**
...... Paperback 0-19-963196-4 **$40.00**
76. **Mammalian Cell Biotechnology** Butler, M. (Ed)
...... Spiralbound hardback 0-19-963207-3 **$60.00**
...... Paperback 0-19-963209-X **$40.00**
75. **Cytokines** Balkwill, F.R. (Ed)
...... Spiralbound hardback 0-19-963218-9 **$64.00**
...... Paperback 0-19-963214-6 **$44.00**
74. **Molecular Neurobiology** Chad, J. & Wheal, H. (Eds)
...... Spiralbound hardback 0-19-963108-5 **$56.00**
...... Paperback 0-19-963109-3 **$36.00**
73. **Directed Mutagenesis** McPherson, M.J. (Ed)
...... Spiralbound hardback 0-19-963141-7 **$55.00**
...... Paperback 0-19-963140-9 **$35.00**
72. **Essential Molecular Biology: Volume I** Brown, T.A. (Ed)
...... Spiralbound hardback 0-19-963110-7 **$65.00**
...... Paperback 0-19-963111-5 **$45.00**
71. **Peptide Hormone Action** Siddle, K. & Hutton, J.C.
...... Spiralbound hardback 0-19-963070-4 **$70.00**
...... Paperback 0-19-963071-2 **$50.00**
70. **Peptide Hormone Secretion** Hutton, J.C. & Siddle, K. (Eds)
...... Spiralbound hardback 0-19-963068-2 **$70.00**
...... Paperback 0-19-963069-0 **$50.00**
69. **Postimplantation Mammalian Embryos** Copp, A.J. & Cockroft, D.L. (Eds)
...... Spiralbound hardback 0-19-963088-7 **$70.00**
...... Paperback 0-19-963089-5 **$50.00**
68. **Receptor-Effector Coupling** Hulme, E.C. (Ed)
...... Spiralbound hardback 0-19-963094-1 **$70.00**
...... Paperback 0-19-963095-X **$45.00**
67. **Gel Electrophoresis of Proteins (2/e)** Hames, B.D. & Rickwood, D. (Eds)
...... Spiralbound hardback 0-19-963074-7 **$75.00**
...... Paperback 0-19-963075-5 **$50.00**
66. **Clinical Immunology** Gooi, H.C. & Chapel, H. (Eds)
...... Spiralbound hardback 0-19-963086-0 **$69.95**
...... Paperback 0-19-963087-9 **$50.00**

65. **Receptor Biochemistry** Hulme, E.C. (Ed)
...... Paperback 0-19-963093-3 **$50.00**
64. **Gel Electrophoresis of Nucleic Acids (2/e)** Rickwood, D. & Hames, B.D. (Eds)
...... Spiralbound hardback 0-19-963082-8 **$75.00**
...... Paperback 0-19-963083-6 **$50.00**
63. **Animal Virus Pathogenesis** Oldstone, M.B.A. (Ed)
...... Spiralbound hardback 0-19-963100-X **$68.00**
...... Paperback 0-19-963101-8 **$40.00**
62. **Flow Cytometry** Ormerod, M.G. (Ed)
...... Paperback 0-19-963053-4 **$50.00**
61. **Radioisotopes in Biology** Slater, R.J. (Ed)
...... Spiralbound hardback 0-19-963080-1 **$75.00**
...... Paperback 0-19-963081-X **$45.00**
60. **Biosensors** Cass, A.E.G. (Ed)
...... Spiralbound hardback 0-19-963046-1 **$65.00**
...... Paperback 0-19-963047-X **$43.00**
59. **Ribosomes and Protein Synthesis** Spedding, G. (Ed)
...... Spiralbound hardback 0-19-963104-2 **$75.00**
...... Paperback 0-19-963105-0 **$45.00**
58. **Liposomes** New, R.R.C. (Ed)
...... Spiralbound hardback 0-19-963076-3 **$70.00**
...... Paperback 0-19-963077-1 **$45.00**
57. **Fermentation** McNeil, B. & Harvey, L.M. (Eds)
...... Spiralbound hardback 0-19-963044-5 **$65.00**
...... Paperback 0-19-963045-3 **$39.00**
56. **Protein Purification Applications** Harris, E.L.V. & Angal, S. (Eds)
...... Spiralbound hardback 0-19-963022-4 **$54.00**
...... Paperback 0-19-963023-2 **$36.00**
55. **Nucleic Acids Sequencing** Howe, C.J. & Ward, E.S. (Eds)
...... Spiralbound hardback 0-19-963056-9 **$59.00**
...... Paperback 0-19-963057-7 **$38.00**
54. **Protein Purification Methods** Harris, E.L.V. & Angal, S. (Eds)
...... Spiralbound hardback 0-19-963002-X **$60.00**
...... Paperback 0-19-963003-8 **$40.00**
53. **Solid Phase Peptide Synthesis** Atherton, E. & Sheppard, R.C.
...... Spiralbound hardback 0-19-963066-6 **$58.00**
...... Paperback 0-19-963067-4 **$39.95**
52. **Medical Bacteriology** Hawkey, P.M. & Lewis, D.A. (Eds)
...... Paperback 0-19-963009-7 **$50.00**
51. **Proteolytic Enzymes** Beynon, R.J. & Bond, J.S. (Eds)
...... Spiralbound hardback 0-19-963058-5 **$60.00**
...... Paperback 0-19-963059-3 **$39.00**
50. **Medical Mycology** Evans, E.G.V. & Richardson, M.D. (Eds)
...... Spiralbound hardback 0-19-963010-0 **$69.95**
...... Paperback 0-19-963011-9 **$50.00**
49. **Computers in Microbiology** Bryant, T.N. & Wimpenny, J.W.T. (Eds)
...... Paperback 0-19-963015-1 **$40.00**
48. **Protein Sequencing** Findlay, J.B.C. & Geisow, M.J. (Eds)
...... Spiralbound hardback 0-19-963012-7 **$56.00**
...... Paperback 0-19-963013-5 **$38.00**
47. **Cell Growth and Division** Baserga, R. (Ed)
...... Spiralbound hardback 0-19-963026-7 **$62.00**
...... Paperback 0-19-963027-5 **$38.00**
46. **Protein Function** Creighton, T.E. (Ed)
...... Spiralbound hardback 0-19-963006-2 **$65.00**
...... Paperback 0-19-963007-0 **$45.00**
45. **Protein Structure** Creighton, T.E. (Ed)
...... Spiralbound hardback 0-19-963000-3 **$65.00**
...... Paperback 0-19-963001-1 **$45.00**
44. **Antibodies: Volume II** Catty, D. (Ed)
...... Spiralbound hardback 0-19-963018-6 **$58.00**
...... Paperback 0-19-963019-4 **$39.00**
43. **HPLC of Macromolecules** Oliver, R.W.A. (Ed)
...... Spiralbound hardback 0-19-963020-8 **$54.00**
...... Paperback 0-19-963021-6 **$45.00**
42. **Light Microscopy in Biology** Lacey, A.J. (Ed)
...... Spiralbound hardback 0-19-963036-4 **$62.00**
...... Paperback 0-19-963037-2 **$38.00**
41. **Plant Molecular Biology** Shaw, C.H. (Ed)
...... Paperback 1-85221-056-7 **$38.00**
40. **Microcomputers in Physiology** Fraser, P.J. (Ed)
...... Spiralbound hardback 1-85221-129-6 **$54.00**
...... Paperback 1-85221-130-X **$36.00**
39. **Genome Analysis** Davies, K.E. (Ed)
...... Spiralbound hardback 1-85221-109-1 **$54.00**
...... Paperback 1-85221-110-5 **$36.00**
38. **Antibodies: Volume I** Catty, D. (Ed)
...... Paperback 0-947946-85-3 **$38.00**
37. **Yeast** Campbell, I. & Duffus, J.H. (Eds)
...... Paperback 0-947946-79-9 **$36.00**

36.	**Mammalian Development** Monk, M. (Ed)		
......	Hardback	1-85221-030-3	**$60.00**
......	Paperback	1-85221-029-X	**$45.00**
35.	**Lymphocytes** Klaus, G.G.B. (Ed)		
......	Hardback	1-85221-018-4	**$54.00**
34.	**Lymphokines and Interferons** Clemens, M.J., Morris, A.G. & others (Eds)		
......	Paperback	1-85221-035-4	**$44.00**
33.	**Mitochondria** Darley-Usmar, V.M., Rickwood, D. & others (Eds)		
......	Hardback	1-85221-034-6	**$65.00**
......	Paperback	1-85221-033-8	**$45.00**
32.	**Prostaglandins and Related Substances** Benedetto, C., McDonald-Gibson, R.G. & others (Eds)		
......	Hardback	1-85221-032-X	**$58.00**
......	Paperback	1-85221-031-1	**$38.00**
31.	**DNA Cloning: Volume III** Glover, D.M. (Ed)		
......	Hardback	1-85221-049-4	**$56.00**
......	Paperback	1-85221-048-6	**$36.00**
30.	**Steroid Hormones** Green, B. & Leake, R.E. (Eds)		
......	Paperback	0-947946-53-5	**$40.00**
29.	**Neurochemistry** Turner, A.J. & Bachelard, H.S. (Eds)		
......	Hardback	1-85221-028-1	**$56.00**
......	Paperback	1-85221-027-3	**$36.00**
28.	**Biological Membranes** Findlay, J.B.C. & Evans, W.H. (Eds)		
......	Hardback	0-947946-84-5	**$54.00**
......	Paperback	0-947946-83-7	**$36.00**
27.	**Nucleic Acid and Protein Sequence Analysis** Bishop, M.J. & Rawlings, C.J. (Eds)		
......	Hardback	1-85221-007-9	**$66.00**
......	Paperback	1-85221-006-0	**$44.00**
26.	**Electron Microscopy in Molecular Biology** Sommerville, J. & Scheer, U. (Eds)		
......	Hardback	0-947946-64-0	**$54.00**
......	Paperback	0-947946-54-3	**$40.00**
24.	**Spectrophotometry and Spectrofluorimetry** Harris, D.A. & Bashford, C.L. (Eds)		
......	Hardback	0-947946-69-1	**$56.00**
......	Paperback	0-947946-46-2	**$39.95**
23.	**Plasmids** Hardy, K.G. (Ed)		
......	Paperback	0-947946-81-0	**$36.00**
22.	**Biochemical Toxicology** Snell, K. & Mullock, B. (Eds)		
......	Paperback	0-947946-52-7	**$40.00**
19.	**Drosophila** Roberts, D.B. (Ed)		
......	Hardback	0-947946-66-7	**$67.50**
......	Paperback	0-947946-45-4	**$46.00**
17.	**Photosynthesis: Energy Transduction** Hipkins, M.F. & Baker, N.R. (Eds)		
......	Hardback	0-947946-63-2	**$54.00**
......	Paperback	0-947946-51-9	**$36.00**
16.	**Human Genetic Diseases** Davies, K.E. (Ed)		
......	Hardback	0-947946-76-4	**$60.00**
......	Paperback	0-94/946-75-6	**$34.00**
14.	**Nucleic Acid Hybridisation** Hames, B.D. & Higgins, S.J. (Eds)		
......	Hardback	0-947946-61-6	**$60.00**
......	Paperback	0-947946-23-3	**$36.00**
12.	**Plant Cell Culture** Dixon, R.A. (Ed)		
......	Paperback	0-947946-22-5	**$36.00**

11a.	**DNA Cloning: Volume I** Glover, D.M. (Ed)		
......	Paperback	0-947946-18-7	**$36.00**
11b.	**DNA Cloning: Volume II** Glover, D.M. (Ed)		
......	Paperback	0-947946-19-5	**$36.00**
10.	**Virology** Mahy, B.W.J. (Ed)		
......	Paperback	0-904147-78-9	**$40.00**
9.	**Affinity Chromatography** Dean, P.D.G., Johnson, W.S. & others (Eds)		
......	Paperback	0-904147-71-1	**$36.00**
7.	**Microcomputers in Biology** Ireland, C.R. & Long, S.P. (Eds)		
......	Paperback	0-904147-57-6	**$36.00**
6.	**Oligonucleotide Synthesis** Gait, M.J. (Ed)		
......	Paperback	0-904147-74-6	**$38.00**
5.	**Transcription and Translation** Hames, B.D. & Higgins, S.J. (Eds)		
......	Paperback	0-904147-52-5	**$38.00**
3.	**Iodinated Density Gradient Media** Rickwood, D. (Ed)		
......	Paperback	0-904147-51-7	**$36.00**

Sets

	Essential Molecular Biology: 2 vol set Brown, T.A. (Ed)		
......	Spiralbound hardback	0-19-963114-X	**$118.00**
......	Paperback	0-19-963115-8	**$78.00**
	Antibodies: 2 vol set Catty, D. (Ed)		
......	Paperback	0-19-963063-1	**$70.00**
	Cellular and Molecular Neurobiology: 2 vol set Chad, J. & Wheal, H. (Eds)		
......	Spiralbound hardback	0-19-963255-3	**$133.00**
......	Paperback	0-19-963254-5	**$79.00**
	Protein Structure and Protein Function: 2 vol set Creighton, T.E. (Ed)		
......	Spiralbound hardback	0-19-963064-X	**$114.00**
......	Paperback	0-19-963065-8	**$80.00**
	DNA Cloning: 2 vol set Glover, D.M. (Ed)		
......	Paperback	1-85221-069-9	**$92.00**
	Molecular Plant Pathology: 2 vol set Gurr, S.J., McPherson, M.J. & others (Eds)		
......	Spiralbound hardback	0-19-963354-1	**$110.00**
......	Paperback	0-19-963353-3	**$75.00**
	Protein Purification Methods, and Protein Purification Applications: 2 vol set Harris, E.L.V. & Angal, S. (Eds)		
......	Spiralbound hardback	0-19-963048-8	**$98.00**
......	Paperback	0-19-963049-6	**$68.00**
	Diagnostic Molecular Pathology: 2 vol set Herrington, C.S. & McGee, J. O'D. (Eds)		
......	Spiralbound hardback	0-19-963241-3	**$105.00**
......	Paperback	0-19-963240-5	**$69.00**
	Receptor Biochemistry; Receptor-Effector Coupling; Receptor-Ligand Interactions: 3 vol set Hulme, E.C. (Ed)		
......	Paperback	0-19-963097-6	**$130.00**
	Human Cytogenetics: (2/e): 2 vol set Rooney, D.E. & Czepulkowski, B.H (Eds)		
......	Hardback	0-19-963314-2	**$130.00**
......	Paperback	0-19-963313-4	**$90.00**
	Peptide Hormone Secretion/Peptide Hormone Action: 2 vol set Siddle, K. & Hutton, J.C. (Eds)		
......	Spiralbound hardback	0-19-963072-0	**$135.00**
......	Paperback	0-19-963073-9	**$90.00**

ORDER FORM for USA and Canada

Qty	ISBN	Author	Title	Amount
			S&H	
	CA and NC residents add appropriate sales tax			
			TOTAL	

Please add shipping and handling: US $2.50 for first book, (US $1.00 each book thereafter)

Name ....

Address ....

....

.... Zip ....

[ ] Please charge $ .... to my credit card
Mastercard/VISA/American Express (circle appropriate card)

Acct. .... Expiry date ....

Signature ....

Credit card account address if different from above:

....

.... Zip ....

[ ] I enclose a cheque for US $............

Mail orders to: Order Dept. Oxford University Press, 2001 Evans Road, Cary, NC 27513